D0874283

Concerted Organic and Bio-organic Mechanisms

NEW DIRECTIONS in ORGANIC and BIOLOGICAL CHEMISTRY

Series Editor: C.W. Rees, CBE, FRS
Imperial College of Science, Technology and Medicine, London, UK

Advisory Editor: Alan R. Katritzky, FRS
University of Florida, Gainesville, Florida

Published Titles and Forthcoming Titles

Activated Metals in Organic Synthesis
Pedro Cintas

The Anomeric Effect
Eusebio Juaristi and Gabriel Cuevas

Aromatic Fluorination
James H. Clark, David Wails, and Tony W. Bastock

Asymmetric Synthetic Methodology
David J. Ager and Michael B. East

Capillary Electrophoresis: Theory and Practice
Patrick Camilleri

C-Glycoside Synthesis
Maarten Postema

Chemical Approaches to the Synthesis of Peptides and Proteins
Paul Lloyd-Williams, Fernando Albericio, and Ernest Giralt

Chiral Sulfur Reagents
M. Mikołajczyk, J. Drabowicz, and P. Kiełbasiński

Chirality and the Biological Activity of Drugs
Roger J. Crossley

Cyclization Reactions
C. Thebtaranonth and Y. Thebtaranonth

Concerted Organic and Bio-organic Mechanisms
Andrew Williams

Dianion Chemistry in Organic Synthesis
Charles M. Thompson

Lewis Acids and Selectivity in Organic Synthesis
M. Santelli and J.-M. Pons

Mannich Bases: Chemistry and Uses
Maurilio Tramontini and Luigi Angiolini

Organozinc Reagents in Organic Synthesis
Ender Erdik

Synthesis Using Vilsmeier Reagents
C. M. Marson and P. R. Giles

Vicarious Nucleophilic Substitution and Related Processes in Organic Synthesis
Mieczyslaw Makosza

Concerted Organic and Bio-organic Mechanisms

Andrew Williams, Ph.D.
University Chemical Laboratory
Canterbury, United Kingdom

CRC Press

Boca Raton   London   New York   Washington, D.C.

Library of Congress Cataloging-in-Publication Data

Williams, Andrew, 1937–
    Concerted organic and bio-organic mechanisms / Andrew Williams.
        p. cm. — (New directions in organic and biological
chemistry)
    Includes bibliographical references and index.
    ISBN 0-8493-9143-1
    1. Organic reaction mechanisms. I. Title. II. Series.
QD502.5.W55 1999
547′.139—dc21   99-42705
                                                               CIP

No claim to original U.S. Government works
International Standard Book Number 0-8493-9143-1
Library of Congress Card Number 99-42705
Printed in the United States of America 1 2 3 4 5 6 7 8 9 0
Printed on acid-free paper

Preface

This book is about the elucidation of concerted mechanisms and the experimental distinction between concerted and stepwise processes in organic and bio-organic reactions. Even though the concept is used extensively in general organic texts, little attention has been devoted to the meaning of a concerted mechanism, to its definition, or to the experimental techniques available for its precise demonstration. The concept of the concerted mechanism was formulated some 90 years ago but it has never been addressed explicitly in a monograph. Studies over the past two decades, most notably from the laboratory of Bill Jencks, have developed precise methods for elucidating concerted mechanisms and have substantially advanced our knowledge over that existing even up to the late 1970's. For this reason we believe that it is useful to gather together the material relating to concerted mechanisms into a single text.

The concept was described by LeBel in 1911 and developed by Lewis in his book, *Valence*, published in 1923 but the evidence distinguishing concerted from stepwise mechanisms was tenuous, even up to the 1960's. There was a massive interest in concerted mechanisms during Ingold's period especially regarding elimination and aliphatic substitution reactions. Subsequent times have provided a critical view and indeed the early experimental basis of the concept, namely the simple observation of inversion of configuration at a chiral centre, is now recognised as *not* being diagnostic of a concerted process (although it is valid that a concerted process *requires* inversion). The diagnosis of concerted mechanisms is inseparable from the demonstration of intermediates which are often expressed in reactions only at very small concentrations; the technique of trapping, which is useful for such low concentrations, is often misinterpreted. We hope in this text to describe some rigorous experimental tools with which the chemist can diagnose, with as much certainty as is feasible, the existence or otherwise of concerted mechanisms.

A reaction with simultaneous bond formation and fission is interpreted as possessing a concerted mechanism. The connection between this interpretation and experimental demonstration is often not clear and for this reason a definition which can be readily tested for a given reaction is necessary before definitive conclusions can be made. The definition of concertedness was formulated in its most precise form by Jencks in 1980 and Dewar in 1984, namely that a concerted mechanism has a single transition state. Armed with this definition, chemists developed methods to exclude stepwise mechanisms and thus diagnose the existence of concerted processes in a number of reactions which were favourable for study. The definition involves a "counting" of transition states. For most of the reactions studied this is not feasible and sub-definitions of concertedness are required, one of which involves study of the relative bond order in the transition structure for bond formation and bond fission.

Advances in understanding lead to further questions; in the case of studies of concertedness these involve borderline mechanisms and those mechanisms where the decomposition of an intermediate can become so fast that it reaches the limits of molecular integrity. It is hoped that this text will encourage the reader to consider these important areas of mechanism.

I am indebted to the following colleagues for their help and advice during the preparation of this text: Paul Bell (Canterbury), Paul Berti (Albert Einstein College of Medicine), Peter Guthrie (Western Ontario), Alvin Hengge (Utah State), Bob Hudson (Canterbury), Howard Maskill (Newcastle), Rory More O'Ferrall (Dublin), Mike Page (Huddersfield), John Richard (Buffalo), Ken Schofield (Exeter), and John Shorter (Hull). I am also grateful to Charles Rees (Imperial College) for encouraging me to write the text and to Christine Andreasen, Felicia Shapiro and Navin Sullivan of CRC Press.

A. Williams, Canterbury, 1999

Table of Contents

Chapter 7
CYCLIC REACTIONS 201

Chapter 8
ENZYME REACTIONS 223

Chapter 1

Definitions

A concerted mechanism occurs when bonds undergo formation and fission in a reaction at the same time. There is a language problem associated with describing the relative and absolute extent of bond formation and fission in the transition structure when a reaction occurs with simultaneous or coupled motions of atoms. Many of the words used to describe this process have their origins in time – *concerted, synchronous, simultaneous*. The *velocity* of a reaction is fundamentally important to our understanding of how processes occur and is also important in the design of industrial chemical processes; it is determined by the free energy of the transition structure which is caused by the change of charge at constituent atoms due to the relative progress of bond formation and fission. Reactions very rarely involve fewer than two *major* bond changes and the question of their relative timing in a mechanism is very important in organic and enzyme reactivity. Indeed, around 1950 Gardner Swain extensively referred to the concept of concertedness with regard to prototropic systems as models of enzymatic processes.[1-3]

Organic and bio-organic reactions have long been documented by a series of curved arrows or hooks (Lapworth[4] and Robinson[5,6]) to represent electron flow during the reaction, and there is a natural tendency, no doubt due to this method of representation, to write them as connected series. We believe that this facility has resulted in the general acceptance of simultaneous bond formation and fission, and it was suggested by Bordwell[7,8] that the acceptance of these mechanisms has depended on their logic and appeal coupled with the attractiveness of using series of curved arrows to simulate electron flow. It has proved very difficult to find "positive" evidence to distinguish between concerted mechanisms and stepwise ones where covalency changes occur in sequence. For the reason that no intermediates have been observed in many cyclical reactions these have sometimes been classed as "no mechanism" reactions to indicate that the reactant is directly converted to product with no intervening molecular species. It is a well-known maxim among mechanistic chemists that the non-observation of an intermediate does *not* exclude it from carrying the major part of the reaction flux of a reaction. The concerted mechanism, like that of many fundamental and now familiar concepts, has been the subject of major confusion over definitions and a definition is necessary which will enable a reaction to be subjected to rigorous tests of mechanism. Even in the classic text on cyclical mechanisms there is no explicit definition of a concerted mechanism (Woodward and Hoffmann, 1970).[9]

1.1 ORIGIN OF THE CONCEPT

The concept of a mechanism where bond changes occur simultaneously was formulated by Le Bel (1911)[10] and Lewis (1923)[11] to explain the Walden inversion (Chapter 4); the introduction of replacing group and expulsion of replaced group are interdependent features of a single concerted process. The concept of the Walden inversion[12] was given a physical interpretation by London (1929)[13,14] and an interpretation based on molecular orbital theory by Olson (1933).[15] Concertedness was also introduced in studies of acid-base catalysed mutarotation of sugars (Lowry, 1925 and 1927)[16,17] where one mechanism considered was that the processes of proton transfer to and from the sugar occur *simultaneously* in the "same electric circuit" (see Chapter 3). The current formal definition of concertedness was already available in the early 1960's because Banthorpe (1963)[18] includes a footnote in his text giving the modern definition.

The terminology of the first recorded observations that bond formation and bond fission could occur in a single step[10,11] did not use today's idiom. The first edition of Hammett's book (1940)[19] simply described the concept in terms of *simultaneous* bonding change. In the early 1950's Gardner Swain[1-3] applied the term *concerted* to the prototropic rearrangements involved in acid-base catalysed mutarotation of sugars and the term only seems to have displaced *simultaneous*, *synchronous* or *at the same time* near the end of the 1950s.[20] The general usage of *concerted* was reinforced by such influential texts as those of Hine (1962)[21], Gould (1962)[22] and Hammett (1970)[23] and by the discussions of the mechanisms led by Woodward and Hoffmann (1970)[9] which extended the term *concerted* to reactions other than that of the mutarotation reaction. *Synchronous* was favoured by British organic chemists as this term is used in Ingold's writing[24,25], that of Bunton (1963)[26] and of Alder, Baker and Brown (1971)[27]. More recent texts such as those of Isaacs (1995),[28] Maskill (1985),[29] Lowry and Richardson (1987)[30] and Page and Williams (1997)[31] conform to the IUPAC recommendations.[32]

1.2   MECHANISM AND ITS DESCRIPTION

The mechanism of a reaction can be described as the structure and energy of a molecule through its progress from reactants to products. Energy is required to transform reactants to products and the energy barrier in a single step reaction is due to a transitional structure which has an existence of only 10 to 100 femtoseconds (Figure 1.1).[33] The collection of reactant molecules is converted into the collection of product molecules through a *transition state*, an assembly of transitional structures, which embraces a collection of transient species having an effectively "normal" thermodynamic distribution of energies even though these structures are not interconvertible within the lifetime during which the arrangement of a given set of atoms resides in the transition state. A Boltzmann distribution of reactant molecules is activated to *effectively* give a Boltzmann distribution of molecules in the transition state. A refined definition of the mechanism of a reaction is *the structures of states on progression from reactant through transition states to product* and this may be used as a mechanistic test to describe any intermediates and all the transition state structures connecting these intermediates, reactants and products.

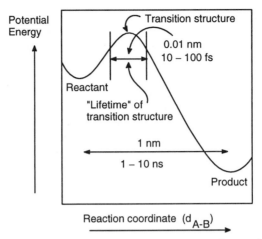

Figure 1.1. Time resolution in a simple reaction involving fission of an A-B bond. The velocity of the system along the reaction coordinate is 10^2 to 10^3 m/sec.

A high level definition of mechanism is the energy surface of a progression of transition states and intermediates as a function of all degrees of freedom. This definition is not attainable at present for reactions in solution although theory and gas phase experimental work have provided information on the energy of single entities as they go through to products for a limited number of reactions.[34,35] The maximum of the potential energy on the reaction coordinate between reactants and products corresponds to the transition state structure. The "transition state" is a quasi-thermodynamic *state* and has a maximum in Gibbs' free energy along the reaction coordinate. This free energy represents the energy of the collection of pseudo-molecules of the transition state distributed amongst the available quantum states of the various degrees of freedom as reflected in the entropy. The maximum in the potential energy along the reaction coordinate is temperature independent whereas the transition state structure may be temperature dependent because of entropy effects. Structure usually refers to potential energy and strictly we should always refer to the "transition state structure" or to the "structure of the activated complex" but common usage abbreviates this simply to the "transition state."

A bimolecular reaction in solution occurs by diffusion of the two reactant molecules through the assembly of solute and solvent molecules and collision to form an *encounter complex* within a solvent cage (Scheme 1.1). If the molecules are charged then the ionic atmosphere adjusts to any changes in the combined charge to give an *active complex* (see Scheme 1.1); reaction may still not be possible until any necessary changes in solvation occur (such as desolvation of lone pairs) to form a *reaction complex* in which bonding changes take place (the "chemical" step). The encounter complex remains essentially intact for the time period of several collisions because of the protecting effect of the solvent surrounding molecules once they have collided. The products of the subsequent reaction could either return to reactants or diffuse into the bulk solvent. Mechanisms as described commonly in undergraduate texts refer to the "chemical" step but it is important to note that the preceding or succeeding "physical" steps can limit the overall rate of reaction especially in system such as enzymes, where the velocity of the "chemical" step has been optimised. A description similar to that for the bimolecular reaction applies to a unimolecular reaction except that formation of the transition state is initiated by energy accumulation in the solvated reactant by collision. Scheme 1.1 gives typical times for reactant molecules *destined* to react. Most encounters do not lead to reaction and only a small fraction of the reaction complexes will have the appropriate transition state solvation in place for reaction to progress.

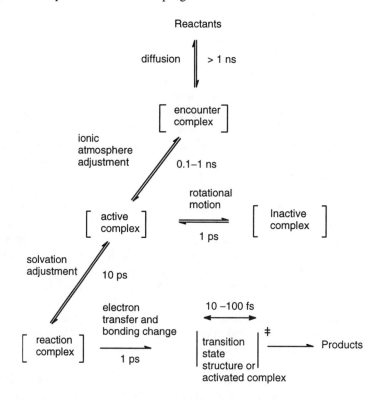

Scheme 1.1. Bimolecular mechanism in solution.

Reaction complexes in enzyme-catalysed reactions are more ordered than those in reactions of simple molecules and constitute thermodynamically favourable species. The enzyme active site provides a special micro-environment for the reaction compared with that for the uncatalysed reaction in bulk solution; however, the majority of encounters of substrate with enzyme molecule are unlikely to result in occupation of the active site.

1.3   THE DEFINITION OF CONCERTEDNESS

Classical descriptions of bond making and breaking usually focus only on the atoms directly involved although many other changes occur during this fundamental process. For example, the first step of a D_N+A_N $(S_N1)^*$ type mechanism nominally involves formation of a 3-coordinate intermediate from a 4-coordinate reactant. The parameter commonly called the reaction coordinate is often represented as the "stretching" of the C-Lg bond. In addition to the C-Lg motion several other relatively minor changes are involved, such as in the R-C bond length and the R-C-R bond angle, and these are often ignored in a description of the reaction coordinate; the degree of "coupling" of these changes does contribute to a detailed description of the mechanism.

The majority of reactions involve at least two major bond-making and bond-fission steps. The few reactions where there is only a single major bonding change are confined to the gas phase and except for reaction of atoms (such as *e.g.* Cl + H) even these involve hybridisation changes.

Reaction mechanisms are often described by potential energy diagrams and these can be two dimensional or three dimensional (Figure 1.2); they are very useful as aids to classification and understanding of mechanisms and refer to the reaction within the common solvation shell of the *reaction complex*. The diagrams should be *multi*-dimensional and it is assumed that a point on the three-dimensional surface represents the energy minimum for all the degrees of freedom not represented in the coordinates for the two major bonding changes (Jencks 1972).[36]

The equations of the curves for these diagrams have been determined empirically and are defined to give the correct slopes at various boundary conditions. The energies at the corners of the three-dimensional diagram are fixed from experimental values and at these coordinates the surface should be at a minimum and therefore be horizontal (Guthrie[38,39] and Lewis[40]). A concerted mechanism is a process where all bond changes occur in a single step without the intervention of an intermediate along the reaction coordinate. This can be expressed as an experimentally verifiable definition of a concerted mechanism. It is simply that a concerted mechanism involves no discrete molecular intermediate between reactants and products (Dewar, 1984[41] and Jencks, 1980[42]). A corollary of the definition is that a concerted mechanism is one with a single transition state and a stepwise process involves more than one transition state on passage to product. Diffusion together of the reactants to form an encounter-complex and diffusion of product from an encounter-complex (see Scheme 1.1) are not parts of the chemical mechanism and in some cases even proton transfer is not considered as a step fundamental to the main reaction. By definition, a fundamental process has no intermediate and must therefore be considered as "concerted." A complex mechanism involving a number of chemical reactions

---

*We shall employ the IUPAC mechanistic nomenclature[37] in this text unless in the author's opinion there is good reason to retain the Ingoldian classification.

in sequence is overall a stepwise process; it can be the case that one or more of the individual steps of this reaction are concerted. It is often a question whether one of these steps, involving more than one substantial bond change, is composed of two steps or is concerted.

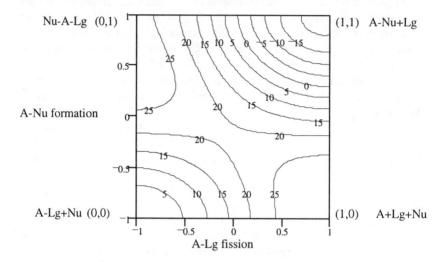

Figure 1.2. Notional contour representation of a More O'Ferrall-Jencks potential energy surface for a simple nucleophilic displacement reaction of A-Lg + Nu →A-Nu + Lg. Numbers on the contours represent energies.

Concerted mechanisms do not necessarily require bond formation and bond fission to be advanced to the same extent in the transition state structures; for example in the transition structure for the A_ND_N mechanism (Scheme 1.2) HO...C and C...Hal are not necessarily "half" bonds. The definition of a concerted mechanism, recommended by the IUPAC, is as follows (Müller, 1994):[32]

"Two or more primitive changes are said to be concerted (or to constitute a concerted process) if they occur within the same elementary reaction. Such changes will normally (though perhaps not inevitably) be *energetically coupled*. In a concerted process the primitive changes may be synchronous or asynchronous. *Energetic coupling* means that the simultaneous (*sic*) progress of the primitive changes involves a transition state of lower energy than that for their successive occurrence."

The term *synchronous* has been used synonymously with *concerted* in the past but there is now a view that it should be applied to a special case of the concerted condition. Müller proposes the following definition:[32]

"A concerted process in which the primitive changes concerned have progressed to the same extent in the transition state is said to be synchronous. The concept does not admit of an exact definition except in the case of concerted processes involving changes in two bonds."

The transition state for a synchronous displacement concerted mechanism can lie at any location on the synchronous diagonal (Figure 1.3). If the transition state is at the central point (B: $\alpha_X = \alpha_Y = 0.5$)[43,44] it is said to be *in balance*. Any other location means that *either* bond formation and bond fission have only partly advanced (A) *or* bond formation and bond fission are both well advanced (C). Concerted mechanisms with transition structures off the synchronous diagonal are *imbalanced*. The location on the tightness diagonal[45] (or disparity mode)[46] of the transition state structure of a concerted identity reaction measures the imbalance between bond formation and bond fission.

Coupling between bond fission and formation enables electronic energy to be transmitted between the bonding changes and hence facilitates a reaction (Bernasconi).[47,48] For this reason a concerted mechanism can sometimes offer a pathway energetically more favourable than its stepwise counterpart (Bordwell).[7,8]

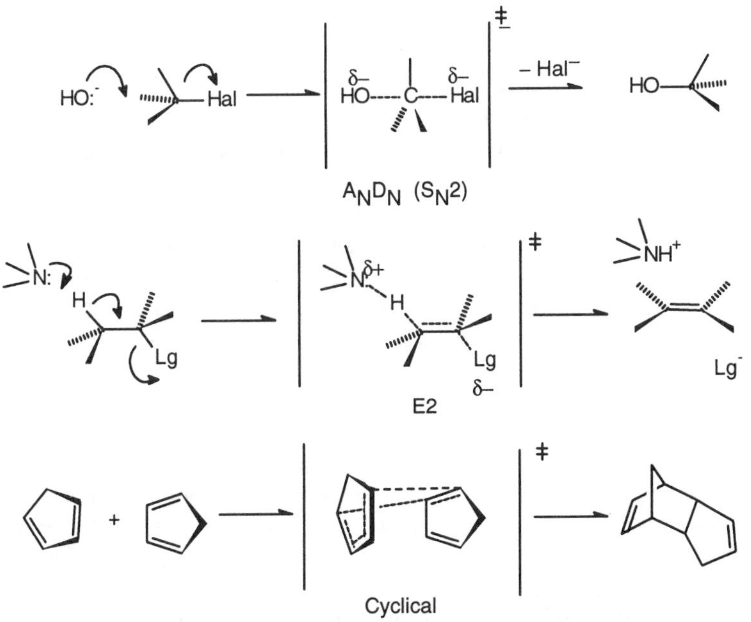

Scheme 1.2. Classical concerted mechanisms.

A stepwise process can lead to the formation of unstable intermediates such as relatively high energy radicals or ions. The concerted mechanism is an obvious route whereby these energetically unfavourable intermediates can be avoided by neutralising the charge or pairing a radical as it is being formed. Related to this is an unfavourable coordination number of an atom which might be expected to result from these processes (in an A_N+D_N reaction at aliphatic carbon the central atom becomes five-coordinate in contravention of the valency rules).

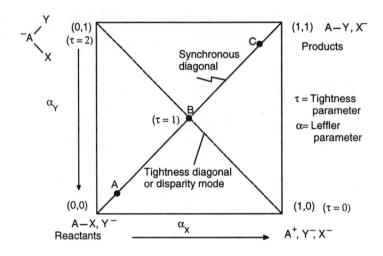

Figure 1.3. More O'Ferrall-Jencks diagram for a displacement reaction; Leffler's $\alpha_X = \beta_X/\beta_{eq}$.

Reactions are favoured which involve least nuclear motion in a given system (Hine)[49,50] because the energy required for small movements of atoms from a stable equilibrium geometry varies approximately as the square of the displacement. The concerted process, however, involves more nuclear motion in a particular step than does a stepwise mechanism because more bonds are being formed or broken thus requiring more atomic movement. The principle of least nuclear motion (Hine[49,50] and Sinnott[51]) is thus a factor which could militate against concertedness. However, the least motion effects are often energetically small and are frequently overridden by other factors. Solvent reorganisation is often a major contributor to the activation energy and it is not easy to apply the principle of least nuclear motion to reactions in which solvent motion is significant.

Improved mechanistic tools and knowledge led to significant understanding of alkyl group transfer and a number of aliphatic nucleophilic substitution reactions originally thought to be concerted are now known to be stepwise.[52,53] Conversely, there are now cases where concertedness has been shown to operate in reactions hitherto thought to be stepwise (acyl group transfer) and due to this renewed interest investigations of concertedness have been subject to considerable scrutiny and critique by Bordwell[7,8] and Williams.[54,55]

An intermediate (Scheme 1.3) does not behave like a normal molecule in the limit where its decomposition has no energy barrier and it is therefore not in equilibrium with reactants and products. The question naturally arises as to "when a barrier is not a barrier." Irrespective of the availability of experimental techniques to detect a very unstable intermediate, an entity is not considered an intermediate if its lifetime is less than h/kT $ca.$ $1.7.10^{-13}$ sec at 25°. However, when intermediates have very short lifetimes, an alternative concerted mechanism may occur which avoids the formation of this high energy intermediate – this is called an *enforced concerted* mechanism by Jencks[42] because concertedness is forced to ensue as the intermediate becomes progressively more reactive. The transition state structure would correspond to that for the formation of the putative intermediate which approximates the structures represented by the NW [0,1] or SE [1,0] corners of the reaction map (Figure 1.3). The stepwise paths (Scheme 1.3) would become concerted when the intermediates $[Y-A-X]^-$ and $[A]^+$ become too reactive to exist; when the barrier is sufficiently low for decomposition of an intermediate for the half life to be in the femtosecond range then there will be concertedness. Barriers which give half lives in the picosecond range (corresponding to rotational motion) will still give rise to stepwise processes. Coupling between bond changes is likely to be minimal because one of the reacting bonds will be almost fully formed in the transition state of an enforced concerted process.

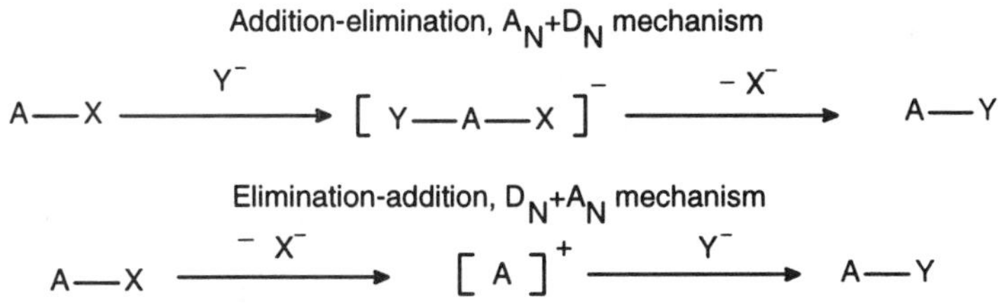

Addition-elimination, $A_N + D_N$ mechanism

$$A-X \xrightarrow{\;\;Y^-\;\;} [\;Y-A-X\;]^- \xrightarrow{\;\;-X^-\;\;} A-Y$$

Elimination-addition, $D_N + A_N$ mechanism

$$A-X \xrightarrow{\;\;-X^-\;\;} [\;A\;]^+ \xrightarrow{\;\;Y^-\;\;} A-Y$$

Scheme 1.3. Some displacement mechanisms with intermediates.

The time scale for the reorganisation of solvation or ionic atmosphere in response to electronic changes is not sufficiently fast (~10ps and 1ns respectively, Scheme 1.1) for them to remain equilibrated. The solvent and ionic atmosphere act like immobile spectators rather than as the coupled rearranging partner usually envisioned; this may be true for reactions where it only takes a very short time to cross the barrier (Bernasconi).[47,48] Some reactions may be viewed as occurring in a pre-organised solvent shell *i.e.* solvent reorganisation occurs *before* bond-making/-breaking in the reactants.

Solvent reorganisation has not hitherto been considered to be a primitive change which could be in concert with, say, bond formation. Coupling between such disparate changes is unlikely; it is possible that interactions between reactant and host within enzyme substrate or host-guest molecular recognition complexes could be coupled with bond formation or fission.

1.4 BOND ORDER AND COUPLING

Adherence to the formal definition of a concerted mechanism as a reaction with a single transition state limits the number of available diagnostic techniques. Reactions with all the major bonding changes substantially advanced in the transition state have concerted mechanisms; conversely it is not true that all concerted mechanisms have substantially advanced major bonding changes. This leads to a *subsidiary definition* that a concerted mechanism has a transition state with substantial bonding changes in all the bonds which suffer fission or formation. This definition is favoured by Saunders[56] and is indicated in Lowe's discussion[57] and in that of Andrist.[58]

Ponec[59,60] introduced a *topological* criterion of concertedness which involves compartmentalising the More O'Ferrall-Jencks diagrams into regions through which the reaction path must travel from reactants to products. The process involving passage through a region denoted as an intermediate (see Figure 1.4) would be a non-concerted mechanism. This *topological* criterion bears a similarity to that described by Lowe.[42] A concerted mechanism results when the transition state lies in a particular area of a potential energy diagram but it is not clear whether a reaction at a borderline on the potential energy surface is concerted or not. The immediate problem with this type of criterion is that the diagrams refer to potential energy whereas experimental techniques are largely based on free energy considerations.

Imbalance is detected when the Leffler parameters (α_X and α_Y, see Figure 1.3)[43,44] that characterise the different bond-forming or bond-breaking processes are not equal. Bernasconi[47,48] formulated the principle that a product stabilising factor whose development at the transition state is late will increase the activation energy and lower the intrinsic rate constant (k_o). Increased imbalance in a concerted mechanism brings the reaction coordinate progressively closer to that for a stepwise process. Thus imbalanced concerted mechanisms need not be more effective than the corresponding stepwise routes. To the extent that there is less effect on the transition state from a weakly advanced primitive step there is less coupling in imbalanced concerted mechanisms than in balanced ones. Synergy between the bonding changes is directly connected to the coupling between them and there is thus no reason to expect increased reactivity simply from a mechanism being concerted.

1.5   POTENTIAL ENERGY OR FREE ENERGY?

The formal definition of concertedness is based on Gibbs free energy and it enables the application of experimental techniques which work on assemblies of molecules (namely most experimental techniques). The parameters which are measured, generally some form of rate constant, are free energies and refer to non-continuous states. This leads to an energy diagram consisting of energy levels corresponding to the free energies of the states.

If it were possible to determine potential energy diagrams the problem would devolve into the question "at which stage does a depression in the maximum in the reaction coordinate turn from a molecule into a transition state?" The question of whether the transition state is located at the maximum in the potential energy surface or at the maximum in free energy has been posed by Albery[61] and Laidler.[62,63] The entropy depends on the spacing of the energy levels in the potential energy surface. The entropic component of the free energy means that the maximum in free energy is *not* the same as that in potential energy (Figure 1.5). The free energy definition of Dewar[41] and Jencks[42] is preferred because it is much easier to measure free energies than potential energies, particularly for solution reactions. These considerations raise the interesting idea that the transition state structure will be temperature dependent. The question of potential energy or free energy has been discussed by Andrist[58] who indicates that potential energy *and* the motion of the system in the surface are the determinants of concertedness.

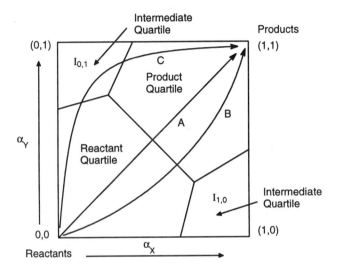

Figure 1.4. *Topological* criterion of concertedness. Coordinates α_X and α_Y represent the bond order of breaking and forming bonds respectively.

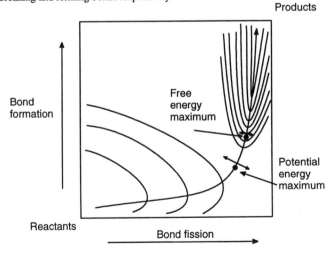

Figure 1.5. Potential energy surface indicating different maxima for free and potential energies. The double-headed arrows represent vibrational motion giving rise to the entropy term.

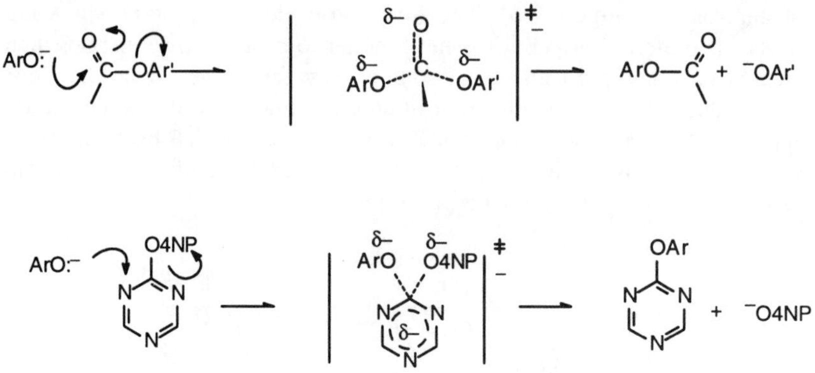

Scheme 1.4. Some concerted organic reaction mechanisms; 4NP = 4-nitrophenyl.

1.6   CLASSES OF CONCERTED REACTIONS

Nucleophilic aliphatic substitution (A_ND_N) - the S_N2 reaction - is the prime example of a classical concerted reaction type. Other classical examples of concerted reactions include pericyclic reactions and base-catalysed eliminations (Scheme 1.2). Some examples of more recently discovered concerted mechanisms are illustrated in Scheme 1.4 and some which remain to be proved as illustrated in Scheme 1.5. Examples of possible concerted enzyme catalysed reactions are shown in Schemes 1.6 and 1.7.

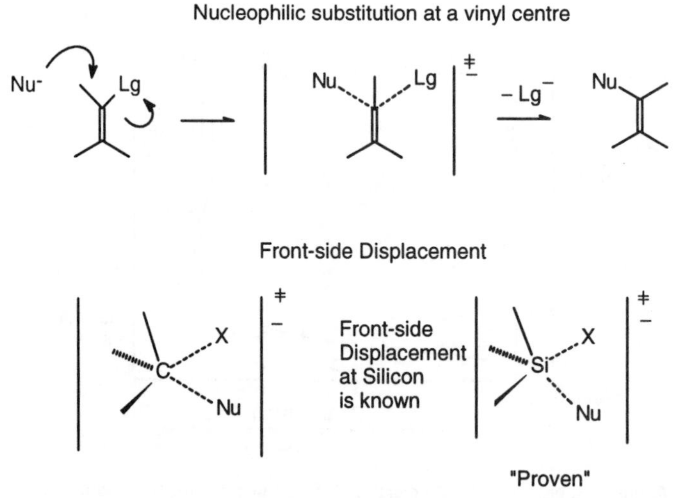

Nucleophilic substitution at a vinyl centre

Front-side Displacement

Scheme 1.5. Some hypothetical concerted organic reaction mechanisms.

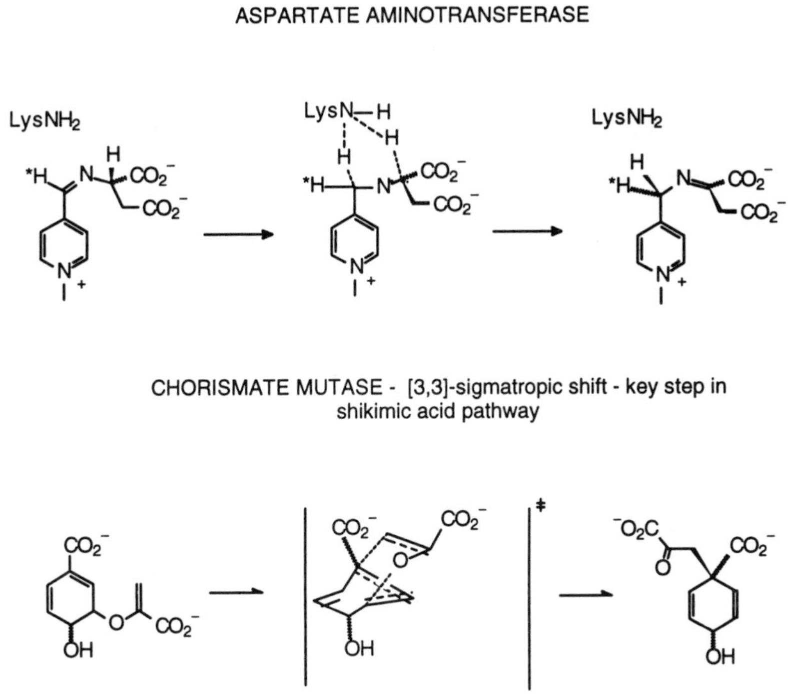

ASPARTATE AMINOTRANSFERASE

CHORISMATE MUTASE - [3,3]-sigmatropic shift - key step in shikimic acid pathway

Scheme 1.6. Some possible concerted enzyme mechanisms with cyclic transition structures.

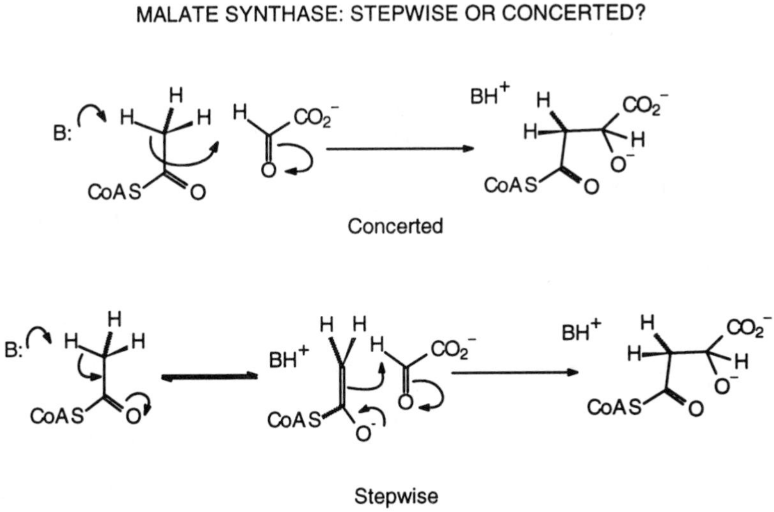

MALATE SYNTHASE: STEPWISE OR CONCERTED?

Concerted

Stepwise

Scheme 1.7. Some possible concerted enzyme mechanisms involving proton transfer.

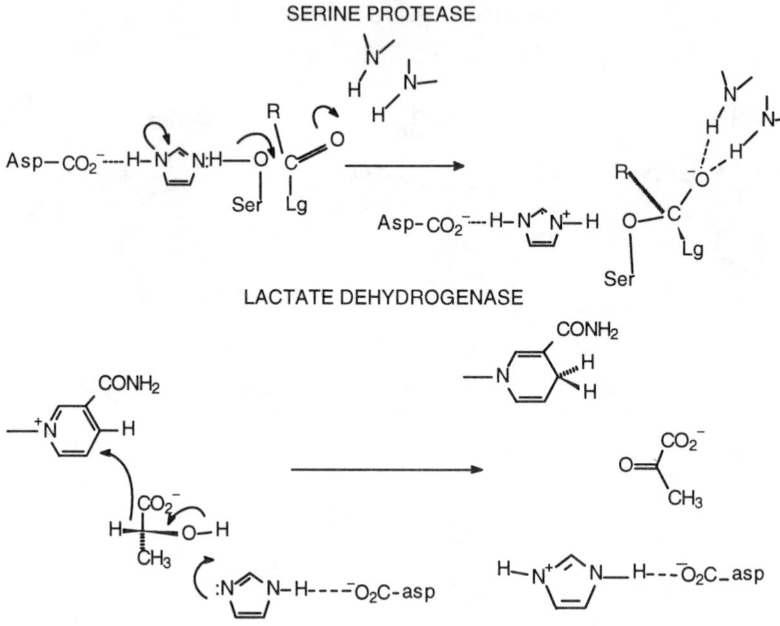

SERINE PROTEASE

LACTATE DEHYDROGENASE

Scheme 1.7 (*continued*). Some possible concerted enzyme mechanisms involving proton transfer.

FURTHER READING

Baggott, J., Molecules caught in the act, *New Scientist*, 17 June 1989.

Baldwin, J. E., and Fleming, R. H., Allene-olefin and allene-allene cycloadditions. Methylene-cyclobutane and 1,2-dimethylenecyclobutane degenerate rearrangements, *Fortsch. Chem. Forschung, 15,* 281, 1970.

Benson, S. W., The range of chemical forces and the rates of chemical reactions, *Accs. Chem. Res., 19,* 335, 1986.

Berson, J. A., Orbital-symmetry-forbidden reactions, *Acss. Chem. Res, 5,* 406, 1972.

Chattaraj, P. K., Cedillo, A., Parr, R. G., and Arnett, E. M., Appraisal of Chemical bond making, bond breaking, and electron transfer in solution in the light of the principle of maximum hardness, *J. Org. Chem., 60,* 4707, 1995.

Evans, M. G., and Polanyi, M., Inertia and driving force of chemical reactions, *Trans. Far. Soc., 34,* 11, 1938.

Fleming, I., *Frontier Orbitals and Organic Chemical Reactions,* John Wiley & Sons Ltd., London, 1976.

Fong, F. K., A successor to transition state theory, *Acss. Chem. Res, 9,* 433, 1976.

Fong, F. K., Relaxation behaviour of equilibrating molecular configurations. Application to the 2-norbornyl problem, *J. Amer. Chem. Soc., 96,* 7638, 1974.

Grunwald, E., Structure-energy relations, reaction mechanisms, and disparity of progress of concerted reaction events, *J. Amer. Chem. Soc., 107,* 125, 1985.

Hoz, S., Is the transition state indeed intermediate between reactants and products? The Michael addition reaction as a case study, *Acss. Chem. Res., 26,* 69, 1993.

Jencks, W. P., How does a reaction choose its mechanism? (Ingold lecture, 1981), *Chem. Soc. Rev. (London), 10,* 345, 1981.

Jencks, W. P., Are structure-reactivity correlations useful? *Bull. Soc. Chim. France,* 218, 1988.

Johnson, C. D., Transition state variation in the Menshutkin reaction? *Tetrahedron Letters, 21,* 2217, 1982.

Klots, C. E., The reaction coordinate and its limitations: an experimental perspective, *Accs. Chem. Res., 21,* 16, 1988.

Marcus, R. A., Skiing the reaction rate slopes, *Science, 256,* 1523, 1992.

McIver, J. W., The structure of transition states: are they symmetric? *Acss. Chem. Res., 7,* 72, 1974.

Page, M. I., Transition-states, standard states and enzymic catalysis, *Int. J. Biochem., 11,* 331, 1980.

Polanyi, J. C., and Zewail, A. H., Direct observation of the transition state, *Accs. Chem. Research, 28,* 119, 1995.

Sheehan, W. F., Along the reaction coordinate, *J. Chem. Ed., 47,* 254, 1970.

Smith, I. W. M., Femtosecond chemistry, *Nature, 343,* 691, 1990.

Smith, I. W. M., Probing the transition state, *Nature, 358,* 279, 1992.

Tedder, J. M., and Nechvatal, A., *Pictorial Orbital Theory,* London, Pitman, 1988.

Truhlar, D. G., Garrett, B. C., and Klippenstein, S. J., Current status of transition state theory, *J. Phys. Chem., 100,* 12771, 1996.

Zewail, A. H., 1988 Laser femtochemistry, *Science, 242,* 1645.

REFERENCES

1. **Swain, C. G.,** General acid-base concerted mechanism for enolisation of acetone, *J. Amer. Chem. Soc., 72,* 4578, 1950.

2. **Swain, C. G., and Brown, J. F.,** Concerted displacements. VII. The mechanism of acid-base catalysis in non-aqueous solvents, *J. Amer. Chem. Soc., 74,* 2534, 1952.

3. **Swain, C. G., and Brown, J. F.,** Concerted reactions. VIII. Polyfunctional catalysis, *J. Amer. Chem. Soc., 74,* 2538, 1952.

4. **Lapworth, A.,** A theoretical derivation of the principle of induced alternate polarity, *J. Chem. Soc., 121,* 416, 1922.

5. **Kermack, W. O., and Robinson, R.,** An explanation of the property of induced polarity of atoms and an interpretation of the theory of partial valencies on an electronic basis, *J. Chem. Soc., 121,* 427, 1922.

6. **Robinson, R.,** *Two Lectures on an "Outline of an Electrochemical (Electronic) Theory of The Course of Organic Reactions,"* Institute of Chemistry of Great Britain and Ireland, London, 1932.

7. **Bordwell, F. G.,** Are nucleophilic bimolecular concerted reactions involving four or more bonds a myth? *Accs. Chem. Res., 3,* 281, 1970.

8. **Bordwell, F. G.,** How common are base-initiated, concerted 1,2-eliminations? *Accs. Chem. Res., 5,* 374, 1972.

9. **Woodward, R. B., and Hoffmann, R.,** The conservation of orbital symmetry, Academic Press, New York, 1970.

10. **Le Bel, J. A.,** A propos de l'inversion optique de Walden, *J. de Chim. Phys., 9,* 323, 1911.

11. **Lewis, G. N.,** *Valence and the Structure of Atoms and Molecules,* The Chemical Catalogue Company, New York, 1923, 113.

12. **Walden, P.,** Sur l'inversion optique des composés organiques (inversion de Walden). *J. de Chim. Phys., 9,* 160, 1911.

13. **London, F.,** Quantum mechanical explanation of activation, *Z. Elektrochemie, 35,* 552, 1929.

14. **London, F.,** The significance of quantum theory for chemistry, *Naturwissenschaften, 17,* 516, 1929.

15. **Olson, A. R.,** The mechanism of substitution reactions, *J. Chem. Phys., 1,* 418, 1933.

16. **Lowry, T. M.,** Studies of dynamic isomerism. Part XVIII. The mechanism of mutarotation, *J. Chem. Soc.,* 1371, 1925.

17. **Lowry, T. M.,** Studies of dynamic isomerism. Part XXV. The mechanism of catalysis by acids and bases, *J. Chem. Soc.,* 2554, 1927.

18. **Banthorpe, D. V.,** *Elimination Reactions*, Elsevier Publishing Co., Amsterdam, 1963, p. 4.

19. **Hammett, L. P.,** *Physical Organic Chemistry*, 1st. edn., McGraw-Hill Book Company, Inc., New York, 1940.

20. **Bartlett, P. D.,** Reaction mechanisms, in *Perspectives in Organic Chemistry*, Todd, A. R. ed., Interscience, New York, 1956, p. 23.

21. **Hine, J.,** *Physical organic chemistry,* McGraw-Hill Book Company, Inc., New York, 1962.

22. **Gould, E. S.,** *Inorganic Reactions and Structure*, revised edition, Holt, Reinhart & Winston, New York, 1962.

23. **Hammett, L. P.,** *Physical Organic Chemistry*, 2nd. edn., McGraw-Hill Book Company, Inc., New York, 1970.

24. **Ingold, C. K.,** *Structure and Mechanism in Organic Chemistry,* 1st edn., G Bell and Sons Ltd. (see also 2nd edn., Cornell University Press, Ithaca, 1969), London, 1953.

25. **Ingold, C. K.,** Principles of an Electronic Theory of Organic Reactions, *Chem. Revs., 15,* 225, 1934.

26. **Bunton, C. A.,** *Nucleophilic Substitution at a Saturated Carbon Atom*, Elsevier Publishing Company, Amsterdam, 1963.

27. **Alder, R. W., Baker, R., and Brown, J. M.,** *Mechanism in Organic Chemistry*, Wiley-Interscience, New York, 1971.

28. **Isaacs, N.,** *Physical Organic Chemistry*, 2nd. edn., Addison Wesley Longman, Harlow, Essex, 1995.

29. **Maskill, H.,** *The Physical Basis of Organic Chemistry*, Oxford University Press, Oxford, 1985.

30. **Lowry, T. H., and Richardson, K. S.,** *Mechanism and Theory in Organic Chemistry,* 3rd. edn., Harper and Row, Publishers, New York, 1987.

31. **Page, M. I., and Williams, A.,** *Organic and Bio-Organic Mechanisms*, Addison Wesley Longman, Harlow, Essex, 1997.

32. **Müller, P.,** Glossary of terms used in physical organic chemistry, *Pure and Applied Chemistry, 66,* 1077, 1994.

33. **Polanyi, J. C., and Zewail, A. H.,** Direct observation of the transition state, *Accs. Chem. Research, 28,* 119, 1995.

34. **Li, C., Ross, P., Szulejko, J. E., and McMahon, T. B.,** High pressure mass spectromtric investigations of the potential energy surfaces of gas phase reactions, *J. Amer. Chem. Soc., 118,* 9360, 1996.

35. **Mills, I. M.,** Potential energy surfaces from vibrational rotational data, *Faraday Discussion, 62,* 7, 1977.

36. **Jencks, W. P.,** General acid-base catalysis of complex reactions in water, *Chem. Revs., 72,* 705, 1972.

37. **Guthrie, R. D., and Jencks, W. P.,** IUPAC recommendations for the representation of reaction mechanisms, *Accs. Chem. Res., 22,* 343, 1989.

38. **Guthrie, J. P.,** Concertedness and E2 elimination reactions. Prediction of transition state position and reaction rates using two-dimensional reaction surfaces based on quadratic and quartic approximations, *Can. J. Chem., 68,* 1643, 1990.

39. **Guthrie, J. P.,** Concerted mechanism for alcoholysis of esters: an examination of the requirements, *J. Amer. Chem. Soc., 113,* 3941, 1991.

40. **Lewis, E. S.,** Linear free energy relationships, in *"Investigation of Rates and Mechanisms of Reactions" Part I,* Bernasconi, C. F., ed., New York, Wiley-Interscience, 1986, p. 871.

41. **Dewar, M. J. S.,** Multibond reactions cannot normally be synchronous, *J. Amer. Chem. Soc., 106,* 209, 1984.

42. **Jencks, W. P.,** When is an intermediate not an intermediate? Enforced mechanisms of general acid-base catalysed, carbocation, carbanion, and ligand exchange reactions, *Accs. Chem. Res.,13,*161, 1980.

43. **Williams, A.,** Effective charge and Leffler's index as mechanistic tools for reactions in solution, *Accs. Chem. Res., 17,* 425, 1984.

44. **Page, M. I., and Williams, A.,** *Organic and Bio-Organic Mechanisms*, Addison Wesley Longman, Harlow, Essex, 1997, Chapter 3.

45. **Grunwald, E.,** Structure-energy relations, reaction mechanisms, and disparity of progress of concerted reaction events, *J. Amer. Chem. Soc., 107*, 125, 1985.

46. **Kreevoy, M. M., and Lee, I. S. H.,** Marcus theory of perpendicular effect on α for hydride transfer between NAD^+ analogues, *J. Amer. Chem. Soc., 106*, 2550, 1984.

47. **Bernasconi, C. F.,** The principle of non-perfect synchronisation, *Adv. Phys. Org. Chem., 27*, 119, 1992.

48. **Bernasconi, C. F.,** The principle of non-perfect synchronisation: more than a qualitative concept? *Acc. Chem. Res., 25*, 9, 1992.

49. **Hine, J.,** The principle of least nuclear motion, *Adv. Phys. Org. Chem., 15*, 1, 1977.

50. **Hine, J.,** The principle of least motion. Application to reactions of resonance stabilised species, *J. Org. Chem., 31*, 1236, 1966.

51. **Sinnott, M. L.,** The principle of least nuclear motion and the theory of stereoelectronic control, *Adv. Phys. Org. Chem., 24*, 113, 1988.

52. **Sneen, R. A.,** Organic ion pairs as intermediates in nucleophilic substitution and elimination reactions, *Accs. Chem. Res., 6*, 46, 1973.

53. **Jencks, W. P.,** How does a reaction choose its mechanism? (Ingold Lecture, 1981), *Chem. Soc. Rev. (London), 10*, 345, 1981.

54. **Williams, A.,** Concerted mechanisms of acyl group transfer reactions in solution, *Accs. Chem. Res., 22*, 387, 1989.

55. **Williams, A.,** The diagnosis of concerted organic mechanisms, *Chem. Soc. Rev., 93*, 1994.

56. **Saunders, W. H.,** Distinguishing between concerted and non-concerted eliminations, *Accs. Chem. Res., 8*, 19, 1976.

57. **Lowe, J. P.,** Is this a concerted reaction? *J. Chem. Ed., 51*, 785, 1974.

58. **Andrist, A. H.,** Concertedness: a function of dynamics or the nature of the potential energy surface? *J. Org. Chem., 38*, 1772, 1973.

59. **Ponec, R., and Strnad, M.,** The least motion principle, concertedness and the mechanisms of pericyclic reactions. A similarity approach, *Coll.Czech.Chem.Commun., 59*, 75, 1994.

60. **Ponec, R.,** Similarity models in the theory of pericyclic reactions, *Topics in current chemistry, 174*, 1, 1995.

61. **Albery, W. J.,** Transition state theory revisited, *Adv. Phys. Org. Chem., 28*, 139, 1993.

62. **Laidler, K. J., and Polanyi, J. C.,** *Progress in Reaction Kinetics*, Longmans, London, 1965, 37.

63. **Laidler, K. J.,** *Theories of Chemical Kinetics*, McGraw-Hill, New York, 1969, p. 76.

Chapter 2

Techniques

The *fundamental* definition of concertedness is the requirement of a single transition state between reactant and product and the *subsidiary* definition is based on the bond order in the bonds undergoing major transformation. The *fundamental* definition needs tools which exclude stepwise paths, whereas measurements of bond order suffice for the *subsidiary* definition. The study of concertedness essentially devolves into one of demonstrating the absence or existence of *intermediates*. The observation of intermediates is clear evidence for a stepwise process but unless there is some other information available the failure to detect an intermediate is not evidence for a concerted mechanism.

2.1 DETECTION OF INTERMEDIATES

2.1.1 Instrumental
A distinction in favour of a stepwise process is manifestly obvious if, during a kinetic study, an intermediate species is observed to accumulate and then decay to give products (Figure 2.1). The nature of the measuring device is not relevant to the argument but is likely to depend on some form of electromagnetic energy absorption or emission. The direct observation of an intermediate depends on a build-up of its concentration to a measurable level and this requires that the decay to product is slow relative to the time resolution of the instrument in use.

The simplest possible case of a stepwise process is shown in Equation (2.1) and this also happens to be one of the most generally applicable mechanisms. The build-up of significant concentrations of the intermediate, B, depends[1] on (a) a favourable equilibrium constant for its formation and (b) a decay of the intermediate slower than its formation. Decay of the intermediate must also be slow enough to permit its kinetic observation. These criteria are embodied in the following requirements: k_2 must be less than $k_1 + k_{-1}$; and k_1 must be greater than k_{-1}. The concentration of B is given as a function of time by Equation (2.2) and it grows to a maximum value $[B_{max}]$ given by Equation (2.3) (Espenson).[2] The maximum concentration of a putative intermediate in the simplest mechanism (Equation (2.1)) could be obtained from estimates of the rate constants by use of linear free energy relationships. This concentration of intermediate could then be assessed by a suitable analytical method, and provided there is confidence in the estimated rate constants the non-observation of an intermediate would be good evidence for excluding a stepwise process. Surprisingly, this direct procedure for excluding an intermediate does not seem to have been employed in studies of concertedness.

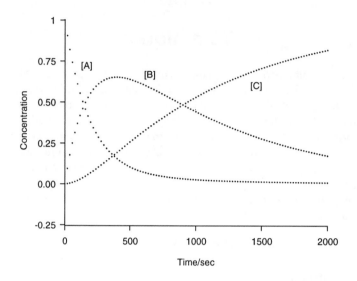

Figure 2.1. Instrumental detection of an intermediate. Lines are drawn from the parameters $A_o =$ 1.00 M, $k_1 = 0.005$ sec^{-1}, $k_2 = 0.001$ sec^{-1}, and $k_{-1} = 2 \ 10^{-4}$ sec^{-1} using Equation (2.2) and equations for [A] and [C] from Espenson.[2]

$$A \underset{k_{-1}}{\overset{k_1}{\rightleftharpoons}} B \overset{k_2}{\longrightarrow} C \tag{2.1}$$

$$[B] = A_o k_1[\exp(-\lambda_1 t) - \exp(-\lambda_2 t)]/(\lambda_2 - \lambda_1) \tag{2.2}$$

where

$$\lambda_1 = 0.5(k_1 + k_{-1} + k_2) + 0.5\{(k_1 + k_{-1} + k_2)^2 - 4k_1 k_2\}^{0.5}$$

and

$$\lambda_2 = 0.5(k_1 + k_{-1} + k_2) - 0.5\{(k_1 + k_{-1} + k_2)^2 - 4k_1 k_2\}^{0.5}$$

$$[B_{max}] = (A_o k_1/\lambda_1)(\lambda_2/\lambda_1)^{\lambda_2/(\lambda_1 - \lambda_2)} \tag{2.3}$$

Recent studies at the very limit of measurements have enabled direct measurements to be made of kinetics in the femtosecond range of time resolution and offer a direct method for determining the lifetimes of extremely reactive putative intermediates (with half lives of *ca.* 50 fs).[3] In principle there is no limit on the value of k_2 which can be measured, but up to now the kinetics of only gas phase reactions have been readily measured.

Intermediates at low concentration levels can be detected by very sensitive instrumental methods such as by spectrofluorimetry or electron spin resonance spectroscopy. The demonstration of such transient species in a reaction does not immediately identify them as carrying the reaction flux; they could be *blind alley* intermediates (*i.e.* intermediates not on the reaction path). The detection of intermediates at such low concentrations requires that they must be demonstrated to carry the reaction flux, and this can only be carried out if the concentration of the intermediate is known and if k_2 is known or calculated. This protocol is required even if the detected intermediate species has a concentration commensurate with that of the reactant at the start of the reaction ($[A_o]$).

2.1.2 Stereochemistry

A cornerstone of techniques for mechanism in organic chemistry is that an *in-line* concerted mechanism for displacement at an aliphatic centre *requires* inversion of configuration at the central carbon atom. It has long been realised that, conversely, the observation of inversion of configuration is *not* evidence for a concerted mechanism because stepwise mechanisms (Scheme 2.1) can be invoked which give the same result provided the carbenium ion is sufficiently reactive. Bordwell pointed out that 90% product formation corresponds to a difference of only 1.4 kcal/mole of activation energy and even 99% only corresponds to 2.7 kcal/mole difference between competing paths.[4] If the bond formation step is fast enough there would be no time for the carbenium ion to diffuse into solution or for the ion-pair complex to reconfigure and this would lead to the observation of either full or partial retention or inversion of configuration; the observation of partial racemisation provides definitive evidence for an intermediate and excludes a concerted process.

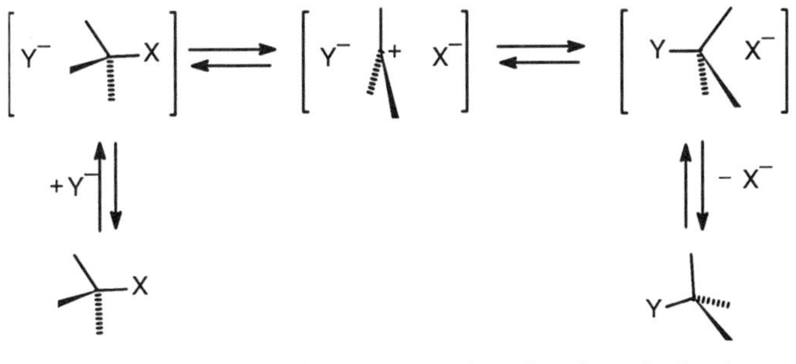

Scheme 2.1. Inversion in a *stepwise* displacement reaction where the carbenium ion reacts faster than it can rotate or diffuse into solvent.

The arguments for substitution at carbon can be applied equally well to substitution at other atomic centres. Stereochemical arguments concerning concertedness classically involve chiral centres but other types of stereochemistry can be employed. Retention or inversion of geometrical configuration as a result of substitution at vinylic centres, for example, are useful tools to discuss concertedness in these reactions (Scheme 2.2).

The stereochemical outcome of a reaction can be employed in favourable cases to measure the lifetime of a putative intermediate. This technique is covered in Section 2.2.2 as a "clocking" device.

Racemisation could occur prior to passage through a chiral reaction path, and this phenomenon could be responsible for a number of as yet undiscovered literature errors. The methoxide ion causes the racemisation of chiral methyl phosphinate esters[5] and reaction of nucleophiles with such species could be accompanied by racemisation caused by the leaving group *released* as the reaction progresses (Scheme 2.3).

Attack by the reagent on the same side of the carbon centre from which the leaving group departs (*front-side attack*) yields retention of configuration in displacement reactions. There are many cases where retention is observed but these may be explained, at least in the case of substitution at carbon, by mechanisms other than concerted front-side attack (Chapter 5). The best documented cases of front-side concerted attack involve substitution at heteroatoms and these are documented in Chapter 6.

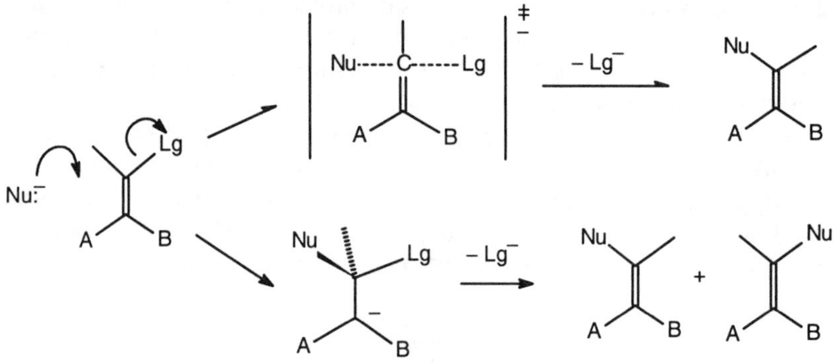

Scheme 2.2. Retention, inversion or racemisation in substitution at a vinyl centre.

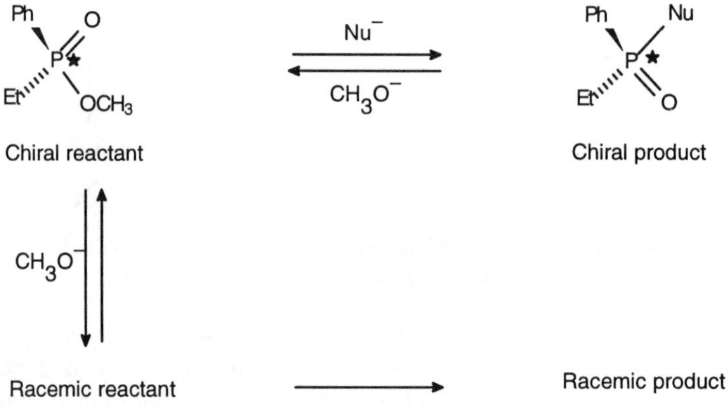

Scheme 2.3. Concerted displacements can yield a racemic product.

2.1.3 Positional isotope exchange

Intermediates can be detected by observation of positional isotope exchange (PIX) in reactions which can be associative or dissociative (Scheme 2.4).

Incorporation of the isotope (B_2) into the original reactant at the position occupied by B_1 is good evidence for an intermediate.

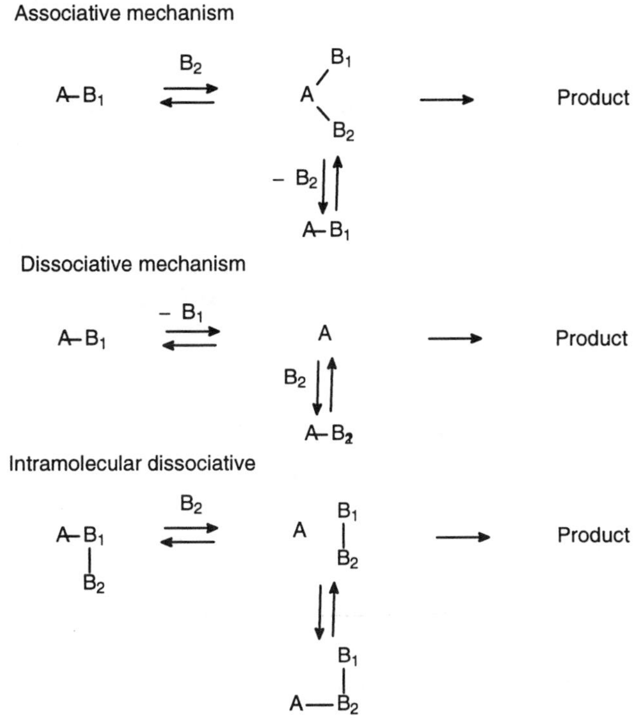

Associative mechanism

A–B_1 $\rightleftharpoons$ $\overset{B_1}{\underset{B_2}{A}}$ $\longrightarrow$ Product

$- B_2 \updownarrow$

A–B_1

Dissociative mechanism

A–B_1 $\overset{- B_1}{\rightleftharpoons}$ A $\longrightarrow$ Product

$B_2 \updownarrow$

A–B_2

Intramolecular dissociative

$\overset{A–B_1}{\underset{B_2}{|}}$ $\overset{B_2}{\underset{\rightleftharpoons}{}}$ $\overset{B_1}{\underset{A \, | \, B_2}{}}$ $\longrightarrow$ Product

$\updownarrow$

$\overset{B_1}{\underset{A—B_2}{|}}$

Scheme 2.4. Positional isotope exchange applied to general associative and dissociative reactions; B_1 and B_2 refer to isotopically substituted groups or atoms.

2.1.4 Trapping

The trapping technique is standard for transient intermediates, and it is exemplified in its general form in Scheme 2.5.

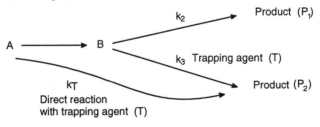

Scheme 2.5. The general trapping experiment.

The observation of trapped product (P_2) could be explained by direct reaction of the trapping agent (T) with reactant (A), and if that were the case then the overall rate constant for the reaction would depend on the concentration of (T). The demonstration of an intermediate by trapping requires that the overall rate constant is independent of concentration of trapping agent, whereas the product ratio $[P_2]/[P_1]$ would be linearly dependent on [T].

2.1.5 Cross-over

Rearrangement reactions (Scheme 2.6) can be considered to involve dissociative stepwise mechanisms. *Cross-over* experiments can indicate if this is the case by carrying out the reaction for a mixture of two reactants (A_1B_1) and (A_2B_2). If the residues (A) and (B) become free then the product would be a mixture B_1A_1, B_2A_1, B_1A_2, and B_2A_2. It is necessary to ensure that the rearrangement rate constants are similar otherwise cross-over will not be observed simply because one reaction is faster than the other. Isotopic substitution or replacement, for example, of methyl by ethyl, are commonly used in these experiments.

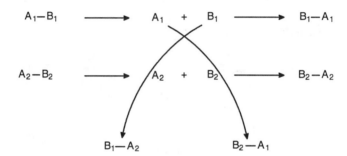

Scheme 2.6. Cross-over criterion for intermediates in a rearrangement reaction.

2.1.6 Non-linear free energy correlations

The build up of measurable concentrations of intermediate does not occur in many cases and the relative concentration is *very* small compared with that of reactant and product. Under these conditions the Bodenstein steady state hypothesis applies and Equation (2.1), the mechanism involving an intermediate, can be solved to give Equation (2.4).[6]

$$\text{Rate} = [A] \, k_1 \, k_2/(k_{-1} + k_2) = k_{\text{overall}} \, [A] \qquad (2.4)$$

Equation (2.4) can be employed to demonstrate the presence of an intermediate if there is some factor such as substituent, concentration or other condition which can vary k_{-1} and k_2 independently allowing the ratio k_{-1}/k_2 to vary from less than to greater than unity. Such behaviour is not predictable but if a change in rate-limiting step occurs there will be a break in a rate-parameter plot (at $k_{-1}/k_2 = 1$) which can be used to demonstrate the presence of an intermediate even though its concentration could be below that for normal instrumental observation. A classical example of this method is the formation of aromatic semicarbazones[7] (Figure 2.2) which was interpreted to be evidence for a tetrahedral intermediate.[8] The individual steps in Equation (2.1) should have linear Hammett dependencies where ρ_n and C_n are the parameters of the equation for k_n. The overall Hammett dependence is given

$$k_{overall} = (10^{(\rho_1 \sigma + c_1)})/(10^{((\rho_{-1} - \rho_2)\sigma + c_{-1} - c_2)} + 1) \qquad (2.5)$$

by Equation (2.5); this fits the data (Figure 2.2) and the value of $\Delta\rho$ is the difference in slope of the two linear portions of the correlation $(\rho_{-1} - \rho_2)$. A similar equation may be written for the Brønsted correlation or any other free energy relationship including those for concentration changes which might affect k_2 but not k_{-1}.

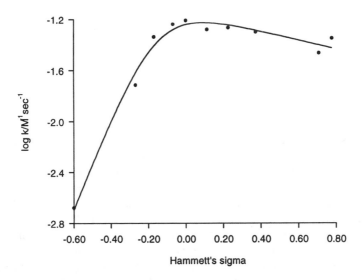

Figure 2.2. Demonstration of a stepwise mechanism for aromatic semicarbazone formation; the theoretical line is obtained from Equation (2.5); data are from *Ref. 7*.

2.2   EXCLUSION OF STEPWISE MECHANISMS

2.2.1 Free energy correlations

Usually there is no way of predicting the value of the σ or pK_a parameters for the condition that $k_{-1} = k_2$ in the stepwise process (Equation 2.1). Whereas the observation of a break in a free energy correlation indicates a stepwise process the converse, the observation of a linear correlation, does *not* exclude an intermediate. A linear plot would result when either the break-point condition does not occur within the range of substrates studied or else the polar substituent effects for the k_{-1} and k_2 rate constants are similar so that the difference in slopes ($\Delta\rho$ or $\Delta\beta$) will be small and a break-point would not be larger than the error in measuring the slope.

Technique of quasi-symmetrical reactions. When B in Equation (2.1) is an intermediate with entering and leaving groups of similar structure, such as a tetrahedral adduct of substituted phenoxide ion and phenyl ester, it is called a *quasi-symmetrical intermediate.* Decomposition of a quasi-symmetrical intermediate to reactants (k_{-1}) and to products (k_2) obey similar linear free energy relationships. A change in rate-limiting step will occur when the intermediate is symmetrical (which occurs when $k_{-1}=k_2$). In the addition-elimination mechanism of Scheme 2.7 the ratio k_{-1}/k_2 becomes unity when entering and leaving groups are identical.

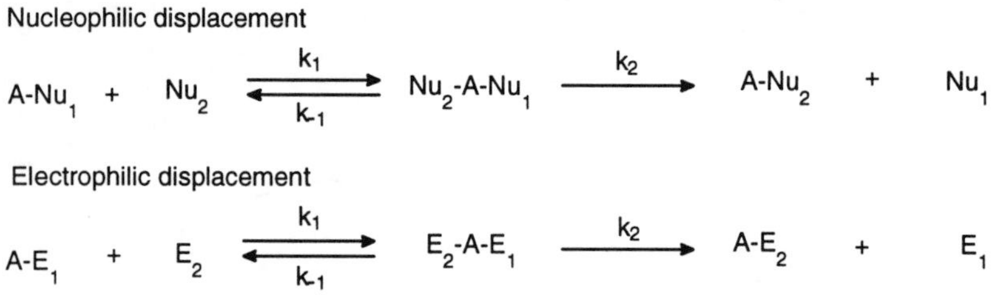

Nucleophilic displacement

$$A\text{-}Nu_1 \; + \; Nu_2 \; \underset{k_{-1}}{\overset{k_1}{\rightleftharpoons}} \; Nu_2\text{-}A\text{-}Nu_1 \; \xrightarrow{k_2} \; A\text{-}Nu_2 \; + \; Nu_1$$

Electrophilic displacement

$$A\text{-}E_1 \; + \; E_2 \; \underset{k_{-1}}{\overset{k_1}{\rightleftharpoons}} \; E_2\text{-}A\text{-}E_1 \; \xrightarrow{k_2} \; A\text{-}E_2 \; + \; E_1$$

Scheme 2.7. Addition-elimination reactions with quasi-symmetrical intermediates.

Thus in a reaction of a series of nucleophiles (Nu_2) with $A\text{-}Nu_1$ a plot of log $k_{overall}$ versus a polarity parameter of the nucleophile (for example the pK_a or σ) will yield a break at the value of the parameter for the nucleofuge Nu_1. A similar result would occur for electrophilic displacement reactions. A free energy correlation as shown in Figure 2.3 illustrates the expected results for an intermediate in a nucleophilic displacement reaction. The technique so far has been applied only to nucleophilic substitution reactions. The absence of a break-point at the required value of the polarity parameter can in principle exclude the intervention of an intermediate and indicate that the mechanism involves only a single transition state. The equation governing the overall rate constant is derived from Equation (2.4)[6] as follows. The Brønsted equations are written generally as Equation (2.6) where k°_n = the value of k_n at the location of the break-point ($\Delta pK_a = 0$).

$$\log k_n/k^{\circ}_n = \beta_n(pK_a - pK_o) = \beta_n\Delta pK_a \qquad (2.6)$$

Thus $\log k_1/k^{\circ}_1 = \beta_1\Delta pK_a$, $\log k_2/k^{\circ}_2 = \beta_2\Delta pK_a$ and $\log k_{-1}/k^{\circ}_{-1} = \beta_{-1}\Delta pK_a$ and Equations (2.7) and (2.8) for $k_{overall}$ follow by substituting into Equation (2.4).[9]

$$k_{overall} = k_1k_2/(k_{-1}+k_2)$$

$$= \{10^{\beta_1\Delta pKa} k^{\circ}_1 10^{\beta_2\Delta pKa} k^{\circ}_2\}/\{10^{\beta_{-1}\Delta pKa} k^{\circ}_{-1} + 10^{\beta_2\Delta pKa} k^{\circ}_2\} \qquad (2.7)$$

$$= 10^{\beta_1\Delta pKa} k^{\circ}_1/\{(k^{\circ}_{-1}/k^{\circ}_2)10^{(\beta_{-1} - \beta_2)\Delta pKa} + 1\} \qquad (2.8)$$

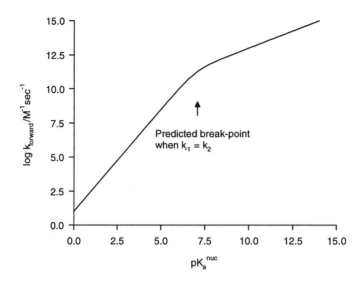

Figure 2.3. Brønsted-type dependence for a nucleophilic displacement reaction involving a quasi-symmetrical intermediate (Scheme 2.7). Predicted break-point is at the pK_a of the leaving group for the symmetrical reaction.

Since $k_{-1} = k_2$ at $\Delta pK_a = 0$ $k^{\circ}_{-1} = k^{\circ}_2$ and gives Equations (2.9 and 2.10)

$$k_{forward.} = 10^{\beta_1 \Delta pKa} k^{\circ}_1 / \{10^{-\Delta\beta\Delta pKa} + 1\} \qquad (2.9)$$

$$k_{reverse} = 10^{\beta_{-2} \Delta pKa} k^{\circ}_{-1} / \{10^{-\Delta\beta\Delta pKa} + 1\} \qquad (2.10)$$

where $\Delta\beta = \beta_2 - \beta_{-1}$. A similar equation can be written for Hammett dependencies with ρ replacing β and σ replacing pK_a. If the investigator is fortunate to possess a system where the reverse reaction can also be followed Equation (2.10) can be derived and fit to the data. Only one such system is at present known (Section 4.3).

The technique of *quasi-symmetrical reactions* relies on a relatively smooth variation of rate with parameter. In practice due attention must be paid to uncertainties in data fitting which arise from microscopic medium effects[10] rather than from experimental error; in general a large number of data points must be obtained to cover a substantial range above and below the value of the parameter for the break-point (ΔpK_a or $\Delta\sigma = 0$).

The technique can be applied to an elimination-addition mechanism (Scheme 2.8).[11-13] The technique distinguishes between the concerted mechanism with the open or exploded transition state and the stepwise process ("B" in Scheme 2.8); the break-point in the free energy correlation occurs at the value of the parameter giving a symmetrical encounter intermediate.

Criteria for observing the break. The sharpness of the break-point in Figure 2.3 depends on the value $\Delta\rho$ (or $\Delta\beta$) which depends on the relative sensitivity of k_{-1} and k_2 to change in the attacking group or leaving group. Changing the leaving group will have a larger effect on k_2 than on k_{-1}, whereas change of attacking group has a larger effect on k_{-1} than on k_2. Thus $\Delta\rho$ (or $\Delta\beta$) is expected to be large in a substitution reaction.

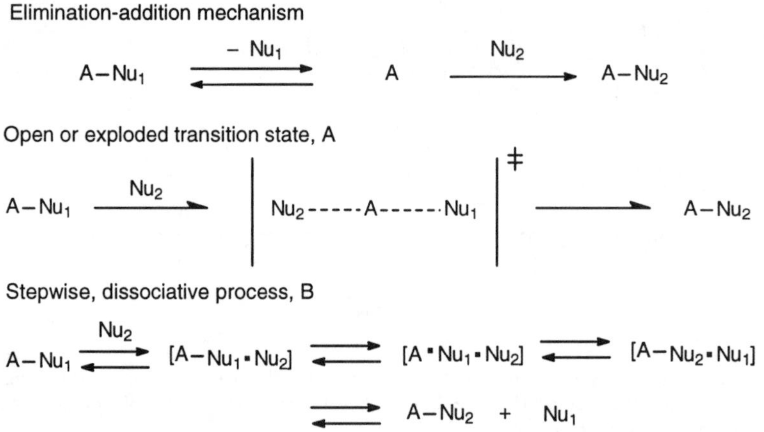

Elimination-addition mechanism

$$A-Nu_1 \; \underset{\longleftarrow}{\overset{-\;Nu_1}{\longrightarrow}} \; A \; \xrightarrow{\;Nu_2\;} \; A-Nu_2$$

Open or exploded transition state, A

$$A-Nu_1 \; \xrightarrow{\;Nu_2\;} \; \left| \; Nu_2 \text{-----} A \text{------} Nu_1 \; \right|^{\ddagger} \; \longrightarrow \; A-Nu_2$$

Stepwise, dissociative process, B

$$A-Nu_1 \; \underset{\longleftarrow}{\overset{Nu_2}{\longrightarrow}} \; [A-Nu_1 \cdot Nu_2] \; \underset{\longleftarrow}{\longrightarrow} \; [A \cdot Nu_1 \cdot Nu_2] \; \underset{\longleftarrow}{\longrightarrow} \; [A-Nu_2 \cdot Nu_1]$$

$$\underset{\longleftarrow}{\longrightarrow} \; A-Nu_2 \; + \; Nu_1$$

Scheme 2.8. Stepwise dissociative and concerted mechanisms for nucleophilic displacement.

The break in a free energy relationship corresponds to a difference in slopes which is measured by $\Delta\rho$ or $\Delta\beta$ and becomes zero in a linear relationship. The distinction between *linear* and *break* depends on the confidence level of the $\Delta\rho$ or $\Delta\beta$ parameter. If the fit has a value $\Delta\rho$ or $\Delta\beta$ no greater than the confidence limit the putative intermediate may not have a significant barrier to decomposition. The value of the substituent effect is related to the change in effective charge and the difference in effective charges between the transition states corresponding to $\Delta\beta$ or $\Delta\rho$ gives an *upper limit* to the change in charge from the intermediate to either forward or return transition states. This value can be compared with the *overall* change in charge for the reaction. A difference of 0.05 was found to be the upper limit of $\Delta\beta$ in phenoxide attack on phenyl acetates; this may be compared with the maximal change of 1.7 units and is too small to accommodate significant barriers for a stable intermediate.[11,14]

It is sometimes possible to estimate a value of $\Delta\beta$ or $\Delta\rho$; in the case of nucleophilic attack at substituted benzoyl group this knowledge leads to the conclusion that even though it may exist no break will be observed because the uncertainty in $\Delta\rho$ or $\Delta\beta$ exceeds the likely absolute value. Scheme 2.9 illustrates the displacement of leaving group by nucleophile at a benzoyl centre. The values of ρ_{-1} and ρ_2 will be approximately the same for structurally similar ligands as the bond fission process of k_{-1} and of k_2 will be affected to essentially the same extent by the substituent X. At the value of σ where $k_{-1} = k_2$ the break-point will be characterised by a $\Delta\rho$ which is very small and is likely to be well within the error limits of its measurement.[15] This situation is not the case in a tetrahedral adduct where the partitioning to reactants and products is not symmetrical (as in the classic case illustrated in Figure 2.2); in this case, the value of $\rho_{-1} - \rho_2$ is substantially positive due to an acid catalysed forward step which has a negative ρ_2 value and a well-defined break is observed. In both unsymmetrical and symmetrical cases the position of the break-point on the σ scale is not predictable (unlike the case when substituent variation is on nucleophile or leaving group).

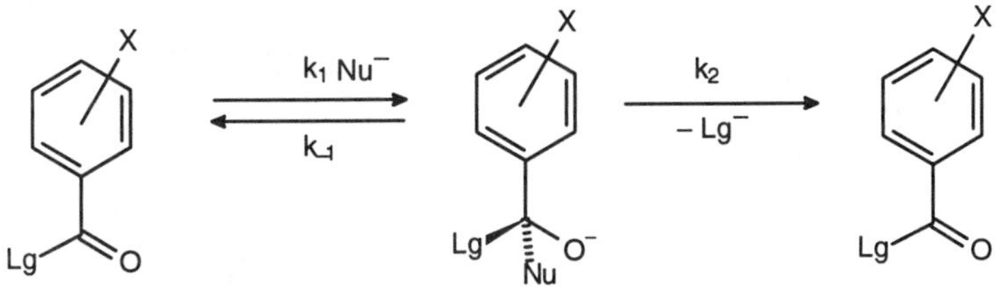

Scheme 2.9. The displacement of a leaving group at a benzoyl centre.

In the case of the reaction of substituted phenoxide ions with 4-nitrophenyl esters values of β_{-1} and β_2 can be calculated from work by Gravitz and Jencks (1974)[16] and of Bernasconi and Leonarduzzi (1982).[17]

2.2.2 Estimation of rate constants

A rate constant in excess of $10^{13}\,sec^{-1}$ (a typical vibration frequency) for decomposition of an intermediate precludes its existence as a discrete molecule although a structure with the same architecture may "exist" on a reaction coordinate.[18] A mechanism involving such a structure on its reaction coordinate would be an *enforced concerted* process. If the rate constant can be predicted from linear free energy relationships to be greater than that for a vibration then the mechanism is required to be concerted.[19] The decomposition of a carbanion species by leaving group departure to form an alkene (Scheme 2.10) is illustrated in Figure 2.4.

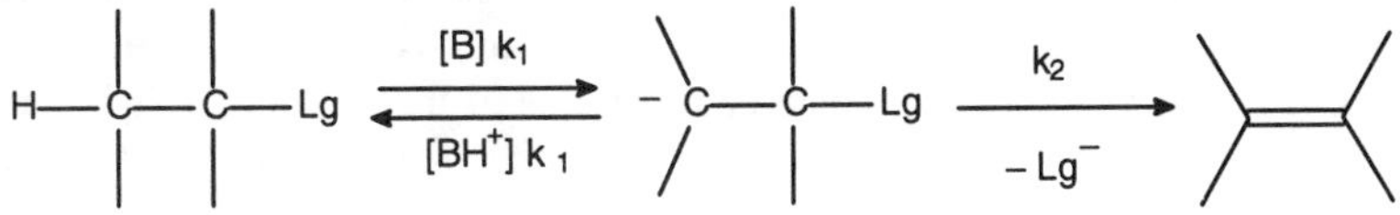

Scheme 2.10. Elimination via a carbanionic intermediate.

The linear free energy relationship of the rate constant against the substituent parameter meets the $10^{13}\ sec^{-1}$ level in the region where the stepwise release of the leaving group is so fast that the carbanion does not have an existence; as soon as it is formed the carbanion collapses to alkene and the mechanism is thus an enforced concerted process. This approach to estimating the pK_a^{lg} at which the mechanism becomes enforced concerted is based on the implicit assumption that the intrinsic barrier for k_2 is zero. The barrier is likely to be small because of the observed high reactivity but if it were not a lower pK_a^{lg} would be needed to drive k_2 to $10^{13} sec^{-1}$.

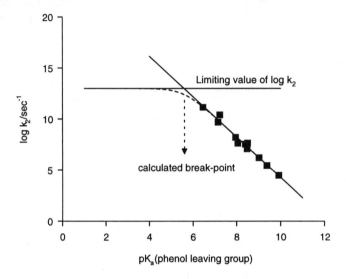

Figure 2.4. A carbanion mechanism changes to enforced concerted for very good leaving groups when the rate constant exceeds $10^{13} sec^{-1}$; data is for the unimolecular decomposition of PhCH⁻ SO₂-OAr to yield a sulphene.[19] The arrow indicates the break-point for zero intrinsic barrier and the dashed curve is from the equation $\log k_2 = 13 - \log(1 + 10^{pKa(lg)}/10^{5.5})$.

"Clocking" devices. Another way of estimating the lifetime of a putative intermediate is to use a "clock" reaction.[19-23] A rotational process involves rate constants close to 10^{12} sec⁻¹ ($t_{0.5}$ ca 10^{-12}sec.)[23] and is thus of use in studying lifetimes in the 10^{-13} sec range for putative intermediates. Scheme 2.11 illustrates the technique with an intermediate [I₁] which can change its conformation to [I₂] with a rate constant k_R. If [I₁] and [I₂] give products P₁ and P₂ which have different structures (for example a difference in stereochemistry) then an estimate can be made of the lifetime of [I₁] from the ratio of [P₁]/[P₂]. In order to assess the rate constants it would be necessary to elucidate the equilibrium constant to be expected between I₁ and I₂ and also the *relative* values of the rate constants for their decomposition. In most of the cases studied $k_1 = k_2$ and the value of k_R is estimated from those for similar conformational changes. The value k_1 is then given by the ratio $k_R P_1/P_2$; due regard needs to be paid to the Curtin-Hammett principle in these

matters.[6] In the case above the value of k_2 (= k_1) is greater than that of k_{-R} (= k_R). If the rate constant k_2 falls below k_{-R} the product ratio becomes $[P_1]/[P_2]$ = $(k_1/k_2)*(k_{-R}/k_R)$ = 1.

The rate of conformational change between two intermediates usually has a high rate constant, and if it can be estimated then the stereochemical outcome of a reaction can be employed to measure the lifetime of an intermediate if the two conformations of the intermediate give products with different configurations (Owens and Berson, 1990).[23]

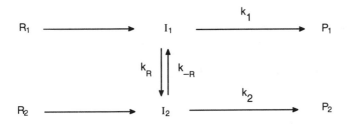

Scheme 2.11. Application of a conformational change (k_R) as a "clock" reaction.

Bauer (1969)[24] indicated that a species ejected from a transition state could be vibrationally activated; if this activation could be detected then a vibrationally activated product with a statistical energy distribution would indicate that it was ejected from a species which had become vibrationally relaxed. Such a species would be an intermediate, and Bauer suggested several systems (Scheme 2.12) where the energy of the excited product could be transferred to carbon dioxide which would then be readily detectable.

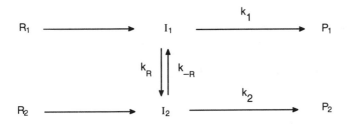

Scheme 2.12. Vibrationally activated extruded molecules as a criterion of concertedness; in the stepwise route the existence of the diradical allows time for the vibrational energy to partition into the cyclobutane fragment.

The technique is exemplified by the pyrolysis of 2,3-diazabicyclo-[2.1.1]hexene which would give a vibrationally activated nitrogen if it were expelled directly from the transition state. The intervention of a diradical intermediate which would survive long enough for the vibrational energy to partition would lead to vibrationally activated bicyclobutane and inactive nitrogen. The technique essentially employs the relaxation time of the partitioning of the vibrational energy as a "clock" against which to measure the lifetime of the intermediate.

2.2.3 Primary isotope effects

Isotope effects constitute a major tool in studies of concertedness.[25-28] A classical application to the electrophilic substitution of benzene illustrates how a concerted process can be excluded. The rate constant was determined by Melander and shown to be independent of deuterium incorporation (Scheme 2.13).[29-31] The absence of a primary deuterium isotope effect in the bimolecular reaction kinetics indicates that the addition step would be rate-limiting in the stepwise process. Even if the H-C bond fission were scarcely advanced in the transition state structure of the concerted mechanism there would be a substantial primary isotope effect because the vibrational zero point energy would be lost. Such a simple exclusion argument should be accompanied by a discussion of the possible retention of zero point energy in the transition state of the putative concerted process. Retention of zero point energy would change the magnitude of a primary kinetic isotope effect depending on the symmetry of the transition state.

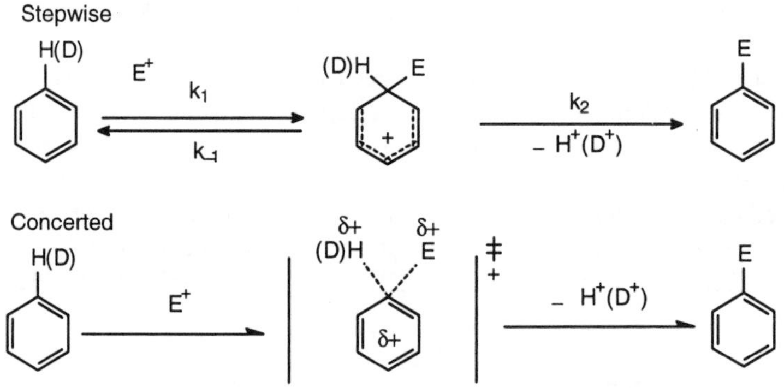

Scheme 2.13. Use of primary isotope effects to exclude a concerted process in aromatic substitution.

If it can be predicted that, in a two-step process, bond formation is rate-limiting then demonstration that bond fission is advanced in the transition state effectively excludes the stepwise mechanism. It is necessary to exclude the possibility that the change in bond order is not due simply to a change in hybridisation as a bond alters in going from reactant to intermediate. Isotope effects are ideal for such experiments; carbonyl acyl group transfer to phenolate ion from 4-nitrophenyl acetate has an ^{15}N-isotope effect of 1.001 ± 0.0002[32] compared with a value of $k^{14/15}$ of 1.002 ± 0.0001 for fission of the O-H bond of 4-nitrophenol.[33] This result is consistent with substantial bond fission in the transition state for a nucleophile where bond formation should be rate-limiting in the putative stepwise process. The ^{15}N-isotope effect of 1.0002 ± 0.0001 for attack of hydroxide ion on the 4-nitrophenyl ester is consistent with a stepwise mechanism. The above example is of considerable interest as the ^{15}N-isotope effect essentially "reports" on the extent of bonding at the *para*-oxygen. Moreover the natural abundance of ^{15}N in the ester serves as label obviating difficult labelling synthesis.[33]

2.2.4 Double isotope fractionation

The change in the overall rate constant caused by variation of the partitioning ratio, k_{-1}/k_2 (Equation (2.1)), can also be affected by changing the isotope affecting, say, the k_2 step.[34,35] This method has the advantage that the overall reaction is not being changed significantly and can therefore be applied to the study of concertedness in enzyme mechanisms which are sensitive to changes in substrate structure. The technique requires that an isotope effect on one bonding change is measured as a function of an isotope change in the other bond and for this reason has been named *the double isotope fractionation test*. Isotopic substitution can fractionally alter the partitioning ratio (k_{-1}/k_2) which is registered by the other isotope effect. The technique is described by consideration of the Claisen condensation which could have stepwise or concerted mechanisms.

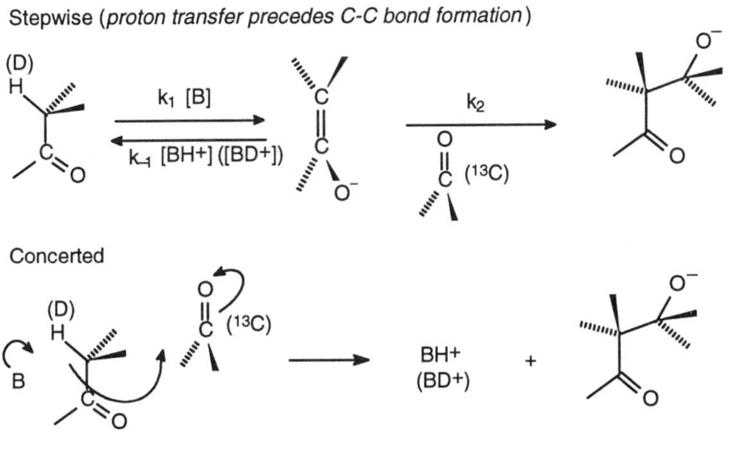

Scheme 2.14. The Claisen condensation.

The stepwise mechanism (Scheme 2.14) could involve (i) rate-limiting proton transfer; (ii) rate-limiting carbon-carbon bond formation or (iii) a mechanism where the partioning ratio (k_{-1}/k_2) is close to unity. The above conditions give rise to a selection of isotope effects respectively (Scheme 2.15): (i) A full $^Hk/^Dk$ effect with no variation on ^{13}C substitution; (ii) a full $^{12}k/^{13}k$ effect with no change on deuterium substitution; (iii) deuterium substitution reduces the ^{13}C isotope effect and (iv) a $^Hk/^Dk$ isotope effect essentially independent of isotopic substitution at the carbon and $^{12}k/^{13}k$ independent of hydrogen/deuterium substitution. The *truth table* (Table 2.1) describes the various diagnostic possibilities.

The double isotope fractionation test also enables the sequence of bonding changes to be determined in stepwise processes and this provides a useful tool in studies of enzyme mechanisms.[34]

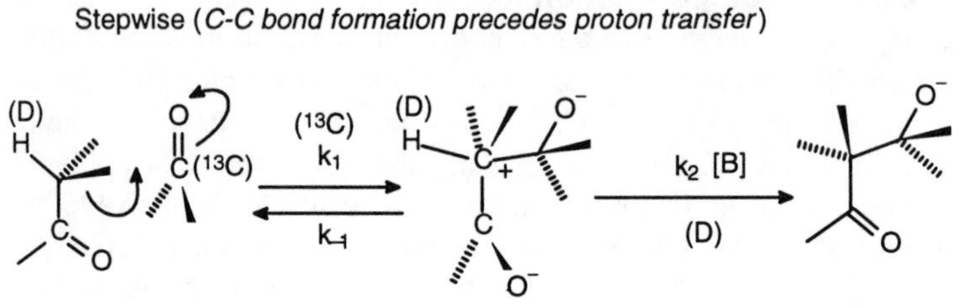

Stepwise (*C-C bond formation precedes proton transfer*)

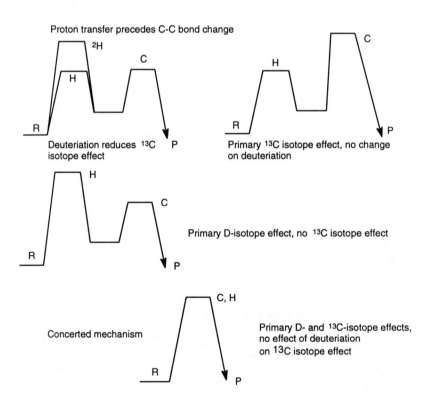

Scheme 2.14 *(continued).* The Claisen condensation (note hypothetical five-valent carbon).

Proton transfer precedes C-C bond change

Deuteriation reduces ^{13}C isotope effect

Primary ^{13}C isotope effect, no change on deuteriation

Primary D-isotope effect, no ^{13}C isotope effect

Concerted mechanism

Primary D- and ^{13}C-isotope effects, no effect of deuteriation on ^{13}C isotope effect

Scheme 2.15. Free energy diagrams for the Claisen condensation and the expected isotope effects; stepwise (*proton transfer precedes C-C formation*) and concerted mechanisms. R = reactant and P = product.

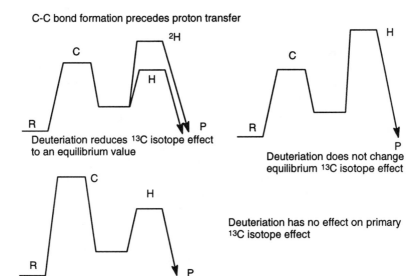

Scheme 2.15 (*continued*). Free energy diagrams for the Claisen condensation and the expected isotope effects; stepwise mechanism (*C-C formation precedes proton transfer*). R = reactant and P = product.

Table 2.1.
Double isotope fractionation effects predicted for various models of the Claisen condensation.[a]

Mechanism		Consequences	
Proton transfer precedes C-C bond making			
Proton transfer rate-limiting	1°	1.00	1.000
C-C bond making rate-limiting	Eq	1.00	1°
Both partially rate-limiting	reduced 1°	<1.00	reduced 1°
C-C bond making precedes proton transfer			
C-C bond making rate-limiting	1.00	1.00	1°
Proton transfer rate-limiting	1°	1.00	Eq
Both partially rate-limiting	reduced 1°	reduces to equilibrium value	1° to equilibrium depending on H-isotope
Concerted	1°	1.00	1°

a) Primary isotope effect ≡ 1°; equilibrium isotope effect ≡ Eq.

2.2.5 Energetics
The major part of the reaction flux would not be carried by a putative intermediate which has a free energy higher than that of the transition state. The energy criterion could be employed to demonstrate concertedness if the

putative intermediates for all possible stepwise processes were too energetic (Scheme 2.16). It is necessary to employ *Gibbs' free energies* in these discussions, and early work by Doering and others[36,37] using the energetic criterion was only capable of employing enthalpies determined from calorimetric studies. Enthalpies may be determined relatively accurately from Benson's group equivalent method[38-40] using tables of standard heats of formation. The estimation of the entropic component was not generally satisfactory but modern work has now rectified that problem and the technique has been extensively developed by Guthrie.[41] Gibbs' free energies of formation of reactive intermediates can be determined easily for aqueous solution and we exemplify the use of the criterion with an example taken from acyl group transfer (Chapter 4) where the tetrahedral adduct and acylium ion are the reactive intermediates. Guthrie studied the thermochemistry of the reaction between 4-nitrophenolate ion with 4-nitrophenyl acetate (Scheme 2.17).[42] The free energies of the two putative intermediates (acylium ion and tetrahedral adduct) can be estimated with reasonable accuracy by techniques pioneered by him.[41] Comparison of the energies with the measured free energy of activation indicate that a stepwise path through either intermediate could not support the observed reaction flux. A concerted mechanism is the only possible alternative.

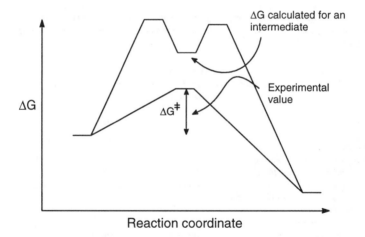

Scheme 2.16. The energy criterion: calculation of the energies of intermediates.

If it can be assumed that the More O'Ferrall-Jencks diagram (Scheme 2.18) includes all the possible mechanistic paths then a surface could be constructed which gives further information about the transition state.[*] Several authors have indicated that the potential energy of the reaction coord-

---

[*]Early attempts to model energy surfaces employed quadratic equations based on Marcus theory but these do not predict the slopes of the surfaces at the edges of the More O'Ferrall-Jencks diagrams.[41-44] The values of $\Delta G^{\ddagger}_{calc}$ are close to those calculated from the quartic equation.

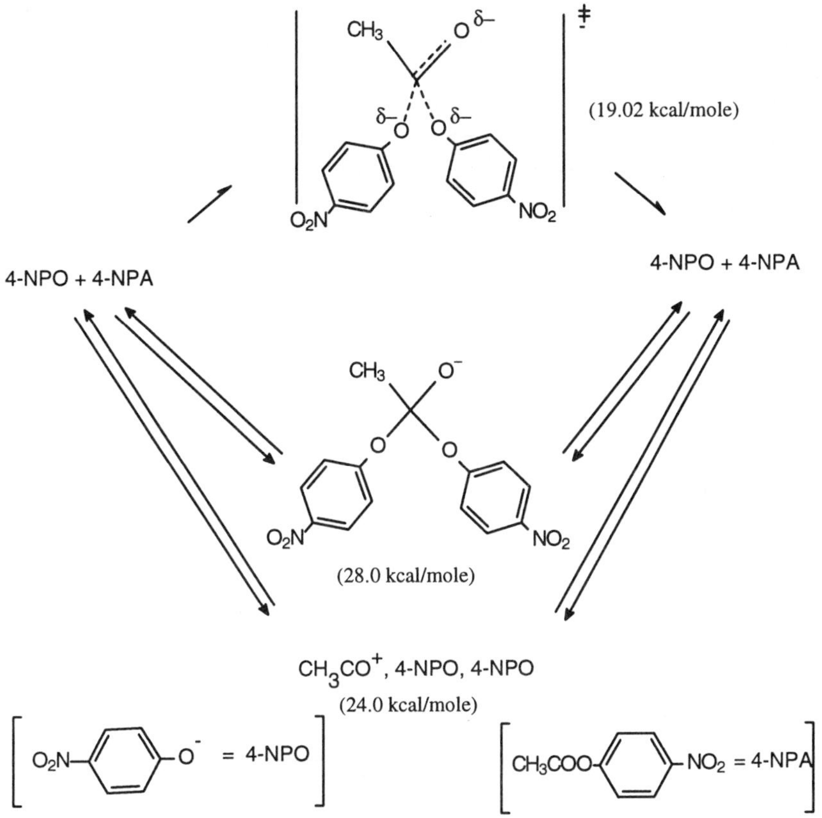

Scheme 2.17. Gibbs' free energies calculated for intermediates in putative stepwise mechanisms for the identity reaction of 4-nitrophenyl acetate with 4-nitrophenolate ion.

-inate of a reaction involving a single bonding change (x) could be modelled by a quartic equation (Equation (2.11)).[42-48]

$$\text{Potential energy (E)} = x^4/4 - (\alpha + 1.0)\, x^3/3 + \alpha.1.0\, x^2/2 + C \qquad (2.11)$$

Equation (2.11) gives curves illustrated in Figure 2.5 and has the correct values of the slope (dE/dx = 0) at reactant (x = 0), transition state (x = α) and product (x = 1).

A two parameter reaction, namely one involving major bond changes "x" and "y" will require an energy *surface* to describe the change in potential energy as a function of x and y and this is described by Equation (2.12) (x and y vary between 0 and 1). Guthrie[42,45] indicated that Equation (2.12) satisfies the various constraints required by such a system. It should be noted that Gibbs' free energy is under discussion and it is possible to utilise Equations (2.11) and (2.12) although the former was originally developed for potential energies. G is a minimum at the four corners of the More O'Ferrall-Jencks diagram and is a maximum at the transition structure. The values of a_1 to a_{10} can be determined from the Gibbs' free energies of the states at the four corners of the diagram and these equations are identified in Scheme 2.19 and refer to the encounter complexes, at energies higher than those of the free reactants by a work term. The surface can be calculated for the hypothetical identity displacement reaction (Figure 2.6) with the values $G_{00} = 0$ (reactant), $G_{11} = 0$ (product), $G_{01} = 25$ (Nu-A-Lg$^-$), $G_{10} = 20$ (Nu, Lg, A) and $G_x = G_y = 1.0$ kcal/mole at 298 K.

$$G = a_1x^2 + a_2y^2 + a_3x^3 + a_4y^3 + a_5x^4 + a_6y^4 + a_7x^2y^3 + a_8x^3y^2$$
$$+ a_9x^3y^3 + a_{10}x^2y^2 \qquad (2.12)$$

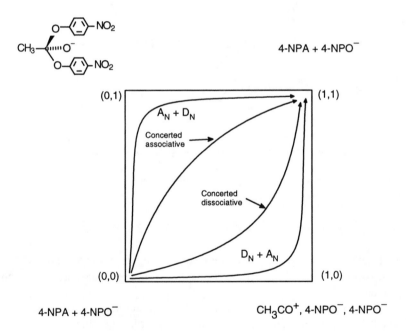

Scheme 2.18. More O'Ferrall-Jencks diagram delineating all possible mechanistic routes for the displacement reaction between 4-nitrophenolate ion(4-NPO) and 4-nitrophenyl acetate (4NPA).

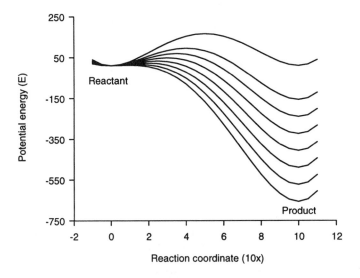

Figure 2.5. Family of curves (Eqn 2.11) of energy versus reaction coordinate, x, for a series of equilibrium constants; potential energy is in arbitrary units and x is the bond order.

$$a_1 = 16G_x + 3(G_{10} - G_{00}) \qquad\qquad a_2 = 16G_y + 3(G_{01} - G_{00})$$

$$a_3 = -32G_x - 2(G_{10} - G_{00}) \qquad\qquad a_4 = -32G_y - 2(G_{01} - G_{00})$$

$$a_7 = a_8 = -6(G_{11} - G_{10} - G_{01} + G_{00}) \qquad a_5 = 16G_x \qquad a_6 = 16G_y$$

$$a_9 = 4(G_{11} - G_{10} - G_{01} + G_{00}) \qquad a_{10} = 9(G_{11} - G_{10} - G_{01} + G_{00})$$

Scheme 2.19. Values of the terms for Equation (2.12) from the experimental energies. $G_{m,n}$ represents corner energies and G_x and G_y are the intrinsic barriers for the "edge" reactions x and y respectively.

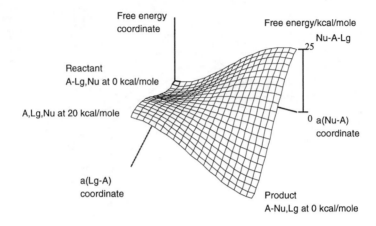

Figure 2.6. Free energy surface calculated for the hypothetical identity displacement reaction between A-Lg and Nu (A-Lg + Nu → A Nu + Lg); free energies of the corners are given.

2.3   TRANSITION STATE STRUCTURE

The subsidiary definition of concertedness requires that a transition state structure corresponds to a concerted mechanism if it involves substantial change in all bonds undergoing formation or fission. The observation of partial bond changes in fission and formation could be employed as evidence for concertedness for those reactions which do not involve enforced concerted mechanisms. If it could be predicted that the first step of the putative stepwise process (Equation (2.1)) were rate-limiting the observation of substantial bond fission by a substituent effect on the bond which is breaking would not be consistent with a stepwise process. Normally, free energy correlations for substituents affecting *both* formation and fission are needed in order to indicate the state of bonding in both bond changes.

2.3.1 Free energy correlations

The Hammond[6] postulate and the reactivity selectivity hypothesis attempt to correlate a change in structure of the transition state with reactivity. As the energy difference between reactant and product increases then the transition state structure should become closer to that of the highest energy state (see Figure 2.5). As the energy of the transition state structure moves towards that of product the reactivity will therefore decrease and selectivity to substituent change will become larger. Substantial controversy has followed these simple ideas due to the observation of a number of systems which, in the event, proved to be special cases. The introduction of extra bonding changes as major contributions to the energy will interfere with the energy if the bonding changes are coupled with each other and it is for this reason that the simple Hammond postulate often breaks down. Such coupling of bond changes, diagnostic of concertedness, is most often seen in cross-correlation effects in free energy relationships. Energy changes in a reaction not associated with bond formation involve effects due to changes in bond angle and bond length; although these have to be considered they are generally assumed to be of second order in comparison with those resulting from bond formation or fission. The transition state structure on one *edge* of the More O'Ferrall-Jencks diagram will comply with Hammond's postulate because the energies of the other two corners (not on this edge) will not substantially affect the position of the transition state. Stepwise mechanisms have transition states on the *edges* of such a diagram.

2.3.2 Coupling between bonding changes

Cross-correlation of substituent effects is probably the most direct experimental revelation of coupling that is known at present. The effect was discussed some time ago by Miller;[49] it was developed by Cordes and Jencks[50-52] and it has had its most recent extensive application by Lee.[53-56] If there is no interaction between two bonding changes, such as reactions occurring at the two ends of a hydrocarbon chain, then there can be little energy transmission and consequently no reason why the two processes should be in concert or indeed cause a rate enhancement by acting synergistically. The coupling between substituent effects can be expressed by a number of equations all based on the same principle. Equation (2.13) derived using Brønsted methodology relates the rate constant to the change in pK_a of both attacking and leaving group (x and y respectively). The cross interaction constant is given by Equation (2.15) which is derived from Equation (2.13).

$$\log k_{xy}/k_{00} = \beta^0_x pK_x + \beta^0_y pK_y + p_{xy}pK_x pK_y \qquad (2.13)$$

$$\beta_x = \partial \log k_{xy}/k_{00} / \partial pK_x = \beta^0_x + p_{xy}pK_y \qquad (2.14)$$

$$p_{xy} = \partial \beta_x / \partial pK_y \qquad (2.15)$$

A similar equation, $p_{yx} = \partial \beta_y / \partial pK_x$, may be derived and, since second derivatives commute, the cross correlation coefficients p_{xy} and p_{yx} are identical. As well as providing evidence for coupling between bonding changes the values of the cross-correlation coefficients provide information about the curvature of the energy surface at the saddle point.[52] A reaction involving two bond changes can be represented by the More O'Ferrall-Jencks diagram (Figure 2.7) where the surface is minimised for the second order energies due to bond angle and bond length changes not associated with bond fission or formation.[57] A stepwise mechanism will follow the edges of such a diagram whereas a concerted mechanism could involve either edges or a path taking a more direct route. The diagram is very similar to that of Ponec's *topological* definition of concertedness (Section 1.4, page 10).[58,59]

The structure of the transition state (in terms of the extent of bonding in the two major bonding changes) can be defined by the Brønsted or Hammett parameter along each edge corresponding to variation of "Lg⁻" and "Nu⁻". Variation of Nu⁻ will have a major effect on the transition state structure measured along the 0,0-0,1 or 1,0-1,1 edges and variation in Lg⁻ will likewise affect the position along the 0,1-1,1 or 0,0-1,0 edges. Effects on the relative energies at the corners of the diagram will move the transition state structure on an edge along the reaction coordinate towards the corner of increasing relative energy according to the Hammond postulate. If the transition state structure lies on the surface away from the edges then the variation in energies of the corners will also move it *perpendicular* to the reaction coordinate *away* from the corner(s) of increasing relative energy and *along* the reaction coordinate *towards* a corner of the diagram undergoing an increase in energy.[60]

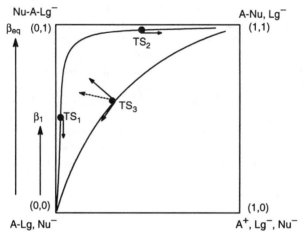

Figure 2.7. Effect of cross-coupling on free energy correlations.

The observation of a "Thornton"[44], "anti-Hammond" or perpendicular effect (movement away from an energy increase) can be useful in diagnosing concertedness as it should only occur when the transition state structure is not on an edge of the More O'Ferrall-Jencks diagram. For the sake of the following arguments we employ Brønsted free energy correlations and a nucleophilic displacement reaction (Figure 2.7) and consider the effects of changing the energies at the corners on the movement of these transition states.

Structure TS₁. The structure TS_1 which corresponds to the transition structure of the rate determining step will give rise to a Brønsted slope of β_1 and an overall β_{eq} value giving a Leffler α_{nuc} for variation of the nucleophile, Nu⁻. If the 0,0 corner becomes less stable than that of the 0,1 corner the transition state structure should move towards 0,0 in accordance with the Hammond postulate; α_{nuc} will decrease as indicated by the arrow at TS_1 in Figure 2.7. Changing the stabilities of 1,1 and 1,0 will not affect the position of the transition structures along the 0,0-0,1 edge *unless* the factors causing the changes *also* affect the stabilities of the 0,0 and 0,1 corners.

Structure TS_2. Similar arguments can be applied to changes in the Leffler α_{lg} caused by effects on the transition structure TS_2 along the top edge of the diagram. If 1,1 becomes less stable than 0,1 T_2 will shift to the right and increase α_{lg} (Figure 2.7).

Structure TS_3. If the transition state of the rate-limiting step is on a reaction coordinate at TS_3 changing the energies of the 1,1 and 0,1 corners could also affect the position of TS_3 along the 0,0-0,1 coordinate if the transition structure is being measured by α_{nuc}. In the case of TS_3 its movement is the resultant of movement along the reaction coordinate towards the more energetic corner (0,0) and perpendicular to the reaction coordinate towards the corner or edge with decreasing energy (0,1). If 1,1 and 0,1 were to decrease in energy TS_3 could move in an "anti-Hammond" direction as shown by the dotted arrow. Such an observation would be good evidence of cross-interaction between changes in Lg^- and Nu^- and a concerted mechanism. It does not follow that a concerted mechanism (with a transition state not at an edge) will necessarily exhibit "anti-Hammond" behaviour. These arguments may be quantified in terms of changes in β_1 as a function of pK_{lg} (p_{xy}).[52]

An enforced concerted process (see section 1.3, page 9) will follow the edges of the More O'Ferrall-Jencks diagram and should not exhibit strong coupling between the bonds undergoing covalency changes.

The element effect[61,62] is a classical mechanistic tool which usually involves comparing the reactivities of a reactant with a series of halogen leaving groups. The basis of the method is that the leaving group should have little effect on the rate constants for a stepwise mechanism with a rate-limiting addition step. A concerted mechanism or a stepwise mechanism with rate-limiting halogen bond fission should be highly sensitive to the identity of the halogen. The variation of halogen directly attached to a reaction centre is rather a gross structural change and it is likely to cause a large perturbation to the system; for this reason a large element effect is not particularly diagnostic and a small element effect must be looked upon with scepticism. The modern equivalent of the element effect, the effect of variation of leaving group substituents within a closely similar set of structures such as substituted phenolate ions or substituted pyridines, is a much better diagnostic tool as it does not substantially perturb the system.

2.3.3 Multiple isotope effects

Isotopic probes can be used to determine the strengths of each major bonding change for the transition state structure and could therefore provide evidence regarding the timing of the bonding changes in the overall mechanism.[63,64] Such studies require a fairly large investment of time and it is necessary to employ theoretical methods to predict isotope effects for given mechanistic hypotheses if multiple isotopes are to be exploited to their optimal extent. The technique is of manifest importance in the solution of mechanisms of cyclical reactions (Chapter 7) which often involve *more* than two major bonding changes.

Multiple isotope effects can be determined with good accuracy and techniques have been reviewed by Saunders[65,66] and Cook.[67] Recently Singleton has reported a rapid and accurate nmr technique, not requiring specially enriched isotopic compositions, for determining multiple isotope effects; this should prove to be very useful in future studies.[68]

The application of theory to the elucidation of concertedness is essentially the calculation of energies of structures within a "grid" of variables (bond length and bond angle) to determine the location of the transition structure or structures within that "grid".[69] Parameters, such as isotope effect or rate constant, can be calculated for the chosen transition state and compared with experimental values. Theory employed in such calculations is of *two types*. The first type involves molecular mechanics, an empirical technique based on known and estimated force constants for bonds; it includes the bond energy-bond order method (BEBO)[70] for estimating force constants for bonds which are only partially formed (as in major bonding changes in a transition state). The second type is quantum mechanics which, in practice, divides into *ab initio* and *semi empirical* molecular orbital theory. Quantum mechanical calculations have essentially the same function as those of molecular mechanics but from a less empirical basis. Whichever method is utilised for determining the energy "grid" it is *predictive* and cannot have the same reliability as experimental conclusions. While theory has no place on its own in *diagnosis* of mechanism it is very powerful when its predictions, for example of isotope effects or rate constants, are compared with experimentally determined values.

Molecular mechanics allows calculation of isotope effects over a range of bonding parameters and the matrix of results is compared with a set of experimental isotope effects.[71,72] The best fit gives the optimal transition state structure within the range of parameters searched. In order to be effective the isotopic substitutions are required to be in atoms participating in bond fission or formation or undergoing a hybridisation change as the reaction proceeds. The reliability of the result depends on the number of isotopic substitutions. The force constants, F_{ii}, may be calculated for partial bonds by making use of Badger's rule (Equation (2.16))[73] where r_i, a, and b are

$$-\ln F_{ii} = (r_i - a)/b \qquad (2.16)$$

respectively the length of the bond, and constants. The force constant is related to the bond order using Badger's equation and the Pauling relationship (Equation (2.17))[74] in Equation (2.18).

$$r_i = r_i^0 - c\ln n_i \qquad (2.17)$$

$$F_{ii} = F_{ii}^0 \, n_i^{c/b} \qquad (2.18)$$

The number of vibrational modes limits the size of molecule which can be treated *via* the calculations because matrices are at least 3n–6. Parts of the molecule more than three bonds removed from the position of the isotopic substitution are assume to have little effect on the calculations. Wolfsberg and Stern have demonstrated that fragments three atoms removed from the isotopically sensitive atoms have very little effect.[75,76] In practice the limit is two bonds and in the case of heavy atom isotope effects atoms more than one bond removed may be omitted from the calculations.

A matrix of calculated isotope effects against bond orders is prepared utilising the Bebovib IV program;[77,78] it is usual to register only two bond orders, that of fission and that of formation, because of the complexities involved if more components are used. The best fit can be determined by simple comparison of the experimental values with the matrix but the criterion (Dev = $[\Sigma_{o-n}[(\text{expt} - \text{calcd})/\text{expt}]^2/(n - 1)]^{0.5})^{72}$ provides a more objective protocol. The aliphatic displacement reaction (Scheme 2.20) provides a convenient example, and Table 2.2 records the matrix of hypothetical results for this reaction. There is a built-in assumption in the application of the bond energy-bond order (BEBO) technique that the bond orders for fission and formation sum to unity (Table 2.2); this equality is not always true but nevertheless the approach has been extremely useful (see Chapter 7) and will probably continue to be used until the molecular orbital technique (below) can be regarded as completely reliable for this type of calculation.

An alternative approach to calculating kinetic isotope effects employs *ab initio* or *semi empirical* molecular orbital theory to determine the vibrational frequencies. The protocol involves determination of the structure of the transition state which is then employed to calculate the frequencies. The force constants may be obtained from *ab initio* calculations and are used to calculate the kinetic isotope effects by use of the Bigeleisen-Mayer formalism (Wolfsberg)[75,76] in the "QUIVER" program employed by Saunders.[79] Paneth[80] reported an alternative program, "ISOEFF", for carrying out similar calculations.

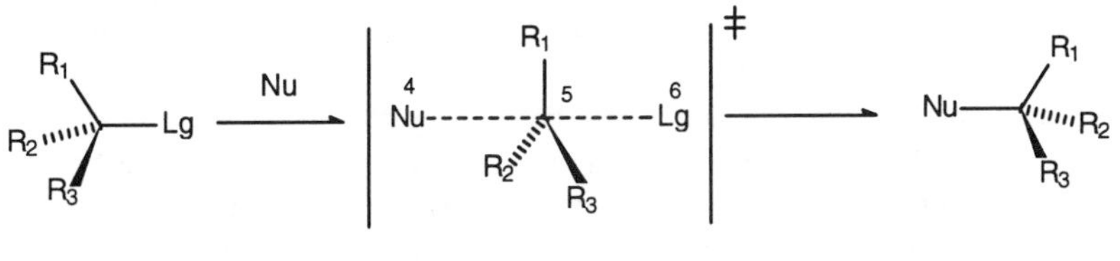

Scheme 2.20. Aliphatic nucleophilic displacement reaction for Table 2.2.

Physical organic chemists familiar with experiments yielding changes in parameters which are substantial are often dismayed with the apparently minute differences seen with heavy atom isotope effects. However, such small effects have even smaller experimental tolerances due to the method of analysis,[81] and studies have been made on systems where a change in rate-limiting step can easily be monitored by the heavy atom isotope effect.

Table 2.2.

Matrix of calculated values of the isotope effects (a_{ij}) for multiple isotopic substitution in a hypothetical aliphatic displacement reaction.[a]

n_{Nu-C}	n_{C-Lg}	R_1	R_2	R_3	Nu (4)	C (5)	Lg (6)	Dev
0.9	0.1	a_{19}	a_{29}	a_{39}	a_{49}	a_{59}	a_{69}	Dev_9
0.8	0.2	a_{18}	a_{28}	a_{38}	a_{48}	a_{58}	a_{68}	Dev_8
0.7	0.3	a_{17}	a_{27}	a_{37}	a_{47}	a_{57}	a_{67}	Dev_7
0.6	0.4	a_{16}	a_{26}	a_{36}	a_{46}	a_{56}	a_{66}	Dev_6
0.5	0.5	a_{15}	a_{25}	a_{35}	a_{45}	a_{55}	a_{65}	Dev_5
0.4	0.6	a_{14}	a_{24}	a_{34}	a_{44}	a_{54}	a_{64}	Dev_4
0.3	0.7	a_{13}	a_{23}	a_{33}	a_{43}	a_{53}	a_{63}	Dev_3
0.2	0.8	a_{12}	a_{22}	a_{32}	a_{42}	a_{52}	a_{62}	Dev_2
0.1	0.9	a_{11}	a_{21}	a_{31}	a_{41}	a_{51}	a_{61}	Dev_1
Expt.		a_{1exp}	a_{2exp}	a_{3exp}	a_{3exp}	a_{5exp}	a_{6exp}	

a) In a_{ij} the subscript "i" refers to the atom substituted (as in Scheme 2.20) and the subscript "j" refers to the bond order of the NuC bond.

2.3.4 Theory

Theory coupled with multiple isotope effects can be used to decide the structure of the transition state and the relative coupling between bond formation and bond fission.[82-84] Provided there is sufficient computer memory available enough coordinate space can be searched to provide reliable information about transition states in the reaction surface. The predictions provide answers to the question of concertedness but still carry the proviso *no different* from that for experimental approaches that models can only be disproved.

In the case of mechanistic studies the free energies of activation calculated for predicted transition states can be compared with experimental values. It is likely that theory will not be in a good position for some time yet to calculate free energies of activation of reactions in solvating solutions; reactions in the gas phase and in non-solvating solutions stand a better chance for success. The attack of the hydroxyl radical on the primary and secondary hydrogens of propane have been simulated and there is relatively good agreement between experimental (18.6 and 19.6 kcal/mole for primary and secondary abstraction respectively) and calculated energies of activation (18.4 and 20.8 kcal/mole respectively) for the AM1-SRP level of theory.[87] The transition state structures for both hydrogen abstractions are calculated to possess O-H and C-H bonds to the migrant hydrogen which are longer than that of a single bond and are in agreement with a concerted process.

Guthrie[88] is attempting, with some success, to calculate rate constants for *solution* reactions using the postulate that the energy for a bond change is a

---

*See *Refs. 85* and *86* for an explanation of symbols used in MO calculations.

quadratic function of bond order to define the reaction surface; the energies of the corners may be obtained from experimental equilibrium constants and distortion energies calculated by MO theory for pairs of species differing only in geometry. Reaction of cyanide ion with formaldehyde and acetaldehyde has $\log k_{obs}$ values of 5.54 and 2.81 respectively. Calculation of distortion energies using AM1 theory gives respectively values of $\log k_{calc}$ of 5.43 and 3.16.

FURTHER READING

Beak, P., Determination of transition state geometries by the endocyclic restriction test: mechanisms of substitution at nonstereogenic atoms, *Accs. Chem. Res., 25,* 215, 1992.

Bernasconi, C. F., ed., *Investigation of Rates and Mechanisms of Reactions, Part 1,* Wiley-Interscience, New York, 1986.

Berti, P. J., Determining transition states from kinetic isotope effects, *Methods in Enzymology, 308,* 355, 1999.

Carpenter, B. K., Intramolecular dynamics for the organic chemist, *Accs. Chem. Res., 25,* 520, 1992.

Connors, K. A., *Chemical Kinetics. The Study of Reaction Rates in Solution,* VCH Verlagsgesellschaft GmbH, Weinheim, 1990.

Cook, P. F. (ed.), *Enzyme Mechanisms from Isotope Effects,* CRC Press, Boca Raton, 1991.

Engel, P. S., He, S. L., Banks, J. T., Ingold, K. U., and Lusztyk, J., Clocking tertiary cyclopropyl carbinyl radical rearrangements, *J. Org. Chem., 62,* 1210, 1997.

Fry, A., Heavy-atom isotope effects in organic reaction mechanism studies, in *Isotope Effects in Chemical Reactions,* Collins, C. J., and Bowman, N. S., eds., New York, 1970, p. 364.

Fry, A., Isotope effect studies of elimination reactions, *Chem. Soc. Rev., 1,* 163, 1972.

Griller, D., and Ingold, K. U., Free-radical clocks, *Accs. Chem. Res., 13,* 317, 1980.

Grunwald, E., Structure-energy relations, reaction mechanism, and disparity of progress of concerted reaction events, *J. Amer. Chem. Soc., 107,* 125, 1985.

Guthrie, J. P., Concerted mechanism for alcoholysis of esters: an examination of the requirements, *J. Amer. Chem. Soc., 113,* 3941, 1991.

Guthrie, J. P., Prediction of rate constants for cyanohydrin formation using equilibrium constants and distortion energies, *J. Amer. Chem. Soc., 120,* 1688, 1998.

Guthrie, J. P., Correlation and prediction of rate constants for organic reactions (Syntex award lecture, 1995), *Can. J. Chem., 74,* 1283, 1996.

Guthrie, J. P., Multidimensional Marcus theory: and analysis of concerted reactions, *J. Amer. Chem. Soc., 118,* 12878, 1996.

Guthrie, J. P., Concertedness and E2 elimination reactions. Prediction of transition state position and reaction rates using two-dimensional reaction surfaces based on quadratic and quartic approximations, *Can. J. Chem., 68,* 1643, 1990.

Hengge, A. C., and Clelland, W. W., Direct measurement of transition state bond cleavage in hydrolysis of phosphate esters of *p*-nitrophenol, *J. Amer. Chem. Soc., 112,* 7421, 1990.

Huisgen, R., Kinetic evidence for reactive intermediates, *Angew. Chem., Int. Ed. Engl., 9,* 751, 1970.

Huskey, W. P., Origins and interpretations of heavy atom isotope effects, in *Enzyme Mechanisms from Isotope Effects,* Cook, P. F., ed., CRC Press, Boca Raton, 1991, p. 37.

Jencks, D. A., and Jencks, W. P., On the characterization of transition states by structure-reactivity coefficients, *J. Amer. Chem. Soc., 99,* 7948, 1977.

Jencks, W. P., A primer for the Bema Hapothle. An empirical approach to the characterization of changing transition state structures, *Chem. Revs., 85,* 511, 1985.

Keeffe, J. R., and Jencks, W. P., Large inverse solvent isotope effects: a simple test for the E1cb mechanism, *J. Amer. Chem. Soc., 103,* 2457, 1981.

Lee, I., Cross-interaction constants and transition state structure in solution, *Adv. Phys. Org. Chem., 27,* 57, 1992.

Lee, I., Mechanistic significance of the magnitude of cross-interaction constants, *J. Phys. Org. Chem.*, 5, 736, 1992.

Melander, L., and Saunders, W. H., *Reaction Rates of Isotopic Molecules*, John Wiley & Sons, New York, 1980.

Newcomb, M., Competition methods and scales for alkyl radical reaction kinetics, *Tetrahedron*, 49, 1151, 1993.

Saunders, W. H., Distinguishing between concerted and non-concerted eliminations, *Accs. Chem. Res.*, 8, 19, 1976.

Willi, A. V., Kinetic carbon and other isotope effects in cleavage and formation of bonds to carbon, in *Isotope Effects in Organic Chemistry*, Buncel, E., and Lee, C. C., eds., Elsevier, New York, 3, 237, 1977.

Williams, A., The diagnosis of concerted organic mechanisms, *Chem. Soc. Rev.*, 93, 1994.

REFERENCES

1. **Bernasconi, C. F., Killion, R. B., Fassberg, J., and Rappoport, Z.,** Kinetics of the reaction of β-methoxy-α -nitrostilbene with thiolate ions - 1st direct observation of the intermediate in a nucleophilic vinylic substitution, *J. Amer. Chem. Soc., 111*, 6862, 1989.

2. **Espenson, J. H.,** *Chemical Kinetics and Reaction Mechanisms*, McGraw-Hill Inc., New York, 1995, p. 75.

3. **Zhong, D., Ahmad, S., and Zewail, A. H.,** Femtosecond elimination reaction dynamics, *J. Amer. Chem. Soc., 119*, 5978, 1997.

4. **Bordwell, F. G.,** Are nucleophilic bimolecular concerted reactions involving four or more bonds a myth?, *Accs. Chem. Res., 3*, 281, 1970.

5. **Hudson, R. F., and Green, M.,** Stereochemistry of displacement reactions at.phosphorus atoms, *Angew. Chem., Int. Ed. Engl.,.2*, 11, 1963.

6. **Page, M. I., and Williams, A.,** *Organic and Bio-Organic Mechanisms*, Addison Wesley Longman, Harlow, Essex, 1997, Chapter 3.

7. **Noyce, D.S., Bottini, A.T., and Smith, S.G.,** Carbonyl Reactions III. The formation of aromatic semicarbazones. A Non-linear rho-sigma correlation, *J. Org. Chem., 23*, 752, 1958.

8. **Anderson, B. M., and Jencks, W. P.,** The effect of structure on reactivity in semicarbazone formation, *J. Amer. Chem. Soc., 82*, 1773, 1960.

9. **Williams, A.,** The diagnosis of concerted organic mechanisms, *Chem. Soc. Rev.*, 93, 1994

10. **Page, M. I., and Williams, A.,** *Organic and Bio-Organic Mechanisms*, Addison Wesley Longman, Harlow, Essex, 1997, p. 121.

11. **Skoog, M. T., and Jencks, W. P.,** Reactions of pyridines and primary amines with N-phosphorylated pyridines, *J. Amer. Chem. Soc., 106*, 7597, 1984.

12. **Bourne, N., and Williams, A.,** Evidence for a single transition state in the transfer of the phosphoryl group ($-PO_2^{-3}$) to nitrogen nucleophiles from pyridine-N-phosphonates, *J. Amer. Chem. Soc., 106*, 7591, 1984.

13. **Bourne, N., Hopkins, A. R., and Williams, A.,** A preassociation concerted mechanism in the transfer of the sulphate group between isoquinoline-N-sulphonate and pyridines, *J. Amer. Chem. Soc., 105*, 3358, 1983.

14. **Dietze, P. E., and Jencks, W. P.,** Swain-Scott correlations for reactions of nucleophilic reagents and solvents with secondary substrates, *J. Amer. Chem. Soc., 108*, 4549, 1986.

15. **Colthurst, M. J., and Williams, A.,** Nucleophilic displacement at the benzoyl centre: a case study of the change in geometry at the carbonyl carbon atom, *J. Chem. Soc., Perkin Trans. 2*, 1493, 1997.

16. **Gravitz, N., and Jencks, W. P.,** Mechanism of general acid-base catalysis of the breakdown and formation of tetrahedral addition compounds for alcohols and a phthalimidium cation. dependence of Brønsted slopes on alcohol acidity, *J. Amer. Chem. Soc., 96*, 507, 1974.

17. **Bernasconi, C. F., and Leonarduzzi, G. D.,** Nucleophilic addition to olefines. 6. Structure-reactivity relationships in the reactions of substituted benzylidene Meldrum's acids with water, hydroxide ion and aryl oxide ions, *J. Amer. Chem. Soc., 104*, 5133, 1982.

18. **Jencks, W. P.,** When is an intermediate not an intermediate? Enforced mechanisms of general acid-base catalysed, carbocation, carbanion, and ligand exchange reactions, *Accs. Chem. Res., 13,* 161, 1980.

19. **Davy, M. B., Douglas, K. T., Loran, J. S., Steltner, A., and Williams, A.,** Elimination-addition mechanisms of acyl group transfer: hydrolysis and aminolysis of arylmethanesulfonates, *J. Amer. Chem. Soc., 99,* 1196, 1977.

20. **Richard, J. P., and Jencks, W. P.,** A simple relationship between carbocation lifetime and reactivity-selectivity relationships for solvolysis of ring-substituted 1-phenylethyl derivatives, *J. Amer. Chem. Soc., 104,* 4689, 1982.

21. **Richard, J. P., and Jencks, W. P.,** Concerted displacement reactions of 1-phenylethyl chlorides, *J. Amer. Chem. Soc., 104,* 4691, 1982.

22. **Richard, J. P., and Jencks, W. P.,** Concerted S_N2 displacement reactions of 1-phenylethyl derivatives, *J. Amer. Chem. Soc., 106,* 1383, 1984.

23. **Owens, K. A., and Berson, J. A.,** Stereochemistry of the thermal acetylenic Cope rearrangement. Experimental test for a 1,4-cyclohexenediyl as a mechanistic intermediate, *J. Amer. Chem. Soc., 112,* 5973, 1990.

24. **Bauer, S. H.,** Operational cxriteria for concerted bond breaking in gas-phase molecular elimination reactions, *J. Amer. Chem. Soc., 91,* 3688, 1969.

25. **Saunders, W. H.,** Distinguishing between concerted and non-concerted eliminations, *Accs. Chem. Res., 8,* 19, 1976.

26. **Willi, A. V.,** Kinetic carbon and other isotope effects in cleavage and formation of bonds to carbon, in *Isotope Effects in Organic Chemistry,* Buncel, E., and Lee, C. C., eds., Elsevier, New York, *3,* 237, 1977.

27. **Fry, A.,** Isotope effect studies of elimination reactions, *Chem. Soc. Rev., 1,* 163, 1972.

28. **Fry, A.,** Heavy-atom isotope effects in organic reaction mechanism studies, In *Isotope Effects in Chemical Reactions,* Collins, C. J., and Bowman, N. S., eds., Van Nostrand Reinhold, New York, 1970, p. 364.

29. **Melander, L.,** Mechanism of nitration of the aromatic nucleus, *Nature, 163,* 599, 1949.

30. **Bonner, T. G., Bowyer, F., and Williams, G.,** Nitration in sulphuric acid. Part IX. The rates of nitration of nitrobenzene and pentadeuterionitrobenzene, *J. Chem. Soc.,* 2650, 1953.

31. **Melander, L.,** The mechanism of electrophilic aromatic substitution. An investigation by means of the effect of isotopic mass on reaction velocity, *Arkiv Kemi, 2,* 213, 1950.

32. **Hengge, A. C., and Hess, R. A.,** Concerted or stepwise mechanisms for acyl transfer reactions of *p*-nitrophenyl acetate? Transition state structures from isotope effects, *J. Amer. Chem. Soc., 116,* 11256, 1994.

33. **Hengge, A. C., and Clelland, W. W.,** Direct measurement of transition state bond cleavage in hydrolysis of phosphate esters of *p*-nitrophenol, *J. Amer. Chem. Soc., 112,* 7421, 1990.

34. **Hermes, J. D., Roeske, C. A., O'Leary, M. H., and Cleland, W. W.,** Use of multiple isotope effects to determine enzyme mechanisms and intrinsic isotope effects. Malic enzyme and glucose-6-phosphate dehydrogenase, *Biochemistry, 21,* 5106, 1982.

35. **Belasco, J. G., Albery, W. J., and Knowles, J. R.,** Double isotope fractionation: test for concertedness and for transition state dominance, *J. Amer. Chem. Soc., 105,* 2475, 1983.

36. **Doering, W. von E., Toscano, V. G., and Beasley, G. H.,** Kinetics of the Cope rearrangement of 1,1-dideuteriohexa-1,5-diene, *Tetrahedron 27,* 5299, 1971.

37. **Doering, W. von E.,** Impact of upwardly revised ΔH^0_f of primary, secondary, and tertiary radicals on mechanistic constructs in thermal reorganisation, *Proc. Natl. Acad. Sci. USA, 78,* 5279, 1981.

38. **Benson, S. W., Cruickshank, F. R., Golden, D. M., Haugen, G. R., O'Neal, H. E., Rodgers, A. S., Shaw, R., and Walsh, R.,** Additivity rules for the estimation of thermochemical properties, *Chem. Rev., 69,* 279, 1969.

39. **Franklin, J. L.,** Prediction of heat and free energies of organic compounds, *Ind. Eng. Chem., 41,* 1070, 1949.

40. **Benson, S. W.,** *Thermochemical Kinetics,* 2nd ed., John Wiley & Sons, New York, 1976.

41. **Guthrie, J. P., and Pike, D. C.,** Hydration of acylimidazoles: tetrahedral intermediates in

acylimidazole hydrolysis and nucleophilic attack by imidazole on esters. The question of concerted mechanisms for acyl transfers, *Can. J. Chem.*, *65*, 1951, 1987.

42. **Guthrie, J. P.,** Concerted mechanism for alcoholysis of esters: an examination of the requirements, *J. Amer. Chem. Soc.*, *113*, 3941, 1991.

43. **Kurz, J. L.,** The relationship of barrier shape to "linear" free energy slopes and curvatures, *Chem. Phys. Lett.*, *57*, 243, 1978.

44. **Thornton, E. R.,** A simple theory for predicting the effects of substituent changes on transition state geometry, *J. Amer. Chem. Soc.*, *89*, 2915, 1967.

45. **Guthrie, J. P.,** Concertedness and E2 elimination reactions. Prediction of transition state position and reaction rates using two-dimensional reaction surfaces based on quadratic and quartic approximations, *Can. J. Chem.*, *68*, 1643, 1990.

46. **le Noble, W. J., Miller, A. R., and Hamann, S. D.,** A simple, empirical function describing the reaction profile, and some applications, *J. Org. Chem.*, *42*, 338, 1977.

47. **Dunn, B. D.,** Pathways of proton transfer in acetal hydrolysis, *Int. J. Chem. Kinetics*, *6*, 143, 1974.

48. **Critchlow, J. E.,** Prediction of transition state configuration in concerted reactions from the energy requirements of the separate processes, *J. Chem. Soc. Faraday Trans. 1*, *68*, 1774, 1972.

49. **Miller, S. I.,** Multiple variation in structure-reactivity correlations, *J. Amer. Chem. Soc.*, *81*, 101, 1959.

50. **Cordes, E. H., and Jencks, W. P.,** General acid catalysis of semicarbazone formation, *J. Amer. Chem. Soc.*, *84*, 4319, 1962.

51. **Lienhard, G. E., and Jencks, W. P.,** Thiol addition to the carbonyl group. Equilibria and kinetics, *J. Amer. Chem. Soc.*, *88*, 3982, 1966.

52. **Jencks, D. A., and Jencks, W. P.,** On the characterization of transition states by structure-reactivity coefficients, *J. Amer. Chem. Soc.*, *99*, 7948, 1977.

53. **Lee, I.,** Cross-interaction constants and transition state structure in solution, *Adv. Phys. Org. Chem.*, *27*, 57, 1992.

54. **Lee, I.,** Mechanistic significance of the magnitude of cross-interaction constants, *J. Phys. Org. Chem.*, *5*, 736, 1992.

55. **Lee, I.,** Characterisation of transition states for reactions in solution by cross-interaction constants, *Chem. Soc. Revs.*, *19*, 317, 1990.

56. **Lee, I., and Sohn, S. C.,** The mechanistic significance of cross-interaction constants ρ_{ij}, *J. Chem. Soc. Chem. Commun.*, 1055, 1986.

57. **Jencks, W. P.,** General acid-base catalysis of complex reactions in water, *Chem. Revs.*, *72*, 705, 1972.

58. **Ponec, R., and Strnad, M.,** The least motion principle, concertedness and the mechanisms of pericyclic reactions - a similarity approach, *Coll. Czech. Chem. Commun.*, *59*, 75, 1994.

59. **Ponec, R.,** Similarity models in the theory of pericyclic reactions, *Topics in current chemistry*, *174*, 1, 1995.

60. **Page, M. I., and Williams, A.,** *Organic and Bio-Organic Mechanisms*, Addison Wesley Longman, Harlow, Essex, 1997, p. 119.

61. **Bird, R. and Stirling, C. J. M.,** Intramolecular reactions. Part XI. Cyclisation of ω-halogenalkyl sulphides: the significance of bromide:chloride ratios in displacement reactions, 1973 *J. Chem. Soc. Perkin Trans. 2*, 1221.

62. **Bunnett, J. F., Garbisch, E. W. Jr., and Pruitt, K. M.,** The "element effect" as a criterion of mechanism in activated aromatic nucleophilic substitution reactions, *J. Amer. Chem. Soc.*, *79*, 385, 1957.

63. **Sühnel, J., and Schowen, R. L.,** Theoretical basis for primary and secondary hydrogen isotope effects, in Enzyme mechanisms from isotope effects, Cook, P. F., ed., CRC Press, Boca Raton, 1991, p. 3.

64. **Huskey, W. P.,** Origins and interpretations of heavy atom isotope effects, in *Enzyme Mechanisms from Isotope Effects*, Cook, P. F., ed., CRC Press, Boca Raton, 1991, p. 37.

65. **Melander, L., and Saunders, W. H.,** *Reaction Rates of Isotopic Molecules*, John Wiley & Sons, New York, 1980.

66. **Saunders, W. H.,** Kinetic isotope effects, in Bernasconi, C. F., ed., *Investigation of Rates*

and Mechanisms of Reactions, Part 1, Wiley-Interscience, New York, 1986, p. 565.

67. **Cook, P. F. (ed.)**, *Enzyme Mechanisms from Isotope Effects*, CRC Press, Boca Raton, 1991.

68. **Singleton, D. A., and Thomas, A. A.**, High-precision simultaneous determination of multiple small kinetic isotope effects at natural abundance, *J. Amer. Chem. Soc., 117*, 9357, 1995.

69. **Rodgers, J., Femec, D. A., and Schowen, R. L.**, Isotopic mapping of transition state structural features associated with enzymic catalysis of methyl transfer, *J. Amer. Chem. Soc., 104*, 3263, 1982.

70. **Dove, J. E.**, Homogeneous gas phase reactions, in Bernasconi, C. F., ed., *Investigation of Rates and Mechanisms of Reactions, Part 1*, Wiley-Interscience, New York, 1986, p. 154.

71. **Kupczyk-Subotkowska, L., Saunders, W. H., and Shine, H. J.**, The Claisen rearrangement of allyl phenyl ether: heavy-atom kinetic isotope effects and bond orders in the transition structure, *J. Amer. Chem. Soc., 110*, 7153, 1988.

72. **Kupczyk-Subotkowska, L., Subotkowski, W., Saunders, W. H., and Shine, H. J.**, Claisen rearrangement of allyl phenyl ether. 1-^{14}C and β-^{14}C Kinetic isotope effects. A clearer view of the transition structure, *J. Amer. Chem. Soc., 114*, 3441, 1992.

73. **Badger, R. M.**, The relation between intermolecular distances and the force constants of diatomic molecules, *J. Chem. Phys., 3*, 710, 1935.

74. **Pauling, L.**, *The Nature of the Chemical Bond*, 3rd ed., Cornell University Press, Ithaca, New York, 1960, p. 239.

75. **Wolfsberg, M., and Stern, M. J.**, The validity of some approximation procedures used in the theoretical calculation of isotope effects, *Pure Appl. Chem., 8*, 225, 1964.

76. **Stern, M. J., and Wolfsberg, M.**, Simplified procedure for the theoretical calculation of isotope effects involving large molecules, *J. Chem. Phys., 45*, 4105, 1966.

77. **Sims, L. B., and Lewis, D. E.**, Bond order methods for calculating isotope effects in organic reactions, in *Isotope Effects in Organic Chemistry*, Buncel, E., and Lee, C. C., eds., Elsevier, New York, *6*, 161, 1984.

78. **Berti, P. J.**, Determining transition states from kinetic isotope effects, *Methods in Enzymology, 308*, 355, 1999.

79. **Saunders, M., Laidig, K. E., and Wolfsberg, M.**, Theoretical calculation of equilibrium isotope effects using ab initio force constants: application to NMR isotopic perturbation studies, *J. Amer. Chem. Soc., 111*, 8989, 1989.

80. **Czyryca, P., and Paneth, P.**, ^{13}C and ^{15}N Kinetic isotope effects on the decarboxylation of 3-carboxybenzisoxazole. Theory *vs.* experiment, *J. Org. Chem., 62*, 7305, 1997.

81. **Saunders, M.**, Kinetic isotope effects, in *Investigation of Rates and Mechanisms of Reactions, Part 1*, Bernasconi, C. F., ed., Wiley-Interscience, New York, 1986, p. 565.

82. **Dewar, M. J., and Ford, G. P.**, Thermal decarboxylation of but-3-enoic acid. MINDO/3 calculations of activation parameters and primary kinetic isotope effects, *J. Amer. Chem. Soc., 99*, 8343, 1977.

83. **Yoo, Y. Y., and Houk, K. N.**, Transition structures and kinetic isotope effects for the Claisen rearrangement, *J. Amer. Chem. Soc., 116*, 12047, 1994.

84. **Chen, J. S., Houk, K. N., and Foot., C. S.**, The nature of the transition structures of triazolinedione ene reactions, *J. Amer. Chem. Soc., 119*, 9852, 1997.

85. **Hirst, D. M.**, *A Computational Approach to Chemistry*, Blackwell Scientific Publications, Oxford, 1990, pp. 29, 46.

86. **Hehre, W. J., Radom, L., Schleyer, P. v. R., and Pople, J. A.**, *Ab Initio Molecular Orbital Theory*, Wiley & Sons Inc., New York, 1986.

87. **Hu, W-P., Rossi, I., Corchado, J. C., and Truhlar, D. G.**, Molecular modeling of combustion kinetics. The abstraction of primary and secondary hydrogens by hydroxyl radical, *J. Phys. Chem. A, 101*, 6911, 1997.

88. **Guthrie, J. P.**, Prediction of rate constants for cyanohydrin formation using equilibrium constants and distortion energies, *J. Amer. Chem. Soc., 120*, 1688, 1998.

Chapter 3

Proton Transfer

3.1   PROTON TRANSFER BETWEEN BASES

The transfer of a proton between bases (Equation 3.1), a special case of a
nucleophilic displacement reaction at a hetero-atom (Chapter 6), is analogous to
the transfer of any other electrophilic group such as the carbonyl or phosphoryl.

$$Nu^- + H\text{-}Lg \rightleftharpoons Nu\text{-}H + Lg^- \qquad (3.1)$$

Invariably the simple transfer of a proton, as in Equation 3.1, is coupled with
other fundamental processes and offers mechanisms by which reactions can be
controlled either by inhibition or catalysis. The proton transfer reaction simply
enables charge to be dispersed from highly unstable cationic or anionic
intermediates thus providing mechanistic routes of lower energy. Organic
synthesis and enzymic catalysis (Chapter 8) are exceptionally important fields
which rely on proton transfer as a means of *enabling* reactions. A large
proportion of reactions employed in synthesis involve proton transfer steps
crucial for their completion, and some classical examples from heterocyclic
chemistry illustrate this point (Scheme 3.1). The other rôle of proton transfer in
control, namely to inhibit a reaction, is important in metabolism but will not be
discussed in this text.

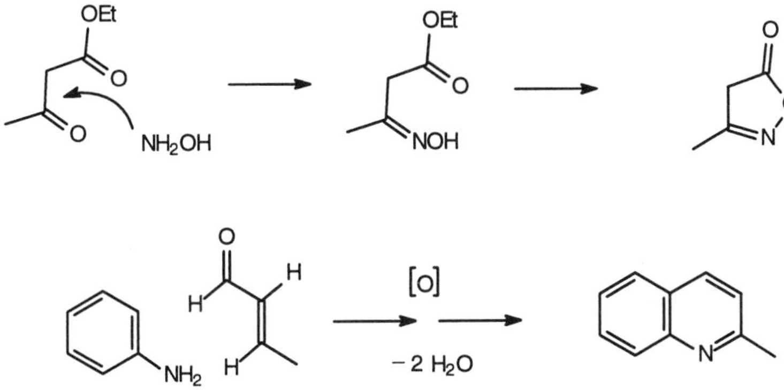

Scheme 3.1. Reactions in heterocyclic syntheses where proton transfer is required.

The elementary proton transfer "step" between nucleophiles has a single
transition structure (**3.1**, Scheme 3.2) where the proton is expressed neither as a
"naked" cation **3.2** nor in the form of an hypervalent hydrogen species **3.3**. It is
not intended that **3.2** and **3.3** be considered as resonance structures of a
resonance hybrid structure. The "naked" proton (**3.2**) and divalent hydrogen
(**3.3**) structures are considered as not unreasonable contributors to the hydrogen-

bond.[1,2] Moreover, hypervalent hydrogen is well known in transition-metal chemistry and in such species as the boron hydrides;[3] the HF_2^- anion is a stable species and "naked" protons are observed in the gas phase. Despite their existence in these systems structures **3.2** and **3.3** are of extremely high energy and this provides a good but not absolute argument for concertedness in the fundamental proton transfer reaction. The "naked" proton has a free energy of reaction with water >100 kcal/mole at 25°. Figure 3.1 illustrates the More O'Ferrall-Jencks diagram for the *two bond* fundamental proton transfer reaction of Equation 3.1.

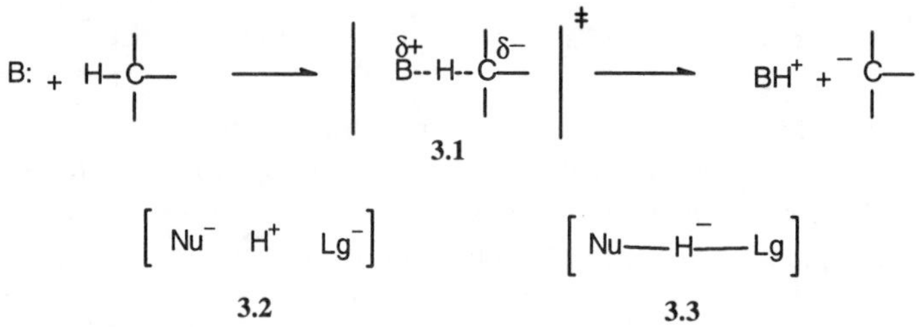

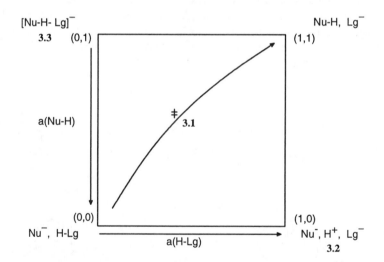

Scheme 3.2. The elementary proton transfer step from carbon acid to base.

Figure 3.1. More O'Ferrall-Jencks diagram for proton transfer between nucleophiles.

Let us consider Scheme 3.3 which represents the most "reasonable" stepwise mechanism for proton transfer from a carbon acid, such as acetone, to a base. The proton transfer from acetone to bases can be readily measured by halogenation. The reactivity is dependent on the strength of the base according to the Brønsted law with exponent $\beta = 0.7$. The reaction also possesses a large primary hydrogen isotope effect. Scheme 3.3 for a stepwise fundamental proton transfer reaction indicates that the observation of general base catalysis is consistent with a rate-limiting step (k_2). However, the isotopically sensitive bond (C-H) has already been broken at this stage so that a primary isotope effect would not be expected. The primary isotope effect is consistent with a rate-limiting k_1 step and although the mechanism of Scheme 3.3 would predict general base catalysis when k_1 is rate-limiting the catalysis would be insensitive to the structure of the base. The value of β would be zero because k_1 does not involve changes in bonding to the base. The experimental observations are consistent with a concerted mechanism with a transition state Structure **3.1** intervening between **3.4** and **3.6** instead of the intermediate **3.5** as in Scheme 3.3. The corresponding case with an addition intermediate involving hypervalent hydrogen (Structure **3.3**) would fit the combined observations of isotope effect and Brønsted data and can only be excluded because of the high energy requirements for such a species. *Ab initio* studies of simple systems predict a concerted transfer of proton between donor and acceptor atoms.[4]

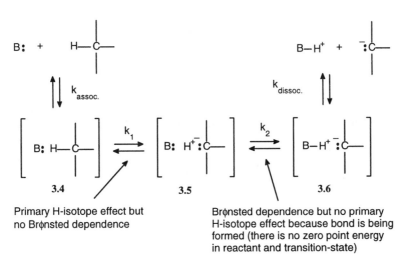

Primary H-isotope effect but no Brønsted dependence

Brønsted dependence but no primary H-isotope effect because bond is being formed (there is no zero point energy in reactant and transition-state)

Scheme 3.3. Stepwise proton transfer involving a putative "naked" proton.

Proton transfer between electronegative base pairs in hydroxylic solvents occurs by relay through one or more intermediate solvent molecules; the reactions are usually fast and limited by diffusion. The process often exhibits an Eigen-type Brønsted correlation[5] and the number of intervening solvent molecules can be determined by proton inventory studies of the number of protons "in-flight".[6] The transfer to carbon is usually much slower and in most cases involves direct transfer. This is attributed to changes in electron delocalisation and rehybridisation accompanied by changes in bond lengths and angles, poor hydrogen-bonding properties of carbon and solvent reorganisation in the rate-determining step.[7]

The coupling of the proton transfer step with other bonding changes introduces the possibility of concerted timing. Two general processes are formulated in Equations (3.2) and (3.3) where proton transfer *enables* a subsequent step k_2. The species X^- and $Y\text{-}H^+$ formed by proton removal or addition possess *extra* energy which can be used to drive the subsequent reaction forward to product. The proton transfer steps as given in Equations (3.2) and (3.3) can be concerted with the subsequent step either when $Y\text{-}H^+$ or X^- do not have a significant molecular lifetime or when concertedness offers a more favourable free energy pathway.

Base catalysis

$$B \quad + \quad H\text{-}X \quad \underset{\xrightarrow{\hspace{2cm}}}{\xleftarrow{\hspace{2cm}}}^{\text{Enabling}} \quad X^- \quad + \quad B\text{-}H^+$$

$$X^- \quad \xrightarrow{\hspace{1cm}}^{k_2} \quad \text{Product} \tag{3.2}$$

Acid catalysis

$$Y \quad + \quad A\text{-}H^+ \quad \underset{\xrightarrow{\hspace{2cm}}}{\xleftarrow{\hspace{2cm}}}^{\text{Enabling}} \quad Y\text{-}H^+ \quad + \quad A$$

$$Y\text{-}H^+ \quad \xrightarrow{\hspace{1cm}}^{k_2} \quad \text{Product} \tag{3.3}$$

A second type of enabling process can be envisaged where substrate on reaction with nucleophile or electrophile yields an unstable species. In order that the reaction proceed to completion a proton transfer step is required (Equations 3.4 and 3.5) to "trap" the reactive intermediate and prevent it from returning to reactant.[8] Such trapping reactions are essentially the *microscopic reverse* of the *enabling* reactions of Equations (3.2) and (3.3) and likewise can exhibit concertedness if the reactive intermediates X^- or $Y\text{-}H^+$ have no significant lifetimes or if the concerted pathway has a more favourable free energy than that of the stepwise route.

$$S \quad \underset{\xrightarrow{\hspace{1cm}}}{\xleftarrow{\hspace{1cm}}} \quad X^- \quad \overset{B\text{-}H^+ \text{ Trapping}}{\underset{\xrightarrow{\hspace{1cm}}}{\xleftarrow{\hspace{1cm}}}} \quad X\text{-}H \quad \xrightarrow{\hspace{1cm}} \quad \text{Product} \tag{3.4}$$

$$S \quad \underset{\xrightarrow{\hspace{1cm}}}{\xleftarrow{\hspace{1cm}}} \quad Y\text{-}H^+ \quad \overset{B \text{ Trapping}}{\underset{\xrightarrow{\hspace{1cm}}}{\xleftarrow{\hspace{1cm}}}} \quad Y \quad \xrightarrow{\hspace{1cm}} \quad \text{Product} \tag{3.5}$$

The trapping and enabling processes described above can be exemplified by reactions by carbonyl and imine addition reactions (Scheme 3.4).

The processes described above can be discussed in terms of the More O'Ferrall-Jencks diagrams (for example Figure 3.2 for Equation (3.2)) although in these reactions there are at least three major bonding changes occurring. In Equation (3.2) there is B-H bond formation and H-X bond fission and one or more bond changes associated with the subsequent reaction of X^-. These three changes are subsumed into two by assuming that the two bond changes associated with proton transfer (the $0,0 \rightarrow 0,1$ or $1,0 \rightarrow 1,1$ axes) are considered as one.

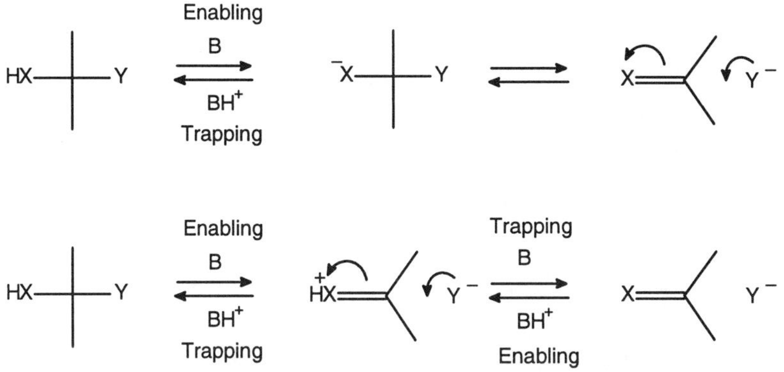

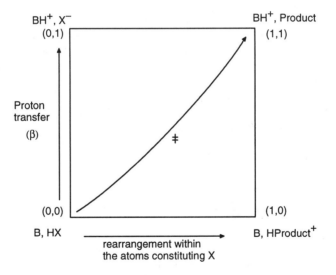

Scheme 3.4 Trapping and enabling by proton transfer exemplified by reactions at carbonyl, imine and thion bonds.

Figure 3.2. More O'Ferrall-Jencks diagram for proton transfer coupled with reaction (Equation (3.2)); the proton transfer coordinate involves two bond changes.

3.2   PROTON TRANSFER AND DISPLACEMENT REACTIONS AT SATURATED CARBON CENTRES

3.2.1   Electrophilic displacement

The halogenation of ketones and aldehydes bearing hydrogen on the carbon atom α to the carbonyl group is one of the earliest reactions to be studied for acid-base catalysis. The rate constant is independent of halogen concentration except at very low concentrations and this is consistent with enolate ion formation being rate-limiting under these conditions (Scheme 3.5). The concerted mechanism is excluded by the absence of a rate law with a term in halogen concentration.

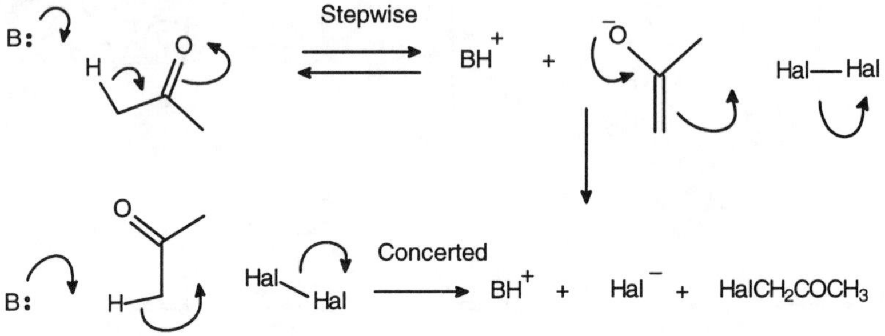

Scheme 3.5. Stepwise or concerted halogenation of acetone.

The halogenation of acetone is more complicated than Scheme 3.5 suggests and the final product includes the haloform ($CHHal_3$); the halogens can also undergo competing reactions with the solvent or base but, in general, the kinetics of enolisation can be readily determined from the halogenation reaction.[9] The formation of the enol can be acid- or base-catalysed and there is also a kinetic term involving both acid and base concentrations ($k_{AB}[A][B]$), which has been investigated thoroughly for the acetate ion-acetic acid pair. Proton inventory studies[10] indicate that there is only one proton "in-flight" in the transition state of the acid- or base-catalysed reaction. The transition state of the acid-catalysed reaction includes a proton which is tightly bound to the carbonyl oxygen (Scheme 3.6).

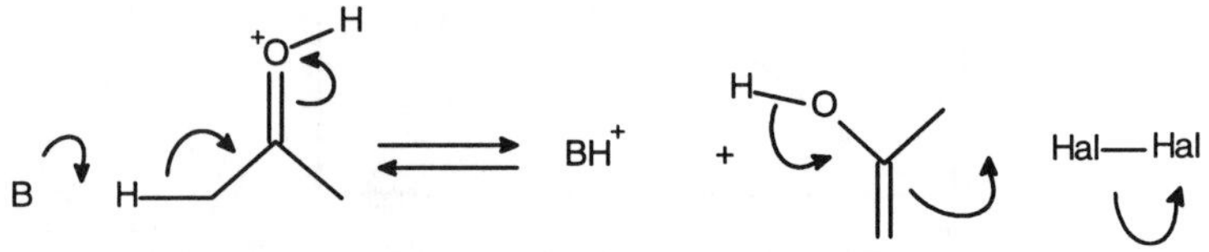

Scheme 3.6. Acid-catalysed halogenation of ketones.

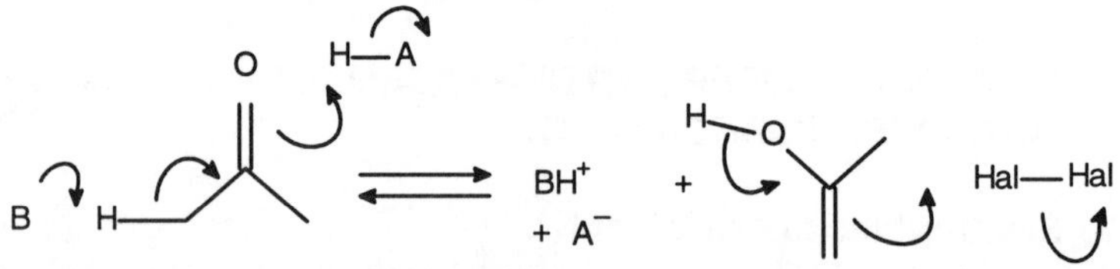

Scheme 3.7. "Push-pull" acid-base catalysed halogenation of ketones.

The third-order term k_{AB} was first observed by Dawson and Spivey (1930).[11] The associated mechanism, described as "push-pull" or "bimolecular", involves removal of the proton from the α-carbon concerted with proton donation by an acid at the carbonyl oxygen (Scheme 3.7). The proportion of reaction flux passing through the third-order term is small[12] in most cases of enolisation. The major part of the flux involves the second-order terms; despite this, the experimental values of k_{AB} have been shown to be well within their error limits. Proton inventory studies on the third-order term[10,13,14] indicate that there are two protons "in-flight" in the transition state. Hegarty and Jencks (1975)[15] addressed the question of whether the intermolecular terms for enolisation of acetone resulted from a concerted push-pull process by use of isotope effect and of Brønsted methodology. The arguments are as follows: the third-order term possesses a normal primary isotope effect of $k^H/k^D = 5.8$ indicating C-H fission in the transition state. The solvent deuterium isotope effect of 2.0 contrasts with the values 1 to 1.25 for k_B and 1.1 to 1.5 for k_{HA} consistent with a proton "in-flight". A larger third-order term for the $Me_3N^+OH\text{-}Me_3N^+O^-$ pair indicates that there is no special effect from the acetate ion-acetic acid pair.

The Brønsted dependence of k_{AB} for acid-base pairs has a slope of 0.15, indicating that the increase in effectiveness of the acid is essentially cancelled by a decrease in effectiveness of the base. If there were no protonation of the carbonyl oxygen in the transition state when the Brønsted value (β_B) for variation of base is expected to be *ca.* 0.88 (β_A, the Brønsted value for variation of acid, is zero). For complete protonation ($\beta_A = 1$) the value of β_B is expected to be 0.45. If it is assumed that β_B is a linear function of β_A (from $\beta_B = 0.88$ at $\beta_A = 0$ to $\beta_B = 0.45$ at $\beta_A = 1$) then the unique set of values that satisfies the Equations $\beta_{AB} = 0.15$ and $\beta_{AB} = \beta_B - \beta_A$ is $\beta_B = 0.66$ at $\beta_A = 0.51$. Thus dependence on the acid in the third-order term is relatively large. Figure 3.3 illustrates the More O'Ferrall-Jencks reaction map for the proton transfer reaction. The synchronous diagonal represents the pathway for $\beta_{AB} = 0$ and the reaction coordinate is offset by a factor giving a β_{AB} of 0.15 where the proton transfer from carbon to base B is slightly in advance of proton transfer from AH^+ to the carbonyl oxygen.

Much larger third-order terms have been measured for the enolisation of aldehydes than for in ketones (Hegarty and Dowling, 1991)[16] and can be explained by the smaller basicity of the oxygen and the higher acidity of the α-proton together with a more stable enol (Keeffe and Kresge, 1990).[17]

3.2.2 Nucleophilic displacement

Proton transfer coupled with nucleophilic attack at saturated centres has been extensively investigated with ketal, acetal and ortho ester hydrolysis.[18] The interest in such studies is that the glycosyl transfer reaction is important in many biological processes. Most of the processes involve an α-carbenium ion intermediate which is stabilised by the electron donating ability of an adjacent oxo group (Scheme 3.8).

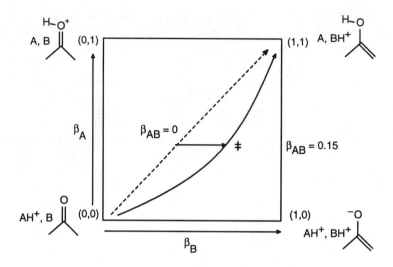

Figure 3.3. Acid-base catalysed enolisation of ketones; the transition state is offset from the synchronous diagonal by $\Delta\beta = 0.15$ (the distance is exaggerated in the diagram).

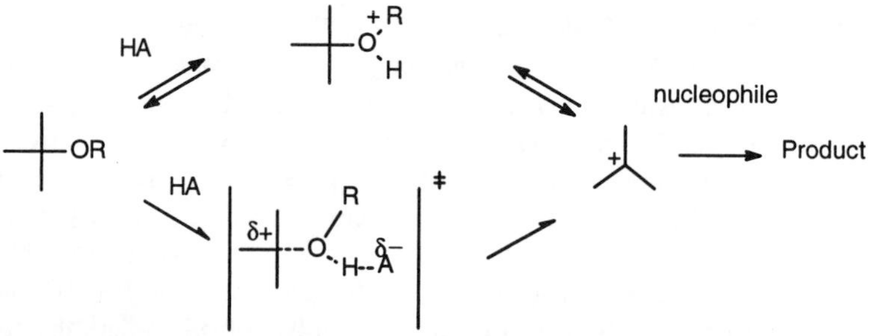

Scheme 3.8. Concerted and stepwise acid-catalysed hydrolysis of acetals, ketals and ortho esters.

A less attractive mechanism could involve the bond formation process concerted with the proton transfer and bond fission giving a three-bond concerted system (Structure **3.7**) The main tools employed in this area have been deuterium isotope effect studies and studies on the effect of structure on reactivity.

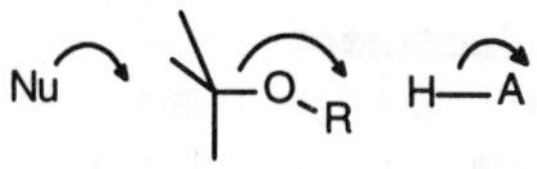

3.7

The observation of general acid as opposed to specific acid catalysis of nucleophilic displacement at ketals, acetals and ortho esters is solvent dependent, and many of the original studies did not appreciate this. The early detection of general acid catalysis employed buffers in dioxane-water mixtures but neglected to consider the influence of specific salt effects (Lahti and Kankaanpera.[19,20] Although general acid catalysis of nucleophilic displacements at carbon bearing ether groups exists, the experimental evaluation of the kinetic parameters needs careful attention to the influence of the salt effects.[21]

3.2.2.1 Intermolecular reactions

The factors leading to general acid catalysis of ether hydrolysis are low basicity of the oxyanion and easy fission of the oxygen-carbon bond which can arise from the stability of the intermediate oxocarbenium ion. General acid catalysis is not generally observed with aliphatic ketals although factors stabilising the carbenium ion (for example Scheme 3.9) lead to general acid catalysis. There are three major electronic changes in the rate-limiting step of concerted general

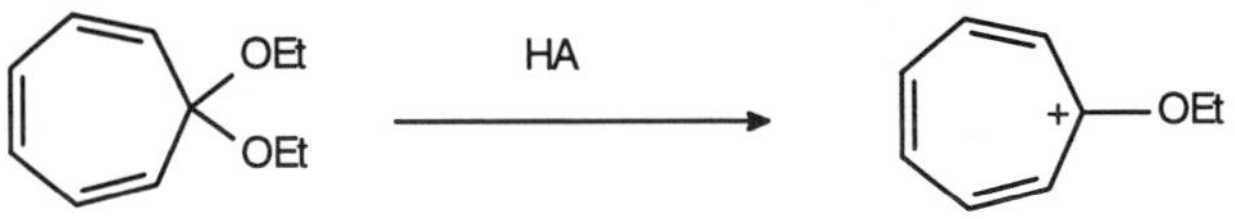

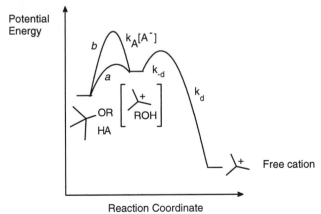

Scheme 3.9. Factors stabilising the oxocarbenium ion promote general acid catalysis.

acid catalysed ether fission: proton transfer to the leaving oxygen, carbon-oxygen bond fission and rehybridisation of the carbenium ion to form the resonance stabilised oxocarbenium ion. The last change is not often considered in simple acetal or ketal hydrolysis and is probably concerted with the C-O bond fission process. Jencks[8] has suggested that Brønsted acid-base catalysis of reversible addition of hydroxylic solvents to carbocations will only be observed when the barrier for addition of alcohols to the cation is larger than for diffusional separation to free ions (case *b*, Figure 3.4). When the barrier is smaller (case *a*, Figure 3.4) the rate-limiting step will be the diffusional process and the Brønsted general acids and bases are not expected to accelerate the transport step (k_d, Figure 3.4).

Figure 3.4. General acid catalysis of reversible addition of hydroxylic solvents to carbocations.

In the hydrolysis of aryl orthocarbonates (Scheme 3.10), the stabilisation of the carbenium ion by resonance with an oxygen atom is not expressed substantially in the transition state of the rate-limiting step.[22,23] The large value for p_{xy} ($\partial\beta_{HA}/\partial\sigma$, Section 2.3.2) for the effect of changing the aryl substituent on the value of the Brønsted exponent for general acid catalysis (~ 0.2) indicates direct coupling between the C-O bond fission and proton donation consistent with a concerted process.

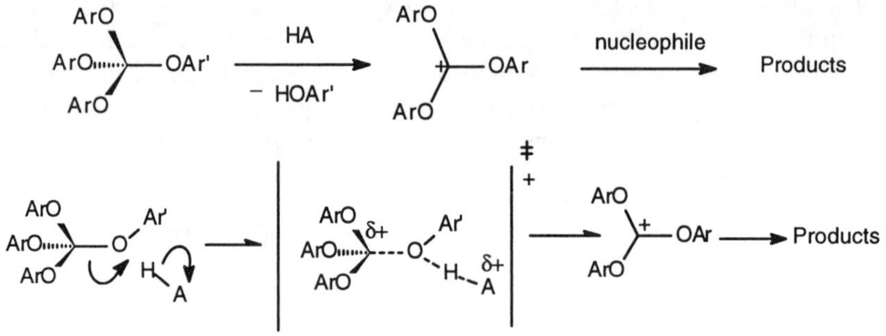

Scheme 3.10. General acid-catalysed hydrolysis of aryl orthocarbonate esters.

The conformation of the initially formed carbenium ion where the positive charge resides on the carbon is not optimal for resonance stabilisation. The tris(aryloxy)carbenium ion has only one reasonably stable conformation (Structure 3.8). Major changes are required for the tris(aryloxy)methyl group to attain the conformation of 3.8 from either the S_4 or D_{2d} conformations. Less conjugation at the transition state is in accord with the large β_{rc} values, greater than expected for the equilibrium formation of the carbenium ion. The visualisation of processes with more than two bonding changes is particularly complicated due to the requirement for more than three dimensions and Kandanarachchi and Sinnott,[23] Grunwald[24] and Guthrie[25] utilise a prism for the three changes. A reaction with three-bond changes, as in proton transfer concerted with a third bond change, is often illustrated with a single coordinate for the two-bond change represented by the proton transfer (Figures 3.2, 3.3, 3.6 and 3.7). Figure 3.8 (see later) is a representation of the energy surface of a reaction with *four* bond changes (two hydrogen-bonds).

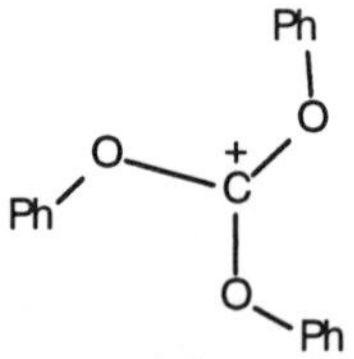

3.8

3.2.2.2 Intramolecular reactions

Proton transfer coupled with nucleophilic attack has been studied in the formation of cyclic sthers (Scheme 3.11). The heavy-atom chlorine isotope effect on the C-Cl bond in the formation of the cyclic ether[26] as well as a substantial β-value for proton transfer (0.36)[26-28] indicate that the proton is "in-flight" in the transition state involving substantial C-Cl bond fission. Assuming a proton transfer is a *single* coordinate there are three major bonding changes which are concerted. The Leffler values of 0.36 and 0.60 for the proton transfer and C-Cl bond fission respectively indicate that there is charge *imbalance* in the transition state involving a build-up of positive charge on the oxygen nucleophile. The β-value of 0.36 is only a little larger than that found in hydrogen-bonding interactions, and it is conceivable that the proton is not in flight in the transition state but that the H-bonding interaction activates the hydroxyl group to nucleophilic attack (Scheme 3.12).

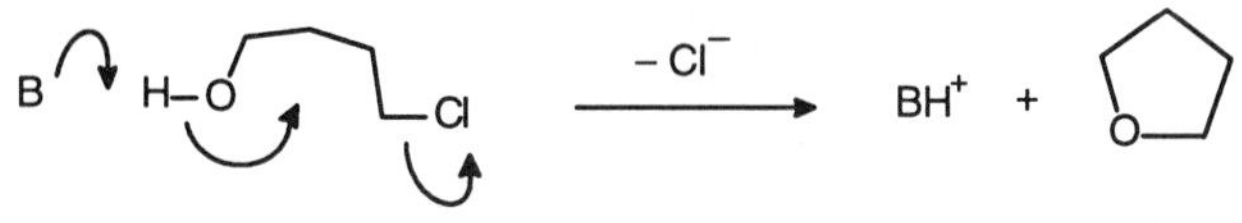

Scheme 3.11. Cyclic ether formation via general base catalysis.

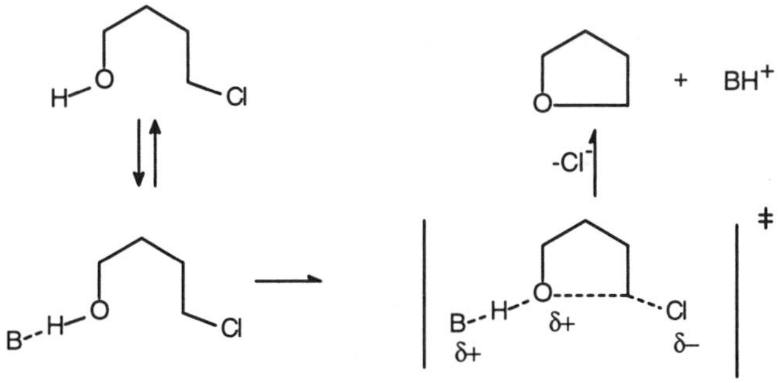

Scheme 3.12. Charge imbalance in the transition state for general base catalysed ether formation.

The reaction of trifluoroethanol with [4(trifluoromethyl)benzyl]methyl(4-nitrophenyl)sulphonium ion is catalysed by buffer bases (Scheme 3.13) and possesses a Brønsted β value of 0.26.[29] The absence of a significant deuterium oxide solvent isotope effect suggests that the proton has not lost zero point energy and is consistent with hydrogen-bonding of the proton by the base catalyst. However, a concerted, coupled mechanism is not excluded because the isotope effect can be decreased by coupling of the motions of the proton and heavy atoms in the transition state.

A further example of proton transfer coupled with nucleophilic attack is a model of methyl transfer enzymes (Scheme 3.13).[30-32] Essentially the same results are obtained with this system, and the Leffler value for base attack of 0.27 and the α-D-isotope effect of 1.17 indicate substantial heavy-atom reorganisation with little proton transfer in the transition state. These results were discussed as proton transfer uncoupled with heavy-atom reorganisation meaning that the reaction coordinate at the transition state is perpendicular to the proton transfer coordinate. There is substantial dynamic coupling of both motions consistent with a concerted process.

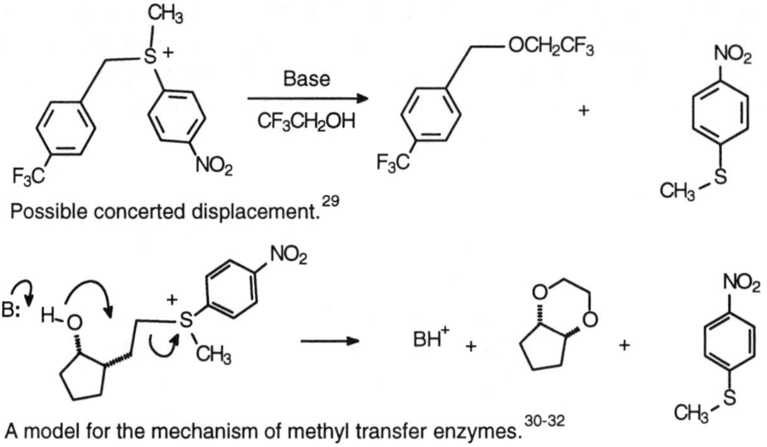

Possible concerted displacement.[29]

A model for the mechanism of methyl transfer enzymes.[30-32]

Scheme 3.13. Displacement of thio ethers from alkyl sulphonium salts.

3.3   PROTON TRANSFER AND DOUBLE BOND CHANGES

3.3.1 Formation of alkenes

The formation of alkenes by elimination can follow a number of mechanistic pathways involving two electron transfers (Scheme 3.14). Elimination mechanisms of alkene formation have been reviewed many times[33-38] and it is useful to refer to Bunnett's early review (1962)[39] which laid the foundations for the mechanistic types seen in elimination reactions.

Bordwell (1972)[40] and Saunders (1976)[41] have critically reviewed the concerted mechanisms in Scheme 3.14 and have shown that while most of the evidence was suspect there are cases where the E2 mechanism is expressed and that most forms of E1cB mechanism are possible as well as the E1 mechanism.

The application of isotope effects to elimination reactions has been in the field of the hydrogen isotope because proton transfer is involved and the cost and techniques for this isotope were favourable compared with those for the other (heavy) atoms. Recent advances have enabled isotope effects to be measured accurately and quickly for heavy atoms (see Chapter 2) and, moreover, the data can be more informative regarding the transition structure than that from the hydrogen isotopes.

9-Fluorenylmethanol undergoes base-catalysed elimination to yield the dibenzofulvene via an E1cB mechanism (steps k_1 and k_2 of Scheme 3.15).

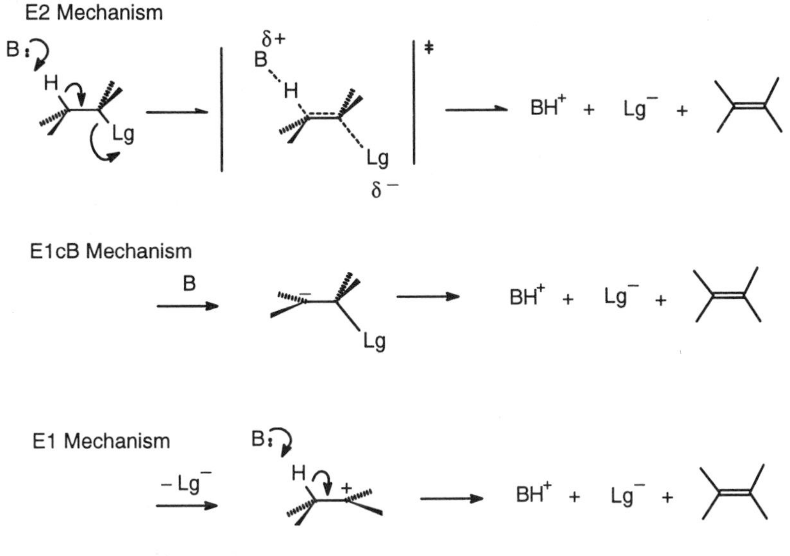

E2 Mechanism

E1cB Mechanism

E1 Mechanism

Scheme 3.14. Mechanisms for formation of alkenes.

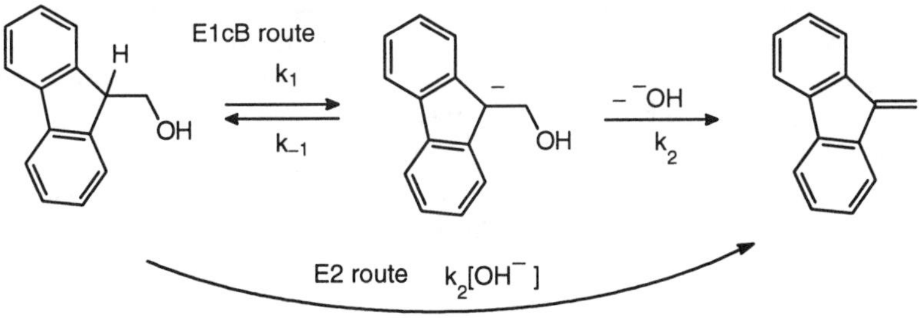

Scheme 3.15. Base-catalysed elimination from 9-fluorenylmethanol to dibenzofulvene.

The rate of tritium exchange in 9-fluorenylmethanol *exceeds* that of elimination and the reaction in D_2O exhibits an induction period, which is due to hydrogen isotope exchange. The inverse deuterium oxide solvent isotope effect is consistent with an E1cB mechanism; the isotope effect on the exchange rate is $k_E{}^H/k_E{}^D = 7.2$.[42]

A further mechanistic tool that can be employed for the E1cB mechanism (in other words, to exclude the concerted E2 process) involves the observation of a large inverse solvent isotope effect. This was employed by Keeffe and Jencks[43,44] for the base-catalysed elimination reactions of 4-nitrophenylethyl alkylammonium ions (Scheme 3.16) which can be accounted for by any one of the mechanisms of Scheme 3.17.

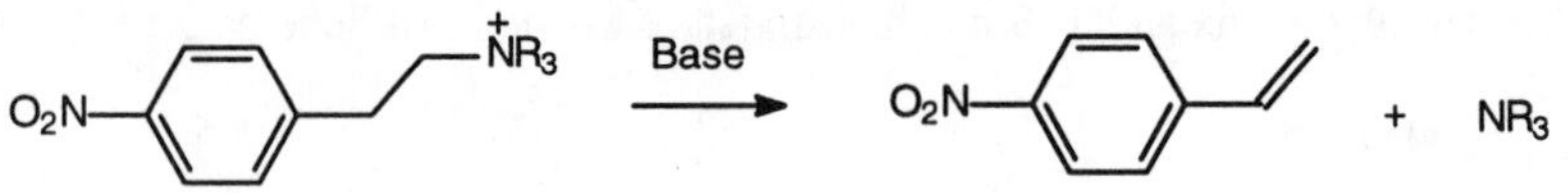

Scheme 3.16. Base-catalysed elimination of 2-(4-nitrophenyl)ethylammonium ion.

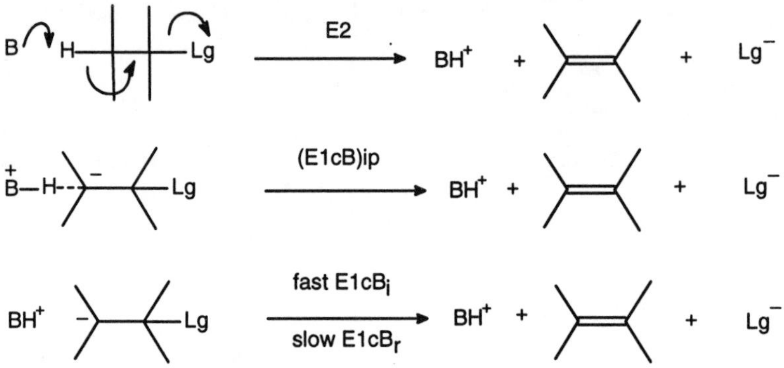

Scheme 3.17. Mechanisms of base-catalysed eliminations.

The inverse solvent isotope effect for *initial* rates of 4-nitrostyrene formation changes with changing buffer concentration (at constant buffer ratio) and is relatively large for high buffer concentration. This variation fits the mechanism of Equation (3.6) which yields the rate law Equation (3.7) and is a sensitive test of the E1cB mechanism when the putative carbanion intermediate suffers partition almost equally distributed between products and reactants. It is essential that initial rates are measured as over a period the exchange of deuterium for hydrogen complicates the kinetics.

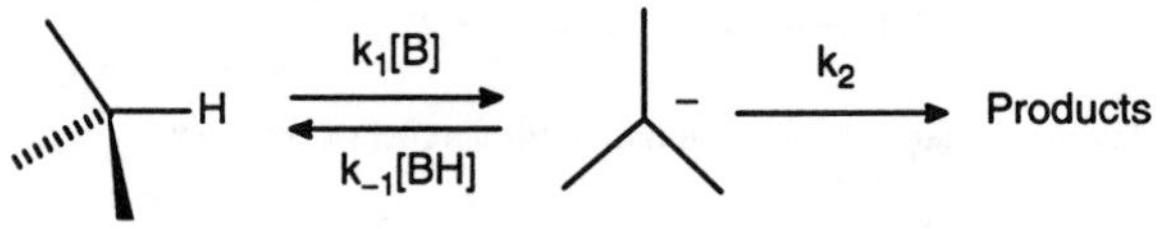

$$\text{Rate} = (k_1[B]k_2)[\text{substrate}]/(k_{-1}[BH] + k_2) \qquad (3.7)$$

The phenyl ethylammonium ions are important in studies of the mechanisms of eliminations because a large number of variables can be studied; these include ring substitution (Hammett ρ values), leaving group effects (element effect and β_{lg}), base strengths (β), kinetic H-isotope effects (primary and secondary), heavy-atom isotope effects, stereochemistry and hydrogen isotope exchange. Bunting and Kanter (1991)[45] showed that variation of the leaving group in the elimination from 4-nitrophenyl ethylpyridinium cation catalysed by hydroxide ion causes a break in the Brønsted plot (Figure 3.5) consistent with a change in rate-limiting step and a stepwise mechanism.

Changing the substituents on the phenyl ring from 4-nitro to less electron attracting groups in the (2-arylethyl)quinuclidinium ions induces an E2 mechanism by transformation from the E1cB process rather than changing between co-existing mechanisms (Gandler and Jencks, 1982).[46] The concerted, E2, mechanisms are diagnosed by demonstrating that both C-H and C-N fission are occurring in the transition state structure. There is no detectable cross-interaction (Section 2.3.2) between the base catalyst and the leaving group for the E1cB elimination (of the 4-nitrophenyl ethyl species) $p_{xy} \sim 0$ $(d\beta_{lg}/dpK_{BH})$. For other species p_{xy} has a positive value of 0.018 corresponding to a change to a less negative β_{lg} as base strength increases. Moreover, β_{lg} becomes less negative as σ increases corresponding to a negative $p_{yy'}$ value of -0.09 $(d\beta_{lg}/d\sigma^-)$. These properties indicate that the reaction coordinate is about $24°$ to the proton transfer coordinate (TS_1, Figure 3.6). In the case of elimination from the 2-arylethyl *bromide* there is more diagonal character to the reaction coordinate at the transition state structure (TS_2, Figure 3.6).

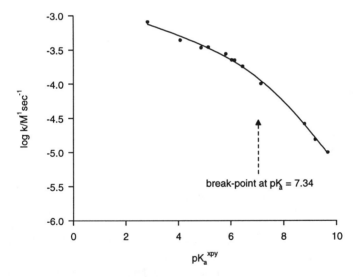

Figure 3.5. E1cB mechanism for the elimination reactions of 4-nitrophenyl ethylpyridinium ions catalysed by hydroxide ion; deprotonation is rate-limiting at pK_a's below the breakpoint. Data are from *Ref. 45* and the line is calculated from: $\log k = -0.531pK_a + 0.18 - \log(1 + 10^{-0.402pKa + 2.93})$.

3.3.1.1 Energy calculations

The 4-nitrophenylethylammonium species eliminate via an E1cB mechanism with a transition state structure for the rate-limiting step which resembles the carbanion intermediate. If the carbanion for other arylethylammonium species had a significant lifetime the reaction path would be expected to have a stepwise mechanism. The observation of E2[44] processes suggests that the carbanion must be too unstable to exist as an intermediate and the mechanism would then be an *enforced* concerted process. Guthrie (1990)[47] was able to estimate the free energies with confidence for the four corners of the More O'Ferrall-Jencks diagram for the elimination reactions of alkyl bromides (Figure 3.7). In the case of the isopropyl and ethyl bromides the observed rate constants for ethoxide ion catalysed elimination have free energies well below those for the two ionic species corresponding to the stepwise routes, and it can therefore be concluded that the mechanism for these substrates is concerted. As expected, the free energy for the trimethylmethyl cation for the tertiary butyl bromide is lower than that for the rate constant in accord with an E1 process.

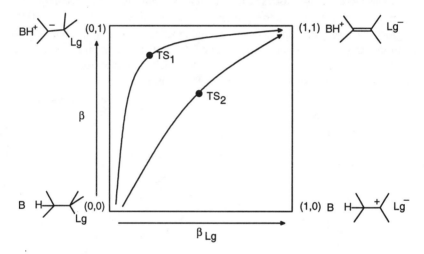

Figure 3.6. More O'Ferrall-Jencks diagram representing the effect of change in leaving group on the elimination mechanism.

The experimental results of Richard (1996)[48] suggest that there is a relatively small difference between the free energy of formation of the tertiary butyl carbocation and that of the transition state for a concerted E1-like elimination mechanism. Calculations may not be sufficiently reliable to make a distinction between E1 and a concerted E1-like mechanism in this case but at least one computational study has provided evidence that the transition state for concerted elimination of tertiary butyl chloride has a lower energy than that for ionisation of the substrate to a carbocation-chloride ion-pair.[49]

3.3.1.2   Lifetimes

Lifetimes of carbanionic intermediates may be estimated by use of the rate constant for the internal return reaction of the carbanion as shown in Scheme 3.18. The values of $k_{-1}(LOH)$ and k_2 may be determined from exchange rate constants (Fishbein and Jencks, 1988)[50,51] $k_{-1}(LOH)/k_2$ and $k_{-1}(LOH)/k_s$ using k_s as a clocking reaction. The value of k_s is taken as $10^{11} sec^{-1}$ which is limited by the reorganisation of water[52-54] and is the rate constant for exchange with bulk solvent of LOH adjacent to the carbanion. The values of k_2 for the expulsion of leaving groups from the carbanion are consistent with a short but significant lifetime. The upper limits for the carbanion of fumaronitrile adducts are near the limit of the bond vibration frequency (2.10^{13}sec^{-1}).

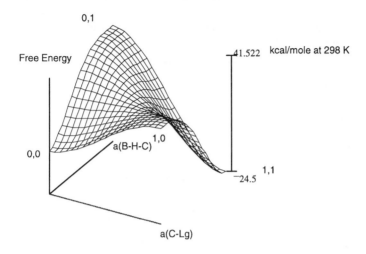

Figure 3.7. Energy contour diagram calculated for the elimination reaction of ethyl bromide; the surface is calculated from data of Guthrie[47] and the coordinates and corners are identified as in Figure 3.6.

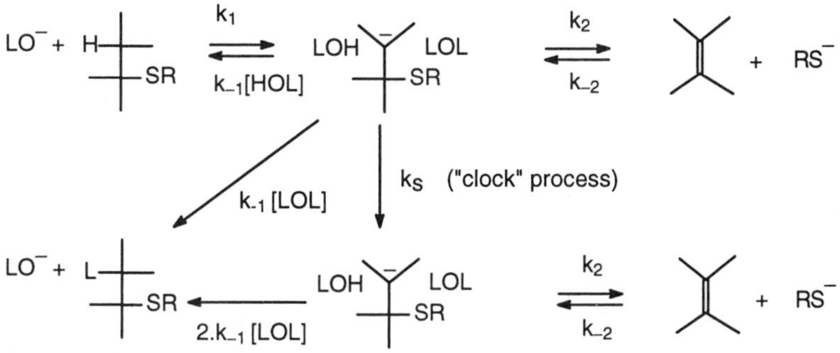

Scheme 3.18. Estimation of the lifetimes of carbanions using a "clock" process.

3.3.2 Reactions at the carbonyl group

Addition of nucleophile to a carbonyl group can be of three types: simple addition (such as in mutarotation), addition followed by departure of a leaving group (a displacement reaction) or addition followed by elimination of the carbonyl oxygen (a condensation reaction) (Scheme 3.19).

3.3.2.1 Mutarotation and enzyme models

The mutarotation reaction of reducing sugars can be monitored by the change in the optical rotation of the solutions, and it was therefore easily measured in the early history of organic reaction mechanisms.[55-57] The reaction involves the intervention of the open-chain aldose (Scheme 3.20). Lowry[55,56] found that both an acid and a base are necessary for the reaction to proceed. Swain[58-60] observed third-order kinetics involving sugar, acid and base concentrations for the reaction in organic solvents and concluded that the mechanism involved a concerted attack of acid and base on the sugar. The alternative view that the mechanism is stepwise was formulated by Pedersen (1934).[61]

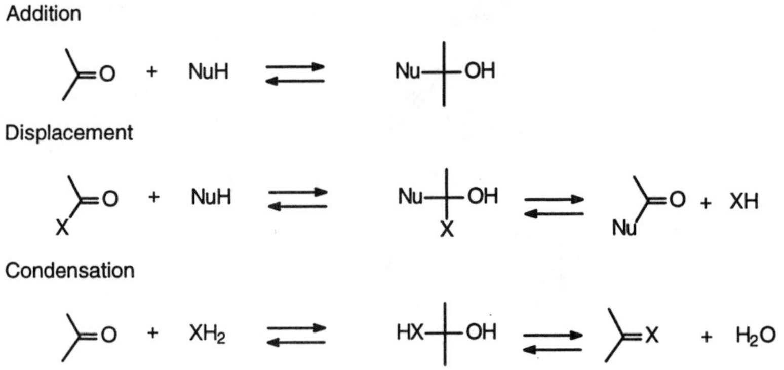

Addition

Displacement

Condensation

Scheme 3.19. Addition-elimination reactions of the carbonyl group.

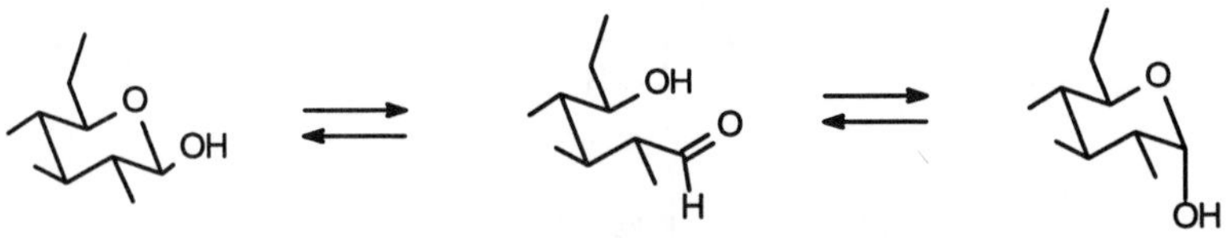

Scheme 3.20. Mutarotation of hexoses.

The observation of third-order kinetics was originally employed to discount the stepwise mechanism. However, a stepwise process would be consistent with third-order kinetics if there were a pre-association process (Scheme 3.21). Reactions carried out in aqueous solution do not exhibit a third-order term and Pocker (1960)[62] showed that the tetra-n-butylammonium cresoxide ion pair is as active in benzene as an ion-pair derived by proton transfer between tertiary amine and phenol.

The termolecular term in the mutarotation of tetramethylglucose was employed to predict that a molecule containing both acid and base functions should act as a superior catalyst to a combination of separate acid and base provided the mechanism were concerted. Swain and Brown[58,59] showed that 2-hydroxypyridine at 0.05M gave a more than 50-fold increase in rate constant for mutarotation than a mixture of phenol and pyridine each at 0.05M concentration. The acceleration must be corrected for the 2-hydroxypyridine's being a weaker base than pyridine and a weaker acid than phenol. It is not possible to compare the third-order rate constant of the pyridine-phenol catalysis directly with the second-order rate constant for catalysis by 2-hydroxypyridine. The evidence now seems to favour a concerted acid-base catalysis for 2-hydroxypyridine but not for the pyridine-phenol mixture.

The tautomeric systems studied by Lowry and Swain were reinvestigated in the late 1960's and early 1970's by Rony[63-66] who extended them to push-pull catalysis in acyl group transfer reactions.

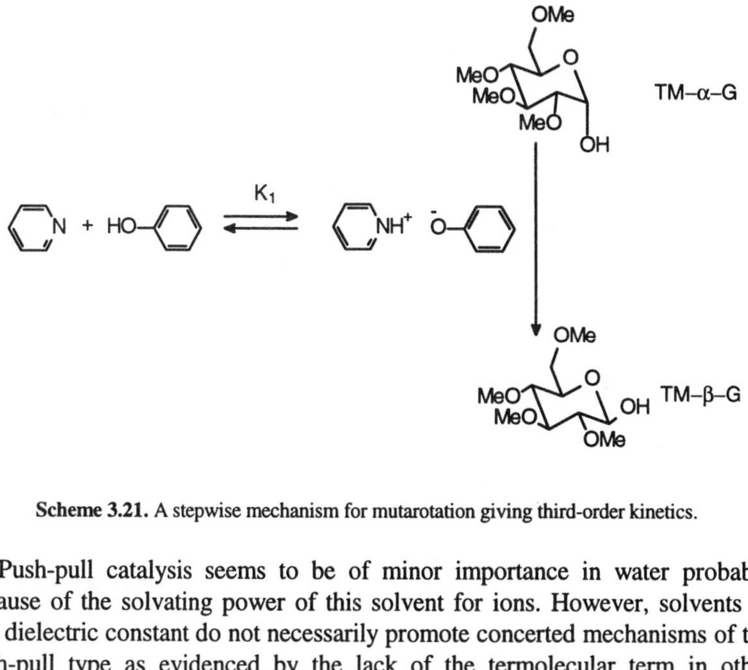

Scheme 3.21. A stepwise mechanism for mutarotation giving third-order kinetics.

Push-pull catalysis seems to be of minor importance in water probably because of the solvating power of this solvent for ions. However, solvents of low dielectric constant do not necessarily promote concerted mechanisms of the push-pull type as evidenced by the lack of the termolecular term in other systems (Bell, 1959).[67]

3.3.2.2 Reactions of carboxylic acid derivatives
General acid catalysis. The addition of nucleophiles to the carbonyl group in carboxylic acid derivatives is fundamental to many processes in biology, and in aqueous solution the reaction is often assisted by acid-base catalysis. The mechanisms for hydrolysis of esters or amides catalysed by acid usually involves a straightforward stepwise process (Scheme 3.22).

Proton transfer concerted with addition at the carbonyl group as well as concerted push-pull mechanisms have been discovered in a number of cases.

The conditions of solvent, leaving group, nucleophile and substituents at the carbonyl function do not always allow the concerted process to occur and in aqueous solutions it was not until the late 1970's that authentic cases were demonstrated. The various mechanisms are shown in Scheme 3.23. Reactions of carboxylic acid derivatives have steps subsequent to the addition process and these may also be catalysed by proton transfer. The overall mechanism for ester aminolysis was proposed by Satterthwait and Jencks (1974)[68] (Scheme 3.24), and it is seen that this comprises proton transfer among the tetrahedral intermediates subsequent to the addition step.

General base catalysis. The methoxyaminolysis of N-acetyltriazole exhibits general base catalysis (k_B) and the Brønsted dependence of the k_B term fits an Eigen-type plot[69] with a break at the pK_a of the donor acid on the zwitterionic intermediate.[70,71]

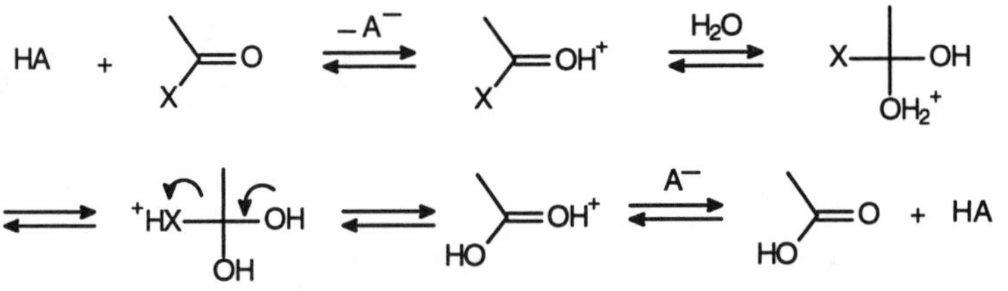

Scheme 3.22. Specific acid-catalysed hydrolysis of amides.

Scheme 3.23. Concerted acid-base catalysed addition to carboxylic acid derivatives.

The reaction of hydrazine with acetylimidazole is catalysed by bases and the curvature exhibited by the Brønsted plot (Page and Jencks, 1972)[72] can be explained simply as a rate-limiting proton transfer reaction for which the limiting rate constants are diffusion processes. A stepwise process is in accord with the *libido* rule that concerted catalysis only occurs when there is a large change in pK_a of the proton donor site where an unfavourable is converted into a favourable proton transfer (Jencks, 1972;[73] Gandour, Maggiora and Schowen, 1974[74]) (Scheme 3.25). When the lifetime of T_- in a similar system is too short for the intermediate to exist, the breakdown of $T_\pm$ catalysed by weak bases will occur through an enforced concerted pathway (Scheme 3.26).

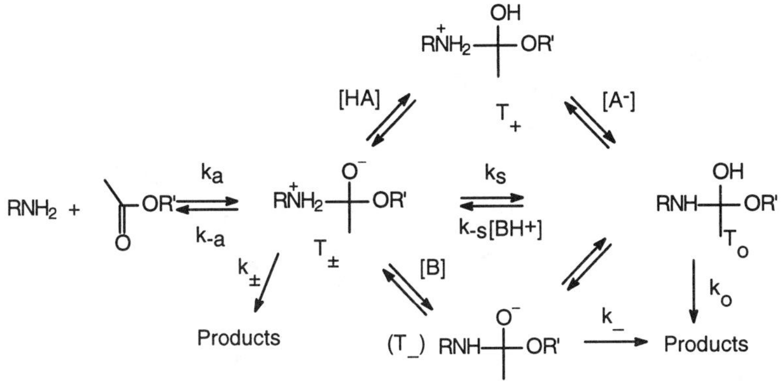

Scheme 3.24. A mechanism for ester aminolysis.

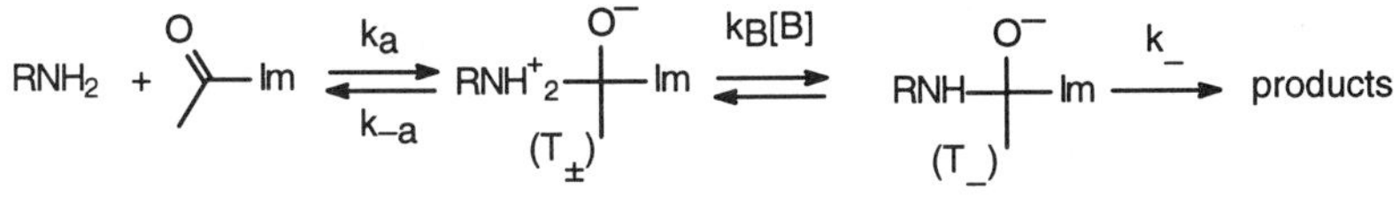

Scheme 3.25. The libido rule indicates that there is a stepwise proton transfer from $T_\pm$ because of the small change in the pK_a of the attacking nitrogen through the reaction.

If the diffusion apart of **3.9** (k_{-d}) is slower than its breakdown to **3.10** (k_a') the lowest energy path for the breakdown of **3.9** will be via **3.10**. By the principle of microscopic reversibility the formation of **3.9** must be through **3.10**, a pre-association (Jencks and Salvesen, 1971)[75] or "spectator" mechanism (Kershner and Schowen, 1971).[76] The base B is not necessary for the formation of $T_\pm$ to go to product as diffusion in of B to form an encounter complex with $T_\pm$ is not fast enough to support it. The k_a' step is rate-limiting when the base is strong and hence $\beta = 0$. If the complex **3.9** has a very short lifetime T^- may be formed directly by the concerted route (k_a'').

General acid catalysis. The general acid-catalysed hydrazinolysis of acetylimidazole possesses a linear Brønsted dependence[72] consistent with a concerted mechanism. In the case of the methoxyaminolysis of acetyltriazole the Brønsted dependence is non-linear for general acid catalysis,[71] indicating a stepwise process. The general acid-catalysed hydrazinolysis of acetylimidazole and general base-catalysed hydrazinolysis of the acetylimidazolium ion are *kinetically ambiguous* mechanisms and the latter mechanism takes the reaction flux of the process. The reaction is an enforced concerted process due to the species **3.11** and **3.12** being too unstable to exist.

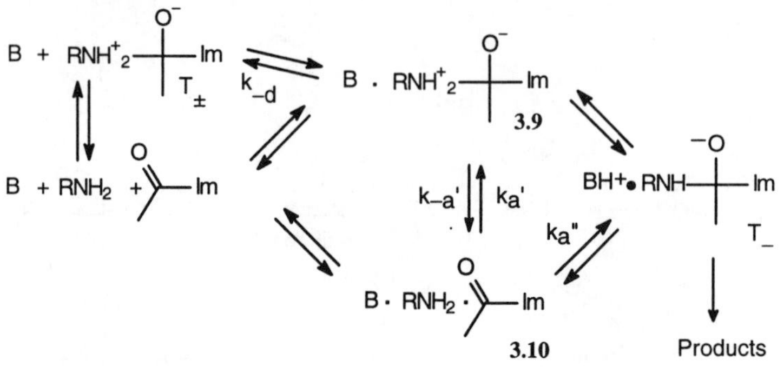

Scheme 3.26. The short lifetime of T_ enforces a concerted pathway for the decomposition of 3.9 in the aminolysis of acylimidazoles.

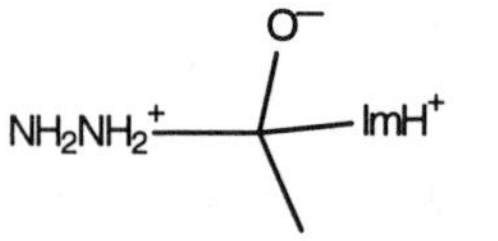

$Bifunctional$ $acid\text{-}base$ $catalysis.$ The bifunctional catalysis observed by Swain[58-60] for mutarotation (see earlier) in non-aqueous solvents provided the inspiration for studies of bifunctional catalysis of acyl group transfer reactions. Examples of concerted, bifunctional acid-base catalysis are rare and it is suspected that catalysis at one end of a system is likely to decrease the importance of catalysis at the other end.[73,77] The most clear-cut examples of bifunctional acid-base catalysis in aqueous solution involve the trapping of unstable intermediates by a stepwise reaction mechanism. Concerted, bifunctional proton transfer has been demonstrated in the methoxyaminolysis of phenyl acetate (Figure 3.8);[78] the catalysts cacodylic acid, bicarbonate ion and the monoanions of phosphates etc. are some 100- to 1000-fold more effective than monofunctional acids of comparable pK_a.

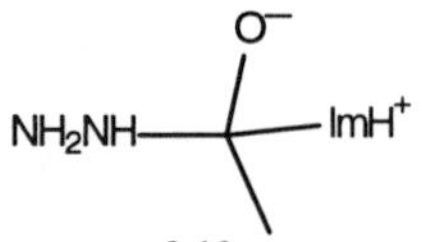

The requirements of efficient bifunctional catalysis are:

(i) $Reaction$ $proceeds$ $through$ an $unstable$ $intermediate$ $which$ has pK_a $values$ $permitting$ $conversion$ to the $stable$ $intermediate$ or $product$ by two $proton$ $transfers$ $after$ $encounter$ $with$ the $catalyst.$

(ii) $Proton$ $transfer$ $with$ $monofunctional$ $catalysts$ $must$ be $slow.$

The enhanced activity of bifunctional catalysts can be explained by two stepwise proton transfers in a "one encounter" mechanism before separation of catalyst or by fully concerted proton transfer after encounter of catalyst with the reactants.[5,79]

Bifunctional concerted catalysis in the methoxyaminolysis of phenyl acetate is in accord with the observed absence of a solvent deuterium isotope effect.[78] The data for bifunctional catalysis shows a higher reactivity than that of monofunctional catalysis and the Brønsted slope (−0.16) indicates that the rate-limiting step is attack of amine with hydrogen-bonding (k'_1)(Scheme 3.27). Downward curvature of the Brønsted line for monofunctional catalysis means that the proton transfer step (k_p) is rate-limiting when the reaction is no longer favourable. The absence of a break in the Brønsted plot for bifunctional catalysis means that proton transfer does not go through k_p but must be by-passed by a faster, concerted mechanism (k_c).

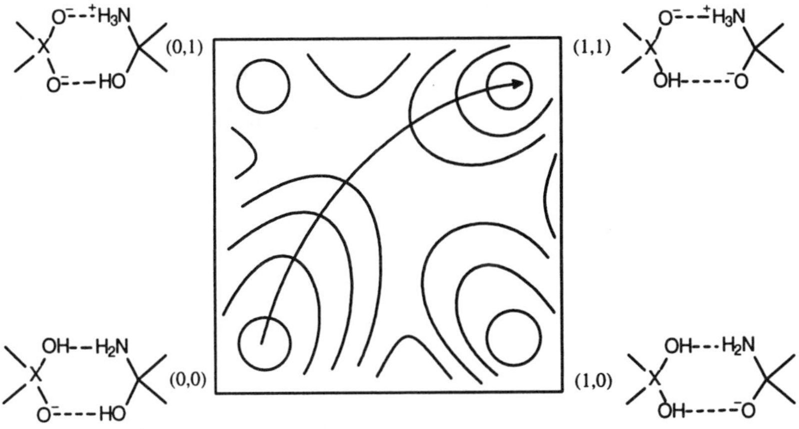

Figure 3.8. Concerted, bifunctional acid-base catalysis of methoxyaminolysis of phenyl acetates; contour lines are notional.

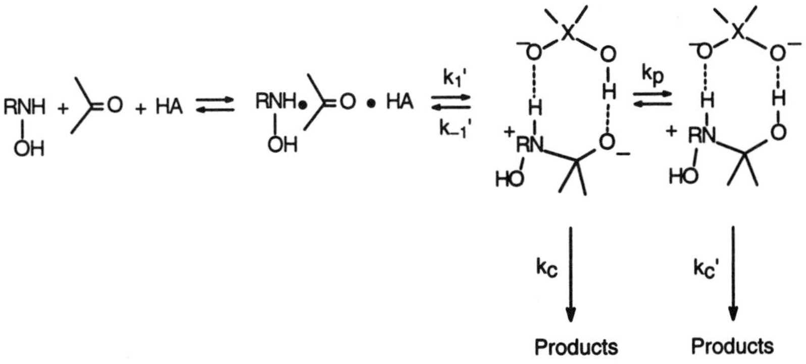

Scheme 3.27. Proton switch mechanism.

Menger (1966)[80] showed that the n-butylaminolysis of 4-nitrophenyl acetate in chlorobenzene involved kinetics second-order in n-butylamine concentration whereas the reaction with benzamidine had only first-order kinetics in benzamidine concentration. It is not possible to compare these rate constants but an upper limit estimate of the kinetic term *first-order* in n-butylamine is some 15,000-fold less than that for benzamidine. A mechanism (Scheme 3.28) was proposed. Anderson, Su, and Watson (1969)[81] showed that the cyclic amidine **3.13** was more reactive than benzamidine toward the 4-nitrophenyl acetate even though a cyclic structure could not be written for this reaction.

A different sort of bifunctional catalysis was observed in the aminolysis of thiol esters in aqueous solution (Anderson, Blackburn, and Murphy, 1972)[82] (Scheme 3.29) whereby the leaving group itself provides the electrophilic assistance at the carbonyl oxygen. Blocking the electrophilic proton by substituting a methyl group causes the reactivity to decrease substantially (Scheme 3.29).

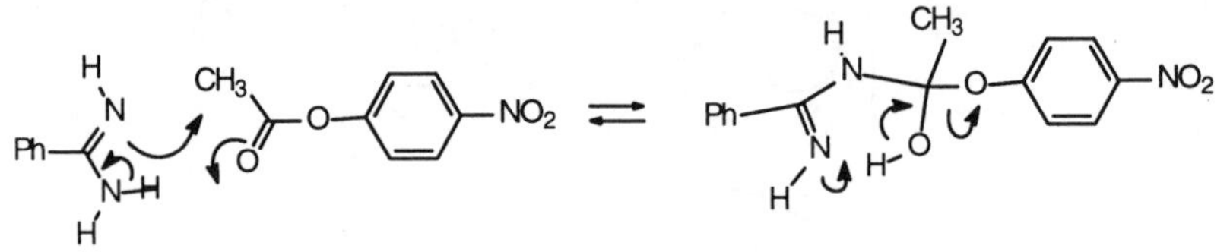

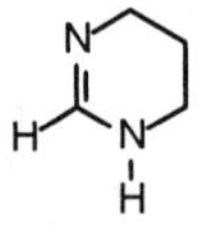

3.13

Scheme 3.28. Bifunctional concerted catalysis of acyl group transfer?

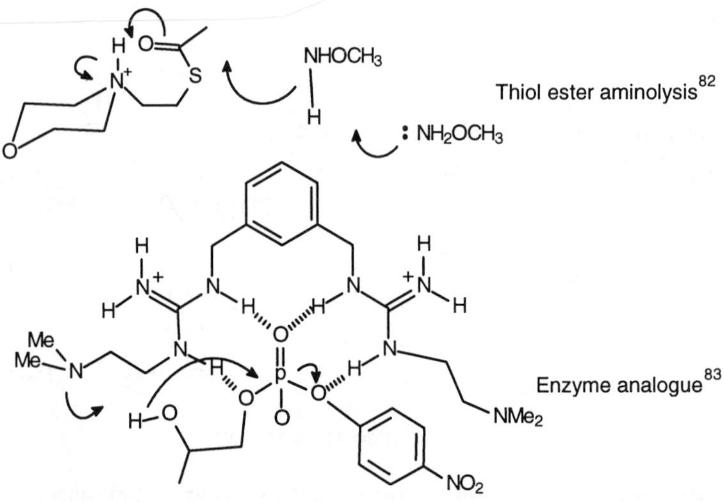

Thiol ester aminolysis[82]

Enzyme analogue[83]

Scheme 3.29. Push-pull mechanisms in acyl group transfer.

The interest in concerted bifunctional catalysis as a source of enzyme activity has given rise to a very active area of research involving the design and synthesis of potential enzyme analogues. We think that design, *de novo*, will not be of much use in the development of such catalysts as our knowledge of the relevant molecular recognition factors, while extensive, is never likely to be sufficiently predictive; it must be remembered that native enzymes locate the substrates very precisely on the enzyme surface. Since enzymes developed by a process of chance it is more reasonable to expect that a *combinatorial* technique will prevail over design. Nevertheless, substantial knowledge will be gained from this sort of work and a recent example is illustrated in Scheme 3.29.[83]

3.3.2.3 Condensation at ketones and aldehydes

Condensation reactions at carbonyl groups involve elimination steps which are essentially the microscopic reverse of the addition process and, as in the reactions of carboxylic acid derivatives, can involve proton transfer between various tetrahedral intermediates. Carbinolamine formation from benzaldehydes and substituted hydrazines is subject to general base catalysis and the Brønsted plot of the rate constant (Figure 3.9) conforms to the Eigen equation[69] consistent with a stepwise proton transfer according to Scheme 3.30.[5,84] The reactivity of the hydroxide ion is some 10-fold larger than that for the diffusion of a regular anion because it transports through water *via* a Grotthus chain mechanism. The observation of such a deviation is diagnostic of the Eigen mechanism and hence of a stepwise process. The experimental observation excludes base acting in

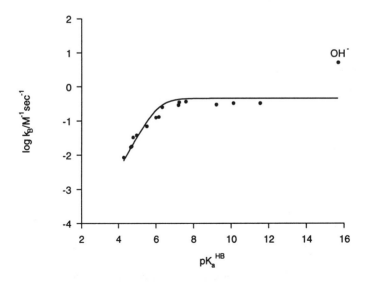

Figure 3.9. Eigen type plot for the general base catalysed of 4-chlorobenzaldehydes with a substituted hydrazine; data from *Refs. 85,86* and line is calculated from $\log k_B = -0.341 - \log (1 + 10^{6.09-pKa})$.

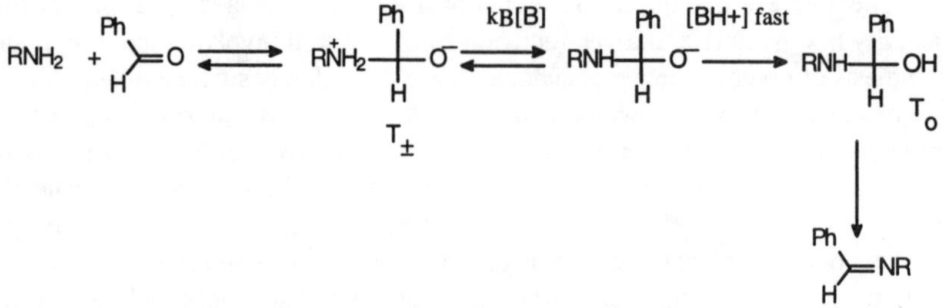

$$RNH_2 + \underset{H}{\overset{Ph}{\diagup}}C{=}O \rightleftharpoons \underset{\underset{T_\pm}{H}}{R\overset{+}{N}H_2{-}\overset{Ph}{\underset{\vert}{C}}{-}O^-} \overset{kB[B]}{\rightleftharpoons} \underset{H}{RNH{-}\overset{Ph}{\underset{\vert}{C}}{-}O^-} \overset{[BH+]\ fast}{\longrightarrow} \underset{H\ \ T_o}{RNH{-}\overset{Ph}{\underset{\vert}{C}}{-}OH}$$

$$\underset{H}{\overset{Ph}{\diagup}}C{=}NR$$

Scheme 3.30. General base catalysed hydrazinolysis of benzaldehydes.

concert with amine attack as in the transition structure **3.14**. Condensation of benzaldehydes and substituted hydrazines involves formation of the carbinolamine assisted by general acid at the carbonyl oxygen as in structure **3.15**; the evidence is that the Brønsted equation for acid catalysed reaction has a small but significant slope (Sayer and Jencks).[85,86] General acid catalysis by acetic acid is an order of magnitude larger than that of general base catalysis by basic amines and thus free $T_\pm$ cannot be an intermediate. The higher reactivity is attributed to a transition state with a high degree of proton transfer from the catalyst.

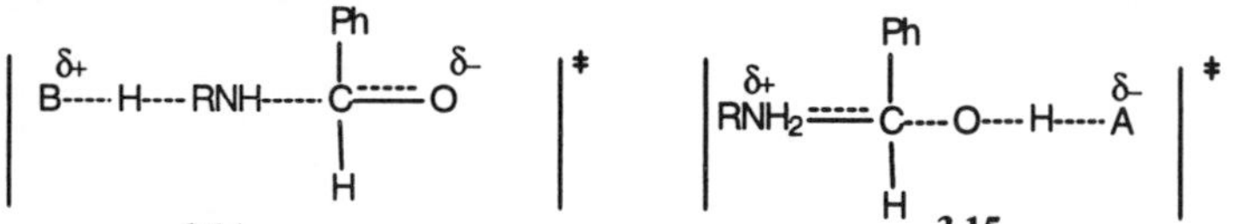

$$\left| \underset{3.14}{\overset{\delta+}{B}{\cdots}H{\cdots}RNH{\cdots}\underset{\underset{H}{\vert}}{\overset{Ph}{\underset{\vert}{C}}}{=\!=}\overset{\delta-}{O}} \right|^{\ddagger} \qquad \left| \underset{3.15}{\overset{\delta+}{RNH_2}{=\!=}\underset{\underset{H}{\vert}}{\overset{Ph}{\underset{\vert}{C}}}{\cdots}O{\cdots}H{\cdots}\overset{\delta-}{A}} \right|^{\ddagger}$$

Transimination[87] is an apparently simple symmetrical reaction that is complicated by the requirement that two protons must be transferred in a stepwise process.

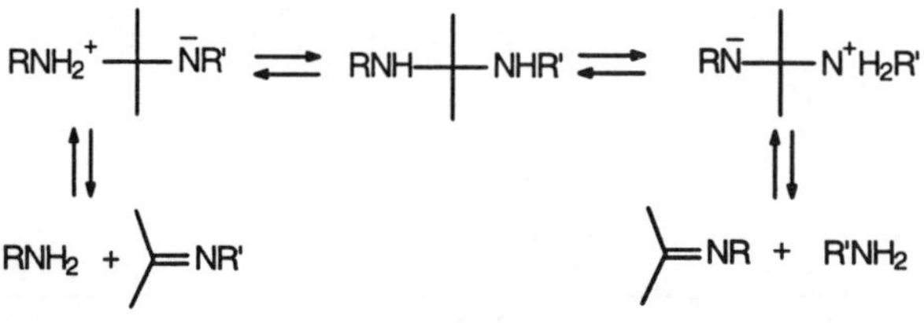

$$RNH_2^+{-}\overset{\vert}{\underset{\vert}{C}}{-}\bar{N}R' \rightleftharpoons RNH{-}\overset{\vert}{\underset{\vert}{C}}{-}NHR' \rightleftharpoons R\bar{N}{-}\overset{\vert}{\underset{\vert}{C}}{-}N^+H_2R'$$

$$RNH_2 + \diagup C{=}NR' \qquad\qquad \diagup C{=}NR + R'NH_2$$

Reactants   Products

Scheme 3.31. Transimination is a symmetrical stepwise reaction.

Base catalysis. A non-linear Brønsted dependence[88] for the base-catalysed reaction of hydroxylamine with benzhydrylidenedimethyl ammonium ion

(Scheme 3.32) is consistent with the trapping mechanism of rate-limiting proton removal from T_+ rather than proton donation to T_0.

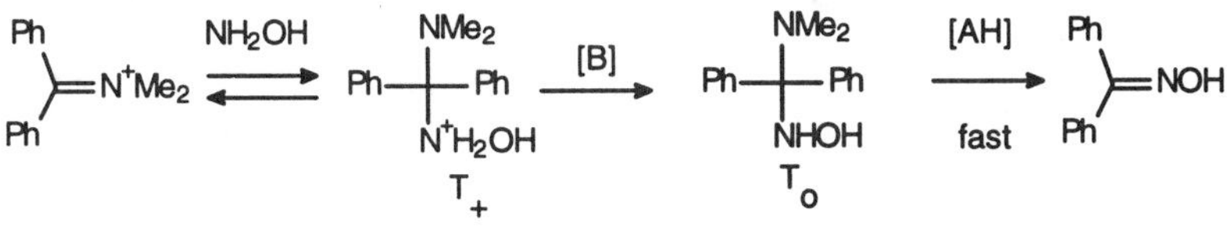

Scheme 3.32. Base-catalysed reaction of hydroxylamine with an iminium ion.

The base-catalysed hydrolysis of the imine is concerted (Kohler, Sandstrom and Cordes);[89] the Brønsted dependence is linear with only a small slope (0.24) and the solvent isotope effect (1.9±0.2) shows no trend or maximum with changing pK_a^{HB}. The small β is consistent with that expected (Structure **3.16**) for the transition state of a concerted mechanism.[90,91]

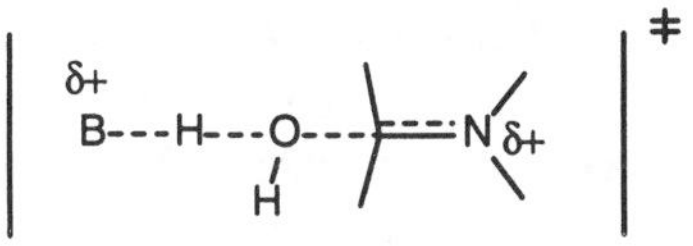

3.16

Acid catalysis. Catalysis of the reaction of hydroxylamine with the 4-methoxy-benzylidene-pyrrolidinium ion exhibits non-linear plots of rate constant versus acid concentration. The absence of a dependence of the rate constant on acid strength with an increased rate constant for the proton (which is due to facilitated Grotthus-type diffusion) is expected for a mechanism which involves trapping an unstable intermediate by an encounter-controlled rate.[92] The observation of a stepwise diffusion-controlled proton transfer rather than a concerted or hydrogen-bonding mechanism is excluded by the "libido" rule[73,74] because proton transfer between catalyst and product is not thermodynamically favourable.

Bifunctional acid-base catalysis. The "proton-switch" mechanism whereby T_1^+ transfers two protons to give T_2^+ via the intermediacy of a bifunctional catalyst (Scheme 3.33) manifests itself in an unexpectedly higher catalytic activity. There is no detectable "proton-switch" mechanism in the hydroxylaminolysis of benzylidenedimethyl-ammonium ion because the intermediate T_1^+ (formed from imine in the T_1^+ conformation shown in **3.17**) cannot isomerise rapidly enough to the conformation where donor and acceptor are on the same side of the molecule (**3.18**).[92] Structure **3.17** is the initially formed intermediate with lone pair antiperiplanar to the entering nucleophile. Catalysis by carboxylic acids in the reaction of hydroxylamine with 4-methoxybenzylidene pyrrolidinium ion was attributed to bifunctional catalysis after rotation/inversion of **3.18** to **3.17**.[92]

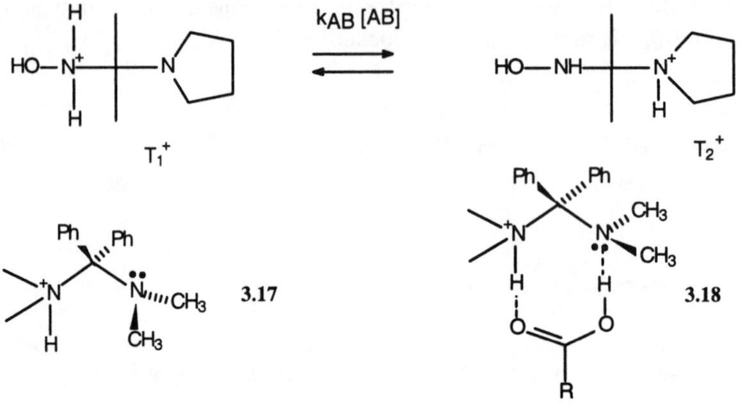

Scheme 3.33. The proton switch process in transimination.

3.3.2.4 Heterocumulenes

The E1cB mechanism of acyl group transfer involves the ionisation of a proton on an atom α to the acyl group followed by expulsion of the leaving group to yield an intermediate, heterocumulene, which then adds water or nucleophile to give the substituted product (Scheme 3.34).[93] The mechanism is general for all acyl groups and is involved in amino-phosphoryl and aminosulphonyl group transfer (Scheme 3.35).[94] General acid base catalysis is not observed for these reactions and the proton transfer is therefore not concerted with departure of the leaving group.

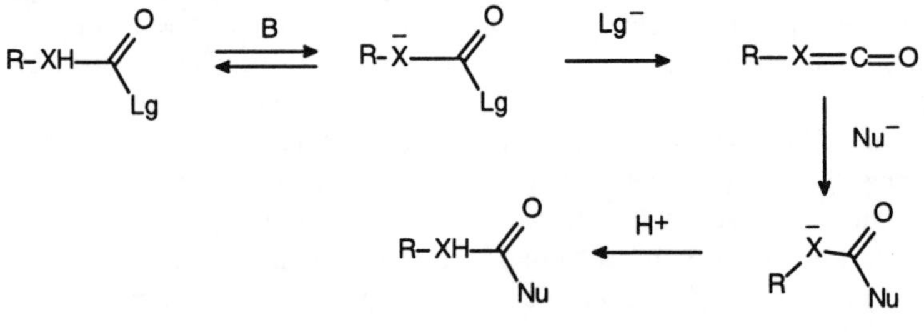

Scheme 3.34. Elimination-addition mechanism of acyl group transfer.

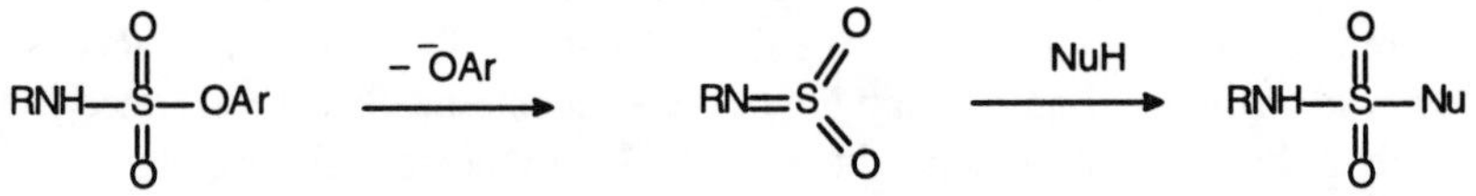

Scheme 3.35. Elimination-addition mechanism of aminosulphonyl group transfer.

Both acid and base catalyses are observed for the aminolysis of isocyanic acid[95,96] and since these exhibit "Eigen"[5,84] type relationships (Figure 3.10) it is evident that in this case the proton transfer is not concerted (Scheme 3.36) with the bond formation step. The deviation of the proton catalysed reactivity from the Eigen-line (B) in Figure 3.10 is diagnostic of the stepwise process involving a diffusion limited step. The stepwise process illustrated for the addition of amines to isocyanic acid involves relatively stable charged intermediates and the estimated pK_a's (10.2 and 1.8, see Figure 3.10) agree with the break-points obtained from Figure 3.10.[95,96] If the stability of the intermediates were to be lowered a concerted proton transfer would become the favourable mechanism involving water or a bifunctional catalyst as a *proton switch*.

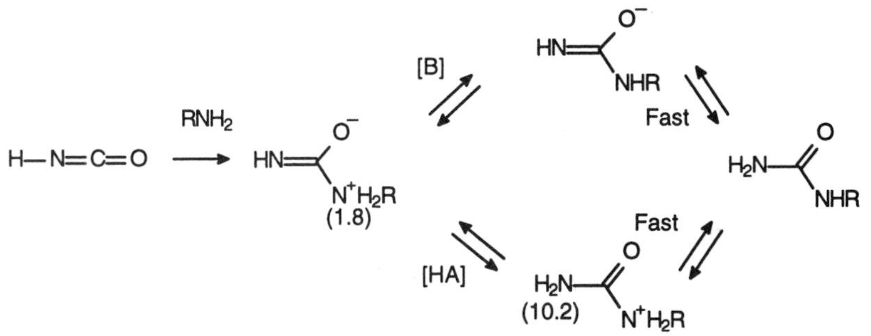

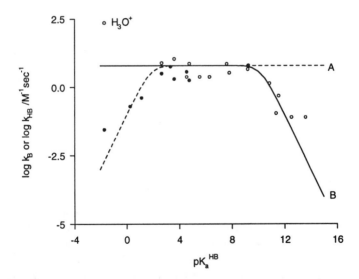

Scheme 3.36. Amine addition to isocyanic acid; numbers in parentheses refer to the pK_a of the indicated acid.

Figure 3.10. Brønsted plots for acid base catalysis of the addition of amines to isocyanic acid; lines A and B respectively are calculated from the equations: $\log k_{HB} = \log 4.0 - \log (1 + 10^{1.8 - pKaHB})$ and $\log k_B = \log 4.0 - \log (1 + 10^{pKaHB - 10.2})$. Data are from *Refs. 95 and 96*.

In the case of the reaction of acetate ion with carbodiimides, analogous to the first step of the well-known peptide coupling protocol, the mechanism is general acid catalysed (Scheme 3.37).[97] The stepwise processes would be circumvented by the concerted mechanism as both intermediates have high energies compared with those of the reactants. In the case of the aminolysis of carbodiimides the zwitterionic intermediate is relatively stable (Scheme 3.38), due to electrostatic interaction and stepwise proton transfer becomes more efficient.[98]

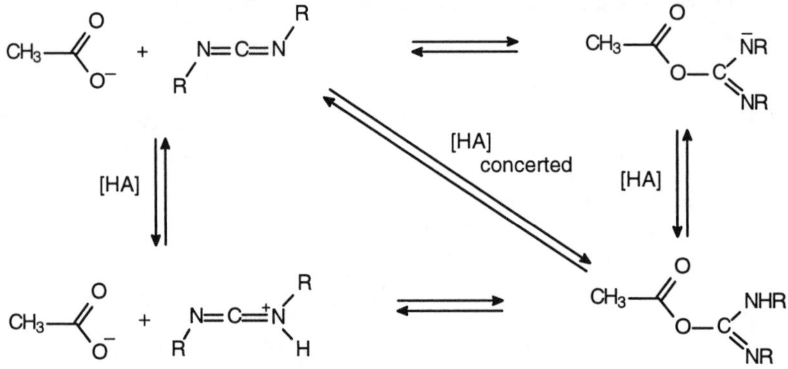

Scheme 3.37. Mechanisms for the acid-catalysed addition of acetate ion to carbodiimides.

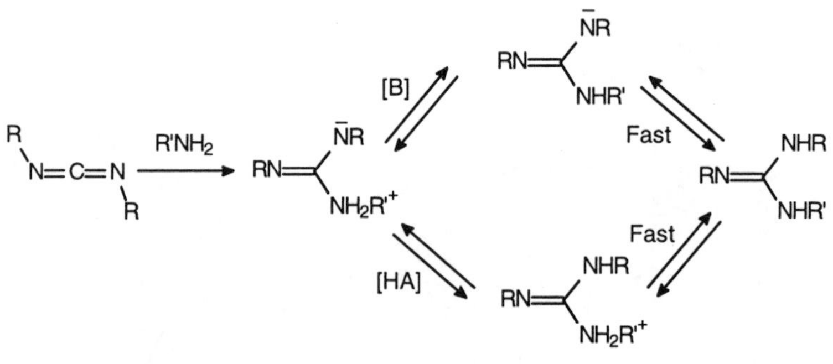

Scheme 3.38. Stepwise mechanisms for the addition of amines to carbodiimides.

Dalby and Jencks (1997)[99] discovered that general bases catalysed the fission of the C-S bond suffered by thiophenyl isothioureas (Scheme 3.39). The reactivity was studied as a function of the base (pK_a) and aryl ring (σ). The value of p_{xy} ($\partial\beta/\partial\sigma lg$) is small and positive (~0.03) for the effect of changing the substituent in the phenyl ring on the Brønsted coefficient for the reverse reaction. A positive value of p_{xy} indicates signficant interaction between the proton donor and the nucleophile in the transition state (Jencks and Jencks, 1977)[100] and the present value is consistent with hydrogen-bonding by the general acid in the transition state. A concerted mechanism enforced by the instability of the intermediate anions **3.19** and **3.20** (Scheme 3.40) cannot, however, be excluded for this reaction. The equilibrium constant for tautomerisation of the nitrile to carbodiimide (Scheme 3.40) is unknown but it is judged to be very unfavourable considering that the equilibrium constant for the analogous tautomerisation of acetonitrile to ketenimine is very low ($K_T = 10^{-22.6}$).[54]

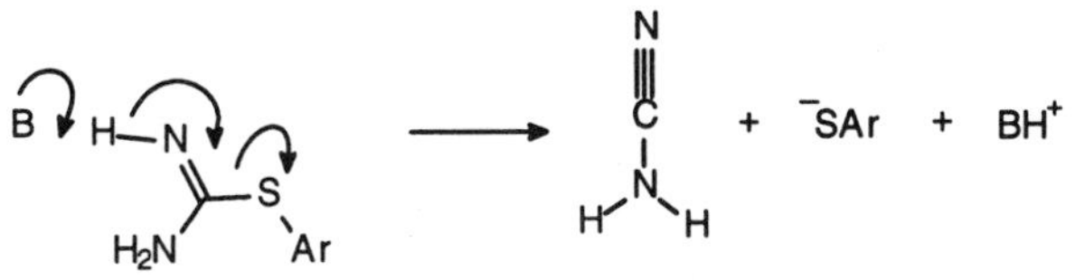

Scheme 3.39. Base-catalysed fission of S-phenylisothioureas.

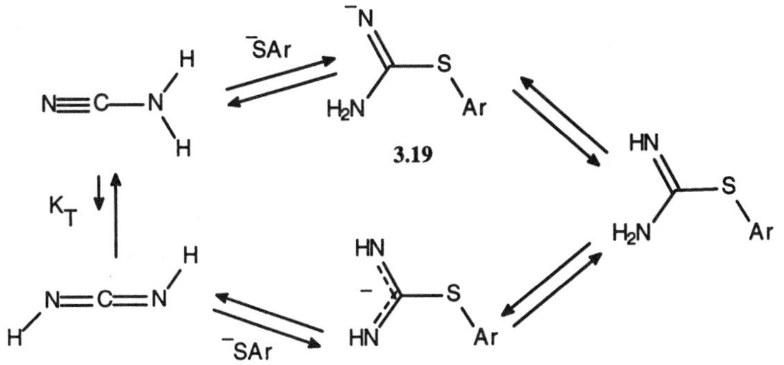

Scheme 3.40. General acid-catalysed addition of thiolate ions to cyanamide.

3.3.2.5 Reactions involving the C=S bond

Sulphene formation. Sulphenes (**3.21**), the carbon analogues of sulphur trioxide,

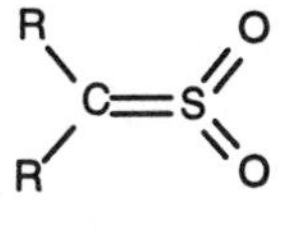

3.21

are reactive intermediates in sulphonyl group ransfer reactions[101-103] and they are formed in base-catalysed reactions from sulphonate derivatives (R_2CHSO_2Lg). The elimination process was shown to be stepwise in the majority of cases *via* the E1cB ($A_{xh}D_H+D_N$) mechanism shown in Equations (3.8 and 3.9). It is possible in the case of the phenylsulphene intermediate to show that the conjugate base **3.22** has a lifetime less than 10^{-13} seconds for certain leaving groups. The pK_a of the phenylmethanesulphonate ester may be determined for a series of aryl esters. Knowledge of the overall value of k_{OH} and that the E1cB mechanism operates enables k_2 to be determined ($k_{OH} = k_2 K_a / K_w$); the Brønsted plot of k_2 is very sensitive to pK_a ($\beta = 2$) and the calculated value of k_2 becomes greater than 10^{13} $M^{-1}sec^{-1}$ at $pK_a \sim 5$. This system is discussed more fully in Chapter 2 and the Brønsted plot is illustrated in Figure 2.4.

$$PhCH_2SO_2OAr \underset{\longleftarrow}{\overset{Base}{\rightleftharpoons}} \quad \overset{\_}{PhCH}-SO_2-OAr \longrightarrow PhCH=SO_2 \qquad (3.8)$$

$$\underset{\textbf{3.22}}{}$$

$$PhCH=SO_2 \overset{H_2O}{\longrightarrow} PhCH_2SO_2-OH \qquad (3.9)$$

3.4 NON-PERFECT SYNCHRONIZATION

3.4.1 The nitroalkane anomaly

Charge imbalance in transition states was first recognised in alkene-forming E2 eliminations by Ingold (1927)[104,105] and the concept was clarified by Bunnett (1962).[39] Imbalance occurs when bonding changes are not synchronous and is reflected in different structure-reactivity coefficients for substituents at different positions within the transition state. The imbalance of charge between the bonding changes is taken up by other bonds and atoms in the reacting system. The most dramatic evidence of imbalance was reported by Bordwell[106] for conflicting Leffler parameters derived for the different bond changes in the formation of pseudo-bases from nitroalkanes (Scheme 3.41). The proton transfer component of a reaction involving bond rehybridisation has been employed as a vehicle for studying the differences in extents of bond changes in a concerted mechanism. The discussion subsequent to Bordwell's discovery indicated that the concept of reaction progress could not be measured by a single parameter when more than one bond was undergoing change. The value of $\alpha_{forward}$ equal to $(\partial logk^X / \partial logK_{eq}{}^X)$ is greater than unity for a number of nitroalkanes indicating that the delocalisation to form the double bond lags behind the proton transfer component in the transition state structure and an imbalance of charge develops on the central carbon atom. It should be emphasised that there are not two steps in the reaction which is concerted *but not synchronous*. The imbalance is illustrated in the More O'Ferrall-Jencks diagram (Figure 3.11). The x coordinate corresponds to the proton transfer step and the y coordinate corresponds to the rehybridisation process. The 0,1 corner is undoubtedly of very high energy and indeed previous authors have hesitated to identify this species (Albery, Kresge, and Bernasconi;[107] Sinnott,[23] and Jencks[99]); the pentacoordinate species is necessary if both vertical coordinates (0,0-0,1 and 1,0-1,1) are to correspond to the same process as is usually required in these diagrams. It is perfectly justified to place a hypothetical structure on a corner of a More O'Ferrall-Jencks diagram; there is precedence that similar, onium-like, structures ($CH_5{}^+$) have an existence (Olah and Rasul[108,109]). Whatever the identity of the structure at the 0,1 coordinates it has a very high energy and the surface will be skewed forcing the transition state structure towards the 1,0 coordinates.

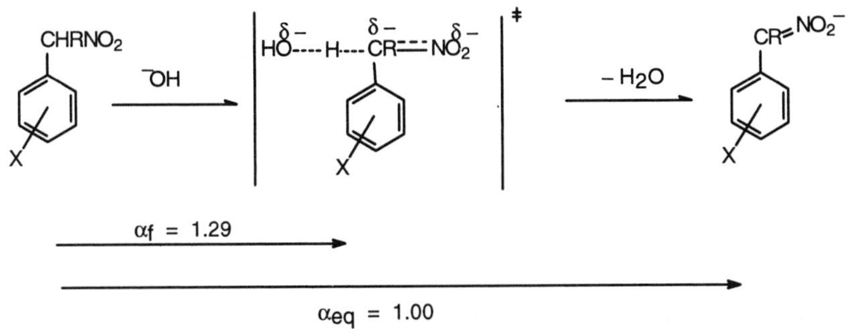

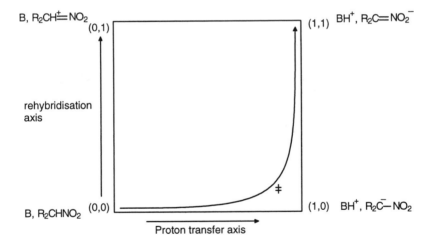

Scheme 3.41. Ionisation of nitroalkanes; $\alpha_f = \alpha_{forward}$ and α_{eq} is the value of the Leffler parameter for the equilibrium reaction (defined as unity).

Figure 3.11. Pseudo-base formation from nitroalkanes.

3.4.2 Product-stabilising factors and the intrinsic barrier

The intrinsic barrier is the Gibbs' free energy of activation for the ergoneutral reaction ($\Delta G^0 = 0$). The concept was introduced by Marcus[110,111] for electron transfer reactions and discussed by Kresge[112,113] and More O'Ferrall.[114] Bernasconi[115,116] proposed that there is a relationship between intrinsic barrier and transition state imbalance and that high intrinsic barriers are associated with lack of synchrony. The principle of *non-perfect synchronisation* is that a product-stabilising factor (*f*) that develops late at the transition state always increases the intrinsic barrier, whereas a product-stabilising factor (*f*) that develops early lowers the intrinsic barrier. Equations (3.10) and (3.11) formalise this principle mathematically. Late development of resonance means that $\lambda_f <$ β_B while stabilisation of the carbanion (in the nitroalkane case by resonance) by the product-stabilising factor implies $\partial \log k_0^R < 0$, ie k_0 is decreased relative to a reference reaction in which the carbanion has no product-stabilising factor.

$$\partial \log k_0^f = (\lambda_f - \beta_B) \, \partial \, \log K_1^f \qquad (3.10)$$

$$\lambda_f = \partial \log k_1^f / \partial \log K_1^f \qquad (3.11)$$

The principle of non-perfect synchronisation is perfectly general and the product-stabilising factor could derive from resonance, hydrogen-bonding interaction, polarisability effects, or solvation.

3.4.3 Hydrogen-bonding and catalysis

Hydrogen-bonding can effect catalysis simply by providing energy to stabilise a complex between catalyst and substrate but without being involved intrinsically in the catalytic mechanism. There would be no effective coupling between hydrogen-bonding involved in binding and the bonding changes of the catalytic reaction.

Another rôle of hydrogen-bonding would be to interact with the catalytic reaction via electronic coupling to transfer energy to assist the bonding changes in a concerted reaction.

Hydrogen-bond interaction could act in carbonyl addition reactions either as an electrophilic component assisting nucleophilic attack or, in the reverse direction, as the hydrogen-bond acceptor assisting departure of a leaving group by enhancing nucleophilicity (Scheme 3.42).

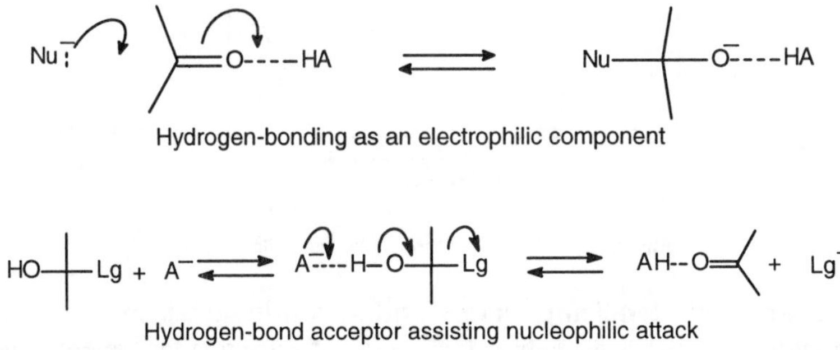

Hydrogen-bonding as an electrophilic component

Hydrogen-bond acceptor assisting nucleophilic attack

Scheme 3.42. Catalysis of carbonyl addition reactions by hydrogen-bonding.

Catalysis by hydrogen-bonding can be considered in connection with the transition from a stepwise pre-association mechanism to a fully concerted process. For example, the stepwise pre-association mechanism for general-acid-catalysed addition of thiol anions to acetaldehyde requires the formation of a complex hydrogen-bonded between the acid catalyst and the carbonyl group (Scheme 3.43). Even if the hydrogen-bond is weak the pre-association mechanism results in acceleration.[117] A change to a weakly basic thiol anion favours stronger hydrogen-bonding and a concerted reaction mechanism should result because the transition state of the return step (k_{-2}) must lie at a lower energy than that of the diffusion step (k_{-HA}).

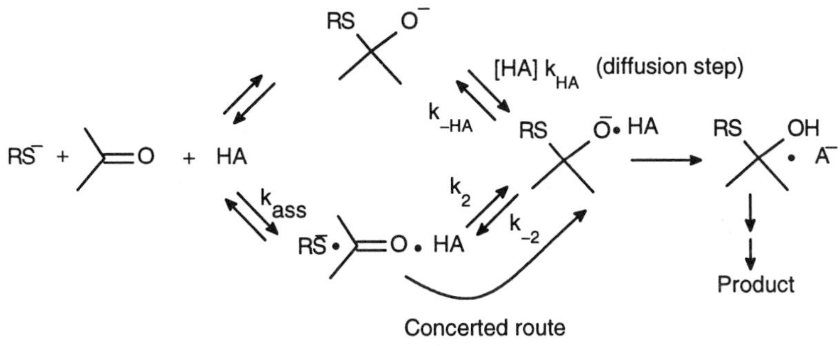

Scheme 3.43. Stepwise mechanisms for general-acid-catalysed addition of thiols to acetaldehyde becomes concerted when $k_{-2} > k_{-HA}$.

FURTHER READING

Ahlberg, P., Janné, K., Löfås, S., Nettelblad, F., and Swahn, L., Bifunctionally catalysed 1,3-proton transfer of a propene by a *sec*-amidine studied by deuterium isotope effects. A stepwise two-proton transfer mechanism, *J. Phys. Chem.*, 2, 429, 1989.

Albery, W. J., A Marcus model for concerted proton transfer, *Disc. Far. Soc.*, 74, 245, 1982.

Albery, W. J., Application of the Marcus relation to concerted proton transfers, *J. Chem. Soc., Farad.Trans. 1*, 78, 1579, 1982.

Baciocchi, E., Base dependence of transition state structure in alkene-forming E2 reactions, *Accs. Chem. Res.*, 12, 430, 1979.

Bernasconi, C. F., The principle of non-perfect synchronisation, *Adv. Phys. Org. Chem.*, 27, 119, 1992.

Bernasconi, C. F., The principle of non-perfect synchronisation: more than a qualitative concept? *Accs. Chem. Res.*, 25, 9, 1992.

Bordwell, F. G., How common are base-initiated, concerted 1,2-eliminations? *Accs. Chem. Res.*, 5, 374, 1972.

Chen, Y., Gai, F., and Petrich, J. W., Solvation of 7-azaindole in alcohols and water. Evidence for concerted, excited-state, double-proton transfer in alcohols, *J. Amer. Chem. Soc.*, 115, 10158, 1993.

Eigen, M., Proton transfer, acid-base catalysis, and enzymatic hydrolysis. Part I: Elementary processes, *Angew. Chem., Int. Ed. Engl.*, 3, 1, 1964.

Ford, W. T., Mechanisms of elimination reactions catalysed by weak bases, *Accs. Chem. Res.*, 6, 410, 1973.

Fry, A., Isotope effect studies of elimination reactions, *Chem. Soc. Rev.*, 1, 163, 1972.

Gandour, R. D., Maggiora, G. M., and Schowen, R. L., Coupling of proton motions in catalytic activated complexes. Model potential-energy surfaces for hydrogen-bond chains, *J. Amer. Chem. Soc.*, 96, 6967, 1974.

Grunwald, E., Reaction mechanism from structure-energy relations. 1. Base-catalysed addition of alcohols to formaldehyde, *J. Amer. Chem. Soc.*, 107, 4710, 1985.

Hibbert, F., Proton transfer from intramolecularly hydrogen-bonded acids, *Accs. Chem. Res.*, *17*, 115, 1984.

Janaway, G. A., Zhong, M., Gatev, G. G., Chabinye, M. L., and Brauman, J. I., [FHNO]⁻ : An intermediate in a spin-forbidden proton transfer reaction, *J. Amer. Chem. Soc.*, *119*, 11697, 1997.

Jencks, W. P., When is an intermediate not an intermediate? Enforced mechanisms of general acid-base catalysed, carbocation, carbanion, and ligand exchange reactions, *Accs. Chem. Res.*, *13*, 161, 1980.

Jencks, D. A., and Jencks, W. P., On the characterization of transition states by structure - reactivity coefficients, *J. Amer. Chem. Soc.*, *99*, 7948, 1977.

Jencks, W. P., Catalysis in chemistry and enzymology, McGraw-Hill, New York, 1969, pp. 196 and 198.

Jencks, W. P., Enforced general acid-base catalysis of complex reactions and its limitations, *Accs. Chem. Res.*, *9*, 425, 1976.

Jencks, W. P., General acid-base catalysis of complex reactions in water, *Chem. Revs.*, *72*, 705, 1972.

Jencks, W. P., Requirements for general acid-base catalysis of complex reactions, *J. Amer. Chem. Soc.*, *94*, 4731, 1972.

Keeffe, J. R., and Kresge, A. J., Catalysis by small molecules in homogeneous solutions, in *Investigations of Rates and Mechanisms of Reactions,* Part 1, 4th edition, Bernasconi, C. F., ed., Interscience, New York, 1986, p. 747.

King, J. F., Return of sulfenes, *Accs. Chem. Res.*, *8*, 10, 1975.

Kreevoy, M. M., and Liang, T. M., Structures and isotopic fractionation factors of complexes A₁HA₂⁻ , *J. Amer. Chem. Soc.*, *102*, 3315, 1980.

Kreevoy, M. M., Liang, T. M., and Chang, K. C., Structures and isotopic fractionation factors of complexes AHA⁻ , *J. Amer. Chem. Soc.*, *99*, 5207, 1977.

Kresge, A. J., What makes proton transfer fast? *Accs. Chem. Res.*, *8*, 354, 1975.

Mähelä, M. J., and Korpela, T. K., Chemical models of enzyme transimination, *Chem. Soc. Rev.*, *12*, 309, 1983.

More O'Ferrall, R. A., Relationship betwen E2 and E1cB mechanisms of ß-elimination, *J. Chem. Soc. (B)*, 274, 1970.

Murrell, J. N., Kettle, S. F. A., and Tedder, J. M., *The Chemical Bond*, John Wiley & Sons, New York, 1985, p. 302.

Parker, V. D., Chao, Y. T., and Zhong, G., Dynamics of proton transfer from radical cations. Addition-elimination or direct proton transfer? *J. Amer. Chem. Soc.*, *119*, 11390, 1997.

Quinn, D. M., and Sutton, L. D., Theoretical basis and mechanistic utility of solvent isotope effects, in *Enzyme Mechanisms from Isotope Effects*, Cook, P. F., ed., CRC Press, Boca Raton, 1991, p. 73.

Saunders, W. H., Distinguishing between concerted and non-concerted eliminations, *Accs. Chem. Res.*, *8*, 19, 1976.

Scheiner, S., Theoretical studies of proton transfers, *Accs. Chem. Res.*, *18*, 174, 1985.

Schowen, R. L., Harmony and dissonance in the concept of proton motions, *Angew. Chem., Int. Ed. Engl.*, *36*, 1434, 1997.

Ta-Shma, R., and Jencks, W. P., How does a reaction change its mechanism? General base catalysis of the addition of alcohols to 1-phenylethyl carbocations, *J. Amer. Chem. Soc.*, *108*, 8040, 1986.

Williams, A., and Douglas, K. T., Elimination-addition mechanisms of acyl transfer reactions, *Chem. Revs.*, *75*, 627, 1975.

REFERENCES

1. **Murrell, J. N., Kettle, S. F. A., and Tedder, J. M.,** *The Chemical Bond*, John Wiley & Sons, New York, 1985, p. 302.
2. **Emsley, J.,** Very strong hydrogen-bonds, *Chem. Soc. Rev.*, 123, 1980.
3. **Greenwood, N. N., and Earnshaw, A.,** *Chemistry of the Elements*, 2nd. edn., Butterworths, London, 1997.

4. **Wolfe, S.,** Marcus theory and the existence or non-existence of transition states in gas phase deprotonation of carbon acids, *Can. J. Chem., 62,* 1465, 1984.

5. **Eigen, M.,** Proton transfer, acid-base catalysis, and enzymatic hydrolysis. Part I: Elementary processes, *Angew. Chem., Int. Ed. Engl., 3,* 1, 1964.

6. **Quinn, D. M., and Sutton, L. D.,** Theoretical basis and mechanistic utility of solvent isotope effects, in *Enzyme Mechanisms from Isotope Effects,* Cook, P. F., ed., CRC Press, Boca Raton, 1991, p. 73.

7. **Keeffe, J. R., and Kresge, A. J.,** Catalysis by small molecules in homogeneous solutions, in *Investigations of Rates and Mechanisms of Reactions, Part 1,* 4th edition, Bernasconi, C. F., ed., Interscience, New York, 1986, p. 747.

8. **Jencks, W. P.,** When is an intermediate not an intermediate? Enforced mechanisms of general acid-base catalysed, carbocation, carbanion, and ligand exchange reactions, *Accs. Chem. Res., 13,* 161, 1980.

9. **Bender, M. L., and Williams, A.,** Ketimine intermediate in amine-catalysed enolisation of acetone, *J. Amer. Chem. Soc., 88,* 2502, 1966.

10. **Albery, W. J., and Gelles, J. S.,** Solvent isotope effects on the enolisation of acetone, *J.Chem. Soc., Farad.Trans. 1, 78,* 1569, 1982.

11. **Dawson, H. M., and Spivey, E.,** Acid and salt effects in catalysed reactions Part XXIV. A study of the catalytic effects produced by acetic acid and acetate buffers under conditions of effectively constant ionic environment, *J. Chem. Soc.,* 2180, 1930.

12. **Bell, R. P., and Jones, P.,** Binary and ternary mechanisms in the iodination of acetone, *J. Chem. Soc.,* 88, 1953.

13. **Albery, W. J.,** A Marcus model for concerted proton transfer, *Disc. Far. Soc., 74,* 245, 1982.

14. **Albery, W. J.,** Application of the Marcus relation to concerted proton transfers, *J. Chem. Soc., Farad. Trans. 1, 78,* 1579, 1982.

15. **Hegarty, A. F., and Jencks, W. P.,** Bifunctional catalysis of the enolisation of acetone, *J. Amer. Chem. Soc., 97,* 7188, 1975.

16. **Hegarty, A. F., and Dowling, J.,** Concerted catalysis in the enolisation of aldehydes, *J. Chem. Soc., Chem. Commun.,* 996, 1991.

17. **Keeffe, J. R., and Kresge, A. J.,** The chemistry of enols. Kinetics and mechanism of enolisation and ketonisation, in *The Chemistry of Enols,* Rappoport, Z., ed., Wiley-Interscience, New York, 1990, Chapter 7.

18. **Fife, T. H.,** General acid catalysis of acetal, ketal and ortho ester hydrolysis, *Accs. Chem. Res., 5,* 264, 1972.

19. **Lahti, M., and Kankaanpera, A.,** A reinvestigation of the general acid catalysis reported for the hydrolysis of ethyl orthoformate in dioxane-water solvents, *Acta Chem. Scand., 24,* 706, 1970.

20. **Lahti, M., and Kankaanpera, A.,** Detection of general acid catalysed hydrolysis in buffer solutions in mixed solvents, *Acta Chem. Scand., 26,* 24, 1972.

21. **Salomaa, P., Kankaanpera, A., and Lahti, M.,** Hydrolysis of acetals and ortho esters. Specific salt effects associated with buffer experiments in mixed solvents, *J. Amer. Chem. Soc., 93,* 2084, 1971.

22. **Kandanarachchi, P., and Sinnott, M. L.,** Hydrolysis of aryl orthocarbonates by general acid-catalysed and spontaneous processes. Characterisation of the water reaction of $(ArO)_3COAr'$ and $(ArO)_3CN_3$, *J. Amer. Chem. Soc., 116,* 5592, 1994.

23. **Kandanarachchi, P., and Sinnott, M. L.,** Hydrolysis of aryl orthocarbonates. Evidence for charge imbalance in the transition state for the general acid-catalysed process, *J. Amer. Chem. Soc., 116,* 5601, 1994.

24. **Grunwald, E.,** Reaction mechanism for structure-energy relations. 2. Acid-catalysed addition of alcohols to formaldehyde, *J. Amer. Chem. Soc., 107,* 4715, 1985.

25. **Guthrie, J. P.,** Multidimensional Marcus theory: and analysis of concerted reactions, *J. Amer. Chem. Soc., 118,* 12878, 1996.

26. **Cromartie, T. H., and Swain, C. G.,** Kinetic and equilibrium chlorine isotope effects in the cyclisation of 2-chloroethanol in protic solvents, *J. Amer. Chem. Soc., 98,* 545, 1976.

27. **Cromartie, T. H., and Swain, C. G.,** Chlorine isotope effects in the cyclisation of chloroalcohols, *J. Amer. Chem. Soc., 97,* 232, 1975.

28. **Swain, C. G., Kuhn, D. A., and Schowen, R. L.,** Effect of structural changes in reactants on the position of hydrogen-bonding hydrogens and solvating molecules in transition states. The mechanism of tetrahydrofuran formation from 4-chlorobutanol, *J. Amer. Chem. Soc., 87,* 1553, 1965.

29. **Dietze, P. E., and Jencks, W. P.,** General-base catalysis of nucleophilic substitution at carbon, *J. Amer. Chem. Soc., 111,* 340, 1989.

30. **Mihel, I., Knipe, J. O., Coward, J. K., and Schowen, R. L.,** α-Deuterium isotope effects and transition state structure in an inytramolecular model system for methyl transfer enzymes, *J. Amer. Chem. Soc., 101,* 4349, 1979.

31. **Coward, J. K., Lok, R., and Tagaki, O.,** General base catalysis in nucleophilic attack at sp^3 carbon of methylase model compounds, *J. Amer. Chem. Soc., 98,* 1057, 1976.

32. **Knipe, J. O., and Coward, J. K.,** Role of buffers in a methylase model reaction. General base catalysis by oxyanions *vs.* nucleophilic dealkylation by amines, *J. Amer. Chem. Soc., 101,* 4339, 1979.

33. **Bartsch, R. A.,** Ionic association in base-promoted β-elimination reactions, *Accs. Chem. Res., 8,* 239, 1975.

34. **Baciocchi, E.,** Base dependence of transition state structure in alkene-forming E2 reactions, *Accs. Chem. Res., 12,* 430, 1979.

35. **Ford, W. T.,** Mechanisms of elimination reactions catalysed by weak bases, *Accs. Chem. Res., 6,* 410, 1973.

36. **Fry, A.,** Isotope effect studies of elimination reactions, *Chem. Soc. Rev., 1,* 163, 1972.

37. **Bartsch, R. A., and Zavada, J.,** Stereochemical and base species dichotomies in olefin-forming E2 eliminations, *J. Amer. Chem. Soc., 80,* 453, 1980.

38. **Saunders, W. H., and Cockerill, A. F.,** *Mechanisms of Elimination Reactions,* Wiley-Interscience, New York, 1973.

39. **Bunnett, J. F.,** The mechanism of bimolecular β-elimination reactions, *Angew. Chem., Int. Ed. Engl., 1,* 225, 1962.

40. **Bordwell, F. G.,** How common are base-initiated, concerted 1,2-eliminations? *Accs. Chem. Res., 5,* 374, 1972.

41. **Saunders, W. H.,** Distinguishing between concerted and non-concerted eliminations, *Accs. Chem. Res., 8,* 19, 1976.

42. **More O'Ferrall, R. A., and Slae, S.,** β-Elimination of 9-fluorenylmethanol in aqueous solution: an E1cB mechanism, *J. Chem. Soc. (B),* 260, 1970.

43. **Keeffe, J. R., and Jencks, W. P.,** Large inverse solvent isotope effects: a simple test for the E1cb mechanism, *J. Amer. Chem. Soc., 103,* 2457, 1981.

44. **Keeffe, J. R., and Jencks, W. P.,** Elimination reactions of N-(2-(p-nitrophenyl)-ethyl)alkylammonium ions by an E1cb mechanism, *J. Amer. Chem. Soc., 105,* 265, 1983.

45. **Bunting, J. W., and Kanter, J. P.,** A change in the rate-determining step in the E1cb reactions of N-(2-(4-nitrophenyl)ethyl)pyridinium cations, *J. Amer. Chem. Soc., 113,* 6950, 1991.

46. **Gandler, J. R., and Jencks, W. P.,** General base catalysis, structure reactivity interactions, and merging of mechanisms for elimination reactions of (2-arylethyl)quinuclidinium ions, *J. Amer. Chem. Soc., 104,* 1937, 1982.

47. **Guthrie, J. P.,** Concertedness and E2 elimination reactions. Prediction of transition state position and reaction rates using two-dimensional reaction surfaces based on quadratic and quartic approximations, *Can. J. Chem., 68,* 1643, 1990.

48. **Toteva, M. M., and Richard, J. P.,** Mechanism for nucleophilic substitution and elimination reactions at tertiary carbon in largely aqueous solutions: lifetime of a simple tertiary carbocation, *J. Amer. Chem. Soc., 118,* 11434, 1996.

49. **Hartsough, D. S., and Merz, M. M.,** Potential of mean force calculations in the S$_N$1 fragmentation of *tert*-butyl chloride, *J. Phys. Chem., 99,* 384, 1995.

50. **Fishbein, J. C., and Jencks, W. P.,** Elimination reactions of ß-cyano thioethers: evidence for a carbanion intermediate and a change in rate-limiting step, *J. Amer. Chem. Soc., 110,* 5075, 1988.

51. **Fishbein, J. C., and Jencks, W. P.,** Elimination reactions of ß-cyano thioethers: internal return and the lifetime of the carbanion intermediate, *J. Amer. Chem. Soc., 110,* 5087, 1988.

52. **Giese. K., Kaatz, U., and Pottel, R.,** Permittivity and dielectric and proton magnetic relaxation of aqueous solutions of the alkali halides, *J. Phys. Chem., 74,* 3718, 1970.

53. **Kaatze, U., Pottel, R., and Schumacher, A.,** Dielectric spectroscopy of 2-butoxyethanol/water mixtures in the complete composition range, *J. Phys. Chem., 96,* 6017, 1992.

54. **Richard, J. P., Williams, G., and Gao, J.,** Experimental and computational determination of the effect of the cyano group on carbon acidity in water, *J. Amer. Chem. Soc., 121,* 715, 1999.

55. **Lowry, T. M.,** Studies of dynamic isomerism. Part XVIII. The mechanism of mutarotation, *J. Chem. Soc.,* 1371, 1925.

56. **Lowry, T. M., and Smith, S. G.,** Studies of dynamic isomersim. Part XXIV. Neutral-salt action in mutarotation, *J. Chem. Soc.,* 2539, 1927.

57. **Brønsted, J. N., and Guggenheim, E. A.,** Contribution to the theory of acid and basic catalysis. The mutarotation of glucose, *J. Amer. Chem. Soc., 49,* 2554, 927.

58. **Swain, C. G.,** General acid-base concerted mechanism for enolisation of acetone, *J. Amer. Chem. Soc., 72,* 4578, 1950.

59. **Swain, C. G., and Brown, J. F.,** Concerted displacement reactions. VIII. Polyfunctional catalysis, *J. Amer. Chem. Soc., 74,* 2538, 1952.

60. **Swain, C. G., and Brown, J. F.,** Concerted displacement reactions. VII. The mechanism of acid-base catalysis in non-aqueous solvents, *J. Amer. Chem. Soc., 74,* 2534, 1952.

61. **Pedersen, K. J.,** The theory of protolytic reactions and prototropic isomerisation, *J. Phys. Chem., 38,* 581, 1934.

62. **Pocker, Y.,** The relative acid dissociation constants of glucose and 2,3,4,6-tetramethyl-O-methylglucose in H_2O and D_2O. The mechanism of mutarotation, *Chemistry and Industry (London),* 968, 1960.

63. **Rony, P. R.,** Polyfunctional catalysis. I. Activation parameters for the mutarotation of tetramethyl-D-glucose in benzene, *J. Amer. Chem. Soc., 90,* 2824, 1968.

64. **Rony, P. R.,** Polyfunctional catalysis. II. General base catalysis of the mutarotation of tetramethyl-D-glucose in benzene and methanol-benzene, *J. Amer. Chem. Soc., 91,* 4244, 1969.

65. **Rony, P. R.,** Polyfunctional catalysis. III tautomeric catalysis, *J. Amer. Chem. Soc., 91,* 6090, 1969.

66. **Rony, P. R., and Neff, R. O.,** Polyfunctional catalysis. IV. Oxy-acid catalysis of the mutarotation of tetramethyl-D-glucose in benzene, *J. Amer. Chem. Soc., 95,* 2896, 1973.

67. **Bell, R. P.,** *The Proton in Chemistry,* 1st. ed., Methuen and Co. Ltd., London, 1959, p. 153.

68. **Satterthwait, A. C., and Jencks, W. P.,** The mechanism of aminolysis of acetate esters, *J. Amer. Chem. Soc., 96,* 7018, 1974.

69. **Eigen, M.,** Proton transfer, acid-base catalysis, and enzymatic hydrolysis. Part I: Elementary processes, *Angew. Chem., Int. Ed. Engl., 3,* 1, 1964.

70. **Fox, J. P., Page, M. I., Satterthwait, A. C., and Jencks, W. P.,** Non-linear Brønsted relationships for general acid-base catalysis of aminolysis reactions, *J. Amer. Chem. Soc., 94,* 4729, 1972.

71. **Fox, J. P., and Jencks, W. P.,** General acid and general base catalysis of the methoxyaminolysis of 1-acetyl-1,2,4-triazole, *J. Amer. Chem. Soc., 96,* 1463, 1974.

72. **Page, M. I., and Jencks, W. P.,** General base and acid catalysis in the hydrazinolysis of acetylimidazole, *J. Amer. Chem. Soc., 94,* 8828, 1972.

73. **Jencks, W. P.,** General acid-base catalysis of complex reactions in water, *Chem. Revs., 72,* 705, 1972.

74. **Gandour, R. D., Maggiora, G. M., and Schowen, R. L.,** Coupling of proton motions in catalytic activated complexes. Model potential-energy surfaces for hydrogen-bond chains, *J. Amer. Chem. Soc., 96,* 6967, 1974.

75. **Jencks, W. P., and Salvesen, K.,** The reaction of thiols with acetylimidazole. Evidence for independent reaction pathways, *J. Amer. Chem. Soc., 93,* 1419, 1971.

76. **Kershner, L. D., and Schowen, R. L.,** Proton transfer and heavy-atom reorganisation in amide hydrolysis. Valence-isomeric transition states, *J. Amer. Chem. Soc., 93,* 2014, 1971.

77. **Jencks, W. P.**, *Catalysis in Chemistry and Enzymology*, McGraw-Hill, New York, 1969, pp.196 and 198.
78. **Cox, M. M., and Jencks, W. P.**, Concerted bifunctional proton transfer and general-base catalysis in the methoxyaminolysis of phenyl acetate *J. Amer. Chem. Soc., 103*, 580, 1981.
79. **Barnett, R. E., and Jencks, W. P.**, Diffusion-controlled proton transfer in intramolecular thiol ester aminolysis and thiazoline hydrolysis, *J. Amer. Chem. Soc., 91*, 2358, 1969.
80. **Menger, F. M.**, The aminolysis and amidinolysis of p-nitrophenyl acetate in chlorobenzene. A facile bifunctional reactivity, *J. Amer. Chem. Soc., 88*, 3081, 1966.
81. **Anderson, H., Su, S., and Watson, J. W.**, Aminolysis Reactions I. Mechanisms of aminolysis and amidinolysis of p-nitrophenyl acetate in chlorobenzene, *J. Amer. Chem. Soc., 91*, 482, 1969.
82. **Anderson, P. E., Blackburn, G. M., and Murphy, S.**, Concerted push-pull catalysis in thiol ester aminolysis, *J. Chem. Soc. Chem. Commun.*, 171, 1972.
83. **Jubian, V., Veronese, A., Dixon, R. P., and Hamilton, A. D.**, Acceleration of a phosphate diester transesterification reaction by bis(alkylguanidinium) receptors containing an appended general base, *Angew. Chem., Int. Ed. Engl., 34*, 1237, 1995.
84. **Eigen, M.**, Kinetics of proton transfer processes, *Disc. Far. Soc., 39*, 7, 1965.
85. **Sayer, J. M., and Jencks, W. P.**, General base catalysis of thiosemicarbazone formation, *J. Amer. Chem. Soc., 91*, 6353, 1969.
86. **Sayer, J. M., and Jencks, W. P.**, Mechanism and catalysis of 2-methyl-3-thiosemicarbazone formation. A second change in rate-determining step and evidence for a stepwise mechanism for proton transfer in a simple carbonyl addition reaction, *J. Amer. Chem. Soc., 95*, 5637, 1973.
87. **Mähelä, M. J., and Korpela, T. K.**, Chemical models of enzyme transimination, *Chem. Soc. Rev., 12*, 309, 1983.
88. **Fischer, H., DeCandis, F. K., Ogden, S. D., and Jencks, W. P.**, Catalysis of transimination by rate-limiting proton transfer to buffer bases, *J. Amer. Chem. Soc., 102*, 1340, 1980.
89. **Koehler, K., Sandstrom, W., and Cordes, E. H.**, Concerning the mechanism of the hydrolysis and aminolysis of Schiff bases, *J. Amer. Chem. Soc., 86*, 2413, 1964.
90. **Funderburk, L. H., Aldwin, L., and Jencks, W. P.**, Mechanisms of general acid and base catalysis of the reactions of water and alcohols with formaldehyde, *J. Amer. Chem. Soc., 100*, 5444, 1978.
91. **Funderburk, L. H., and Jencks, W. P.**, Structure-reactivity coefficients for general acid catalysi of semicarbazone formation, *J. Amer. Chem. Soc., 100*, 6708, 1978.
92. **Hogg, J. L., Jencks, D. A., and Jencks, W. P.**, Catalysis of transimination through trapping by acids and bases, *J. Amer. Chem. Soc., 99*, 4772, 1977.
93. **Williams, A., and Douglas, K. T.**, Elimination-addition mechanisms of acyl transfer reactions, *Chem. Revs., 75*, 627, 1975.
94. **Williams, A., and Douglas, K. T.**, Hydrolysis of aryl N-methanesulphonates: evidence consistent with an E1cB mechanism, *J. Chem. Soc. Perkin Trans. 2*, 1727, 1974.
95. **Williams, A., and Jencks, W. P.**, Urea synthesis from amines and cyanic acid: kinetic evidence for a zwitterionic intermediate, *J. Chem. Soc., Perkin Trans. 2*, 1753, 1974.
96. **Williams, A., and Jencks, W. P.**, Acid and base catalysis of urea synthesis: nonlinear Brønsted plots consistent with a diffusion-controlled proton-transfer mechanism and the reactions of imidazole and N-methylimidazole with cyanic acid, *J. Chem. Soc., Perkin Trans. 2*, 1760, 1974.
97. **Ibrahim, I. T., and Williams, A.**, Direct spectrophotometric obsevation of an O-acylisourea intermediate: concerted general acid catalysis in the reaction of acetate ion with a water-soluble carbodiimide, *J. Chem. Soc. Perkin Trans. 2*, 1459, 1982.
98. **Ibrahim, I. T., and Williams, A.**, A stepwise mechanism in the reaction of amines with carbodiimides to form guanidines, *J. Chem. Soc. Perkin Trans. 2*, 1455, 1982.
99. **Dalby, K. N., and Jencks, W. P.**, General acid catalysis of the reversible addition of thiolate anions to cyanamide, *J. Chem. Soc. Perkin Trans. 2*, 1555, 1997.
100. **Jencks, D. A., and Jencks, W. P.**, On the characterization of transition states by structure-reactivity coefficients, *J. Amer. Chem. Soc., 99*, 7948, 1977.
101. **King, J. F.**, Return of sulfenes, *Accs. Chem. Res., 8*, 10, 1975.

102. **Davy, M. B., Douglas, K. T., Loran, J. S., Steltner, A., and Williams, A.**, Elimination-addition mechanisms of acyl group transfer: hydrolysis and aminolysis of arylmethane-sulfonates, *J. Amer. Chem. Soc.*, *99*, 1196, 1977.

103. **Kice, J. C.**, Mechanisms and reactivity of organic oxyacids of sulphur and their anhydrides, *Adv. Phys. Org. Chem.*, *17*, 65, 1980.

104. **Ingold, C. K.**, The mechanism of olefine elimination (Faraday Lecture), *Proc. Chem. Soc.*, 265, 1962.

105. **Hanhart, W., and Ingold, C. K.**, The nature of the alternating effect in carbon chains. Part XVIII. Mechanism of exhaustive methylation and its relation to anomalous hydrolysis, *J. Chem. Soc.*, 997, 1927.

106. **Bordwell, F. G., Boyle, W. J., Hautala, J. A., and Lee, K. C.**, Brønsted coefficients larger than 1 and less than 0 for proton removal from carbon acids, *J. Amer. Chem. Soc.*, *91*, 4002, 1969.

107. **Albery, W. J., Kresge, A. J., and Bernasconi, C. F.**, Marcus-Grunwald theory and the nitroalkane anomaly, *J. Phys. Org. Chem.*, *1*, 29, 1997.

108. **Olah, G. A., and Rasul, G.**, From Kekulé's tetravalent methane to five-, six-, and seven-coordinate protonated methanes, *Accs. Chem. Res.*, *30*, 245, 1997.

109. **Olah, G. A.**, Super electrophiles, *Angew. Chem., Int. Ed. Engl.*, *105*, 767, 1993.

110. **Marcus, R. A.**, Theoretical relations among rate constants, barriers, and Brønsted slopes of chemical reactions, *J. Phys. Chem.*, *72*, 891, 1968.

111. **Marcus, R. A.**, Unusual slopes of free energy plots in kinetics, *J. Amer. Chem. Soc.*, *91*, 7224, 1969.

112. **Kresge, A. J.**, What makes proton transfer fast? *Accs. Chem. Res.*, *8*, 354, 1975.

113. **Kresge, A. J.**, The Brønsted relation and recent developments, *Chem. Soc. Rev.*, *2*, 475, 1973.

114. **More O'Ferrall, R. A.**, Substrate isotope effects, in *Proton Transfer Reactions*, Caldin, E. F., and Gold, V., eds., Chapman & Hall, London, 1975, p. 201.

115. **Bernasconi, C. F.**, The principle of non-perfect synchronisation, *Adv. Phys. Org. Chem.*, *27*, 119, 1992.

116. **Bernasconi, C. F.**, The principle of non-perfect synchronisation: more than a qualitative concept? *Accs. Chem. Res.*, *25*, 9, 1992.

117. **Gilbert, H. F., and Jencks, W. P.**, Mechanisms for enforced general-acid catalysis of the addition of thiol anions to acetaldehyde, *J. Amer. Chem. Soc.*, *99*, 7931, 1977.

Chapter 4

Nucleophilic Displacements at Unsaturated Carbon

Up to about 1950 nucleophilic displacement reactions at unsaturated carbon were considered to possess concerted or stepwise mechanisms. Dewar[1] advanced the concerted mechanism and Ingold[2,3] the stepwise process of ester hydrolysis. Bender's inspirational experiment with isotope exchange[4] initiated a period during which the stepwise process was preferred; most experiments were confirmatory of Bender's hypothesis although it was recognised that concerted mechanisms might prevail for bimolecular reactions under certain conditions (Kivinen[5,6] and Jencks[7]).

A classical kinetic isotope experiment excluding an S_E2 mechanism was also carried out by Melander[8] contemporaneously with that of Bender's and the existence of a Wheland intermediate[8-10] has been considered mandatory since that time. Because of these influential experiments there has been little credence given to concerted mechanisms in *electrophilic* aromatic substitution. The demonstration and identification of Meisenheimer complexes have ensured that, until recently, the stepwise (A_N+D_N) mechanism is the preferred mechanism for *nucleophilic* aromatic substitution.

Theoretical considerations have been invoked to exclude the possibility of concerted mechanisms in nucleophilic substitution reactions at trigonal carbon.[11] An in-line displacement process is energetically favourable in substitution at aliphatic carbon where the HOMO of the attacking nucleophile overlaps with the LUMO of the bond to the leaving group (Salem).[12] Front-side attack whereby the stereochemistry of the reaction involves retention of configuration involves highly repulsive interactions with the LUMO of the departing group in aliphatic displacements (Chapter 5). Repulsive interactions can become much less important in displacements at the trigonal carbon and recent molecular orbital calculations[13] have reversed the notion that concerted mechanisms be excluded.

The mechanism of displacement at the trigonal centre of the vinyl group analogous to the classical S_N2 process is feasible as has been shown theoretically[13] and has been given the term S_N2-vin mechanism.[14]

The positional isotopic exchange experiment of Bender was so influential that concerted processes of ester hydrolysis were not considered as possible until the mid 1980's. Evidence consistent with concerted displacement at acyl centres slowly accumulated until the new technique of *quasi symmetrical reactions* was discovered (Chapter 2) which excluded a stepwise carbonyl group transfer between weakly basic nucleophiles. The possibility of concerted substitution at carbonyl centres is now established and requires a re-examination of the question of concertedness in other trigonal carbon centres.

4.1　Displacements at the carbonyl group

The transfer of an acyl group between acceptor and donor atoms forms the basis of many *synthetic* routes in organic chemistry and mainly involves nucleophiles as acceptors and donors. Electrophilic displacement mechanisms are known and are exemplified by the action of thiamine (Scheme 4.1). However, electrophilic displacements are usually a combination of elementary steps (as in the thiamine case). The fundamental reaction (Scheme 4.2) where electrophile replaces an electrophile by a one or two step process is not known in acyl group chemistry probably due to the difficulty in bringing an electrophile (E$^+$) up to the electrophilic carbonyl carbon atom.

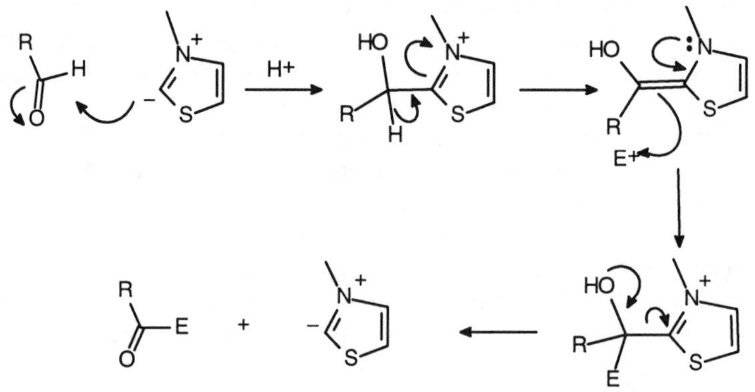

Scheme 4.1. Action of thiamine, an overall *electrophilic* displacement process.

4.1.1　Mechanisms

Mechanisms available for transfer of the acyl group can be discussed in the context of group transfer; these are available for the transfer of a *general* group and bear a strong relationship to other group transfers such as alkyl substitution. Scheme 4.2 illustrates the fundamental mechanisms that are available for displacements at the carbonyl group; these mechanisms are not equally important. Radical mechanisms have been shown to occur as indicated in Scheme 4.3[15-21] but no questions have been addressed regarding concertedness. The system shown in Scheme 4.3 could be subject to experiments of the Berson type (Chapter 2) estimating the lifetime of the adduct and also to chemically induced dynamic nuclear polarisation studies, but so far as we are aware such studies have not yet been attempted.

The heterolytic associative mechanism for carbonyl group transfer reactions involves the formation of a tetrahedral intermediate and conversion of trigonal carbon to a tetrahedral configuration occurs because: (i) the carbonyl carbon-oxygen π bond is usually weaker than the bond to the leaving group (C-Lg); (ii) four-coordinate carbon presents no unusual steric or electronic problems; (iii) solvation of the negative charge density on the oxygen helps to stabilise the tetrahedral intermediate; (iv) the intermediate usually has a significant lifetime due to barriers to expulsion of the incoming and leaving groups. The reaction is driven by the energy of formation of the Nu-C bond. Many carbonyl group transfer reactions appear to occur through an associative pathway even with acylating agents with good leaving groups.

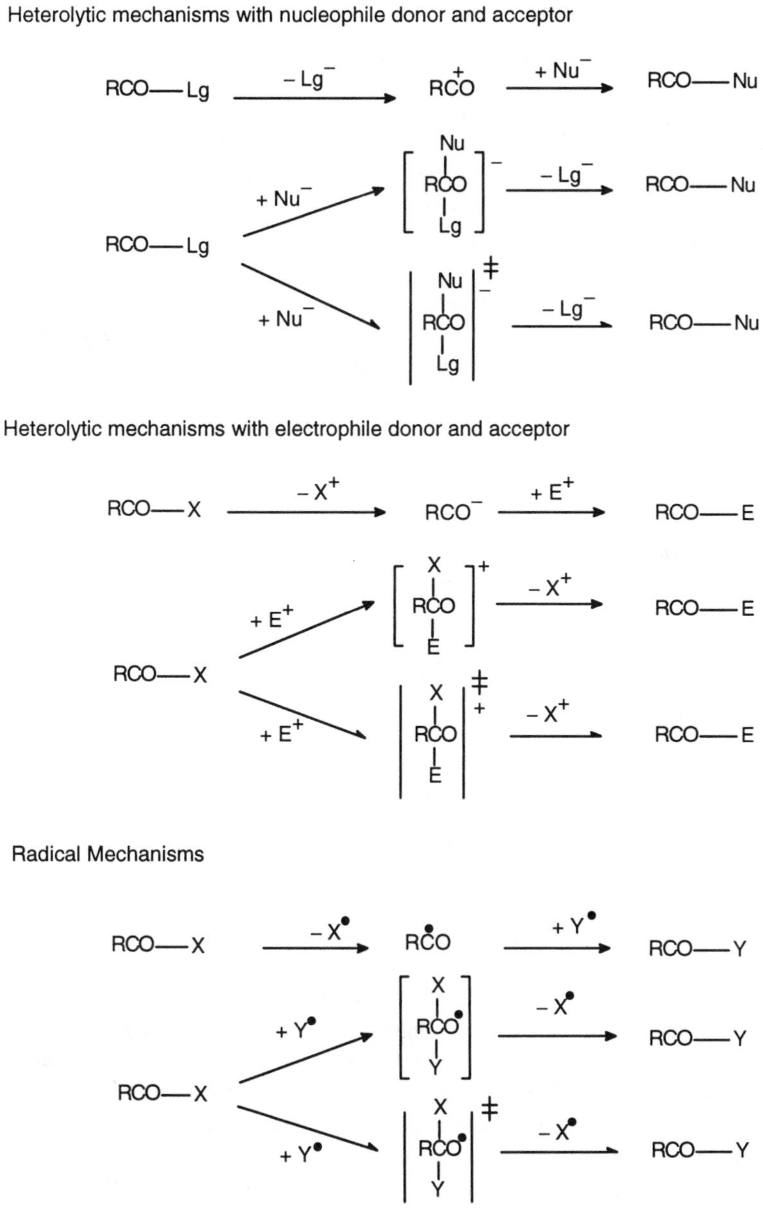

Heterolytic mechanisms with nucleophile donor and acceptor

Heterolytic mechanisms with electrophile donor and acceptor

Radical Mechanisms

Scheme 4.2. Mechanisms for acyl group transfer - radical and heterolytic.

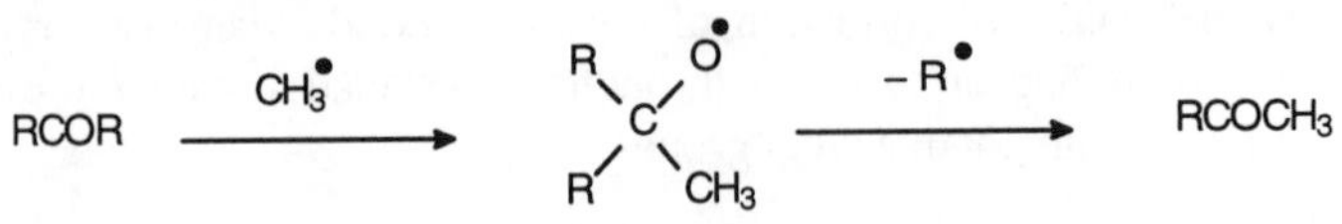

Scheme 4.3. Radical substitution at trigonal carbon.

The nature of the leaving group, nucleophile and the energy of the putative intermediate predicates whether or not a concerted mechanism will ensue in acyl group transfer. Weakly basic nucleophile and leaving group together with an unstable intermediate will encourage the mechanism to be concerted. This is given quantitative expression in Section 4.1.7.

4.1.2 Demonstration of intermediates

The observation of exchange between the ester carbonyl oxygen and ^{18}O-labelled water during the hydrolysis of esters (Bender 1951)[4] is consistent with an associative mechanism (Scheme 4.4); unless the proton transfer step is fast compared with the other steps no exchange would occur.

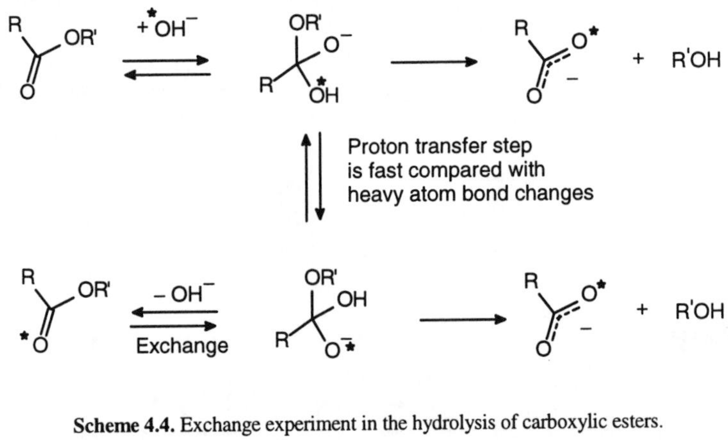

Scheme 4.4. Exchange experiment in the hydrolysis of carboxylic esters.

Detection of the addition intermediate is generally difficult because of its reactivity and many laboratories have searched for stable examples. If the potential leaving group is very unstable then the tetrahedral intermediate may accumulate. The well-known adduct from water and trichloro-acetaldehyde cannot go forward to products because the initial leaving groups would be relatively unstable hydride or trichloromethyl anions. Neither of these unfavourable steps are circumvented by concerted processes (Scheme 4.5). Observable *reactive* tetrahedral intermediates which may be generated by very fast reactions (Scheme 4.5) have been thoroughly investigated.[22,23]

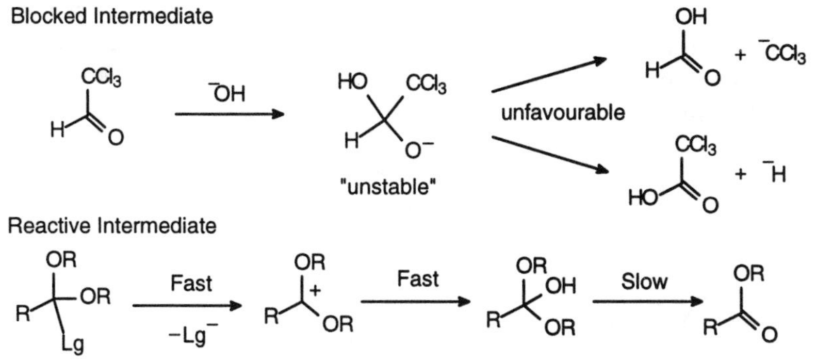

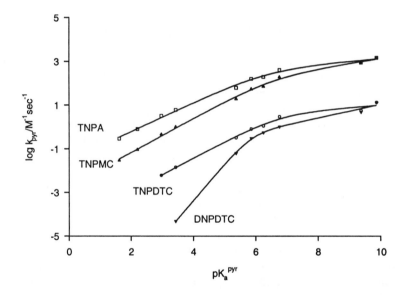

Scheme 4.5. Observation of tetrahedral intermediates.

Figure 4.1. Demonstration of intermediates in pyridinolysis reactions of some esters: DNPDTC = O-ethyl 2,4-dinitrophenyl dithiocarbonate;[24] TNPDTC = O-ethyl 2,4,6-trinitrophenyl dithiocarbonate;[24] TNPA = 2,4,6-trinitrophenyl acetate;[27] TNPMC = 2,4,6-trinitrophenylmethylcarbonate.[27] The lines are calculated from Equation (2.5) using fitted parameters.

Intermediates have been demonstrated extensively in the pyridinolysis of aryl esters, and Figure 4.1 illustrates the application of the classical non-linear free energy relationship for a number of such reactions.[24-30] Break-points ($k_{-1} = k_2$) occur at different pK_a^{pyr} values for the reactions illustrated.

The acylium ion (RCO^+) of the heterolytic dissociative mechanism has long been accepted as an intermediate in gas phase reactions[31-39] and is likely to be very reactive in nucleophilic solvents. The intervention of an acylium ion intermediate in solution reactions depends on stabilising factors and on the leaving group. Scheme 4.6 illustrates some examples of intermediates which have been detected.[31,36,40,41]

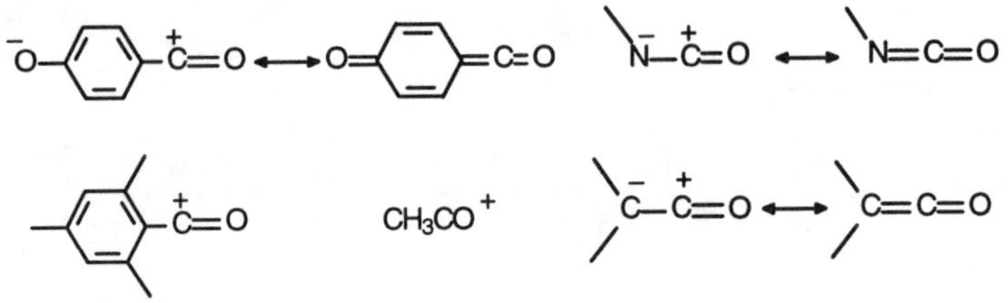

Scheme 4.6. Some acylium ion intermediates and analogues.

Ion cyclotron resonance spectrometry detected an addition intermediate $CH_3CCl_2O^-$ in reaction of CH_3COCl with Cl^- (Asubiojo, and Brauman, 1979);[42] the electron affinity of the corresponding CH_3CCl_2O radical is much higher than expected for an alkoxy radical and this was advanced as evidence that the adduct is a loose ion-molecule complex. Later work (Han, and Brauman, 1987)[43] showed that the two chlorines in the $CF_3CCl_2O^-$ adduct are not equivalent because scrambling of isotopic chlorines does not occur in the final product; this is consistent with an asymmetric adduct but not consistent with a tetrahedral, oxyanion, intermediate where the chlorines are identical.

4.1.3 Free energy correlations

Jencks and Gilchrist[7] observed that the sensitivity to substituents on the nucleophile in acyl group transfer is not very much different from that for the leaving group consistent with coupling between bond fission and bond formation. The difference between nucleophile and nucleofuge type was one of the factors making the interpretation difficult. Any residual differences in donor and acceptor can be factored out for the phenolysis of oxazolinones because it is possible to obtain both Leffler parameters α_{lg} and α_{nuc} for this reaction. Scheme 4.7 illustrates the Leffler parameter map for the reaction.[44] The α_{lg} of 0.36 is large and similar to that for α_{nuc} (0.42) indicating substantial coupling between bond formation and bond fission and providing good *prima facie* evidence for a concerted mechanism. The large Leffler parameter (0.36) for the ring oxygen in the addition of the aryl oxide ion to the oxazolinone indicates a large change in bonding to the carbonyl carbon not predicted from the stepwise mechanism (Scheme 4.7).

A possible transition state for the concerted mechanism of nucleophilic substitution at acyl centres has planar geometry which would become tetrahedral depending on the relative advancement of bond formation and fission to entering and leaving groups respectively (Scheme 4.8). When the nucleophile and leaving groups are not identical non-symmetrical structures will be formed.

As the donor and acceptor groups become weaker nucleophiles there would be a transition from bond angle $\theta = 109°$ through to $\theta = 180°$.

Concerted mechanism

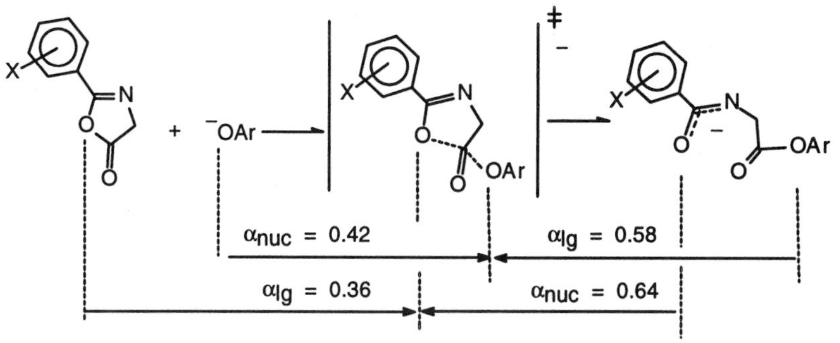

Mechanism involving an addition intermediate

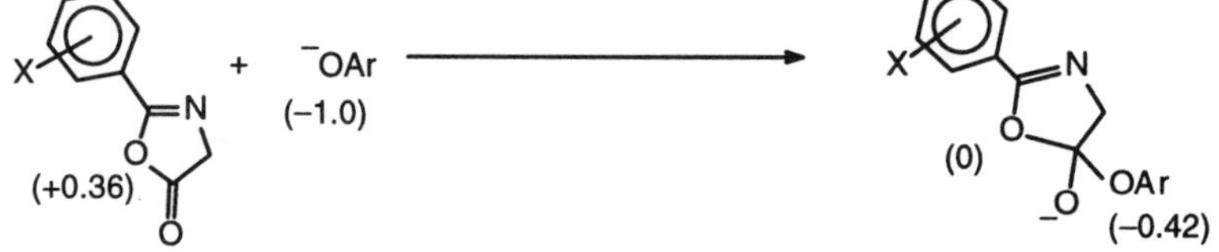

Scheme 4.7. Strong coupling between bond formation and fission in the phenolysis of oxazolinones; a large Leffler parameter (0.36) for the endocyclic C-O fission is not consistent with the stepwise process.

The reaction of substituted pyridines with the N-methoxycarbonyl-isoquinolinium ion to yield N-methoxycarbonyl-pyridinium ions has a linear Brønsted correlation over a range of pK_a values greater than and less than that of isoquinoline[45] (Figure 4.2). This result is not consistent with a stepwise process involving a tetrahedral intermediate because this would require a change in rate-limiting step to occur at the pK_a of isoquinoline and give rise to a break-point at that place. The data can be analysed according to the Equation (2.9) and the value $\Delta\beta$ of zero indicates that there is no charge difference between the nitrogen on the substituted pyridine for both transition states of the putative stepwise process. This means that the transition states are identical and this can only be so if the reaction is concerted. Structure-reactivity studies on the *quasi-symmetrical* reaction enable a full effective charge map to be constructed (Scheme 4.9). The values of the Leffler α_{nuc} and α_{lg} indicate substantial coupling between the two major bonding changes.

Intermediates

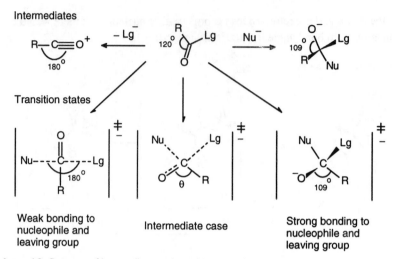

Transition states

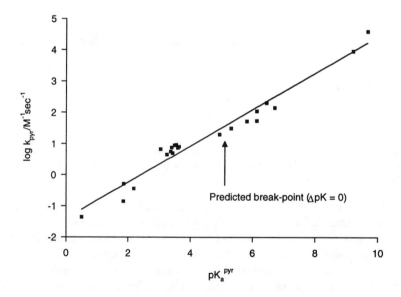

| Weak bonding to nucleophile and leaving group | Intermediate case | Strong bonding to nucleophile and leaving group |

Scheme 4.8. Geometry of intermediates and transition states in acyl group transfer between similar nucleophiles and leaving groups.

Figure 4.2. The quasi-symmetrical reaction of pyridines with N-methoxycarbonyl isoquinolinium ion (Data are from *Ref. 45*); line is the best fit linear regression to the data.

The value of β_{nuc} is approximately 0.6 for attack of substituted pyridines on the isoquinolinium ion when the second step of the *putative* two-step mechanism (decomposition of the tetrahedral intermediate) would be rate-limiting ($pK_{nuc} < pK_{lg}$). Since k_{xpy} under these conditions is k_1k_2/k_{-1} the $\beta_{eq(1)}$ for *formation* of the intermediate (k_1/k_{-1}) must be less than 0.6. Since the overall β_{eq} (for formation of products from reactants) is 1.6 the equilibrium constant for formation of product from intermediate must have a $\beta_{eq(2)} > 1.0$. The overall β_{eq} (1.6) is the sum of $\beta_{eq(1)}$ and $\beta_{eq(2)}$ for the two consecutive steps. It is not conceivable that the second step is more sensitive to substituent on the attacking pyridine where the substituent is directly linked to the changing bond. These conclusions are not consistent with a stepwise process but fit a concerted mechanism.

Putative stepwise mechanism (variation of substituent on nucleophile)

Effective charge map (variation of substituent on both nucleophile and leaving group)

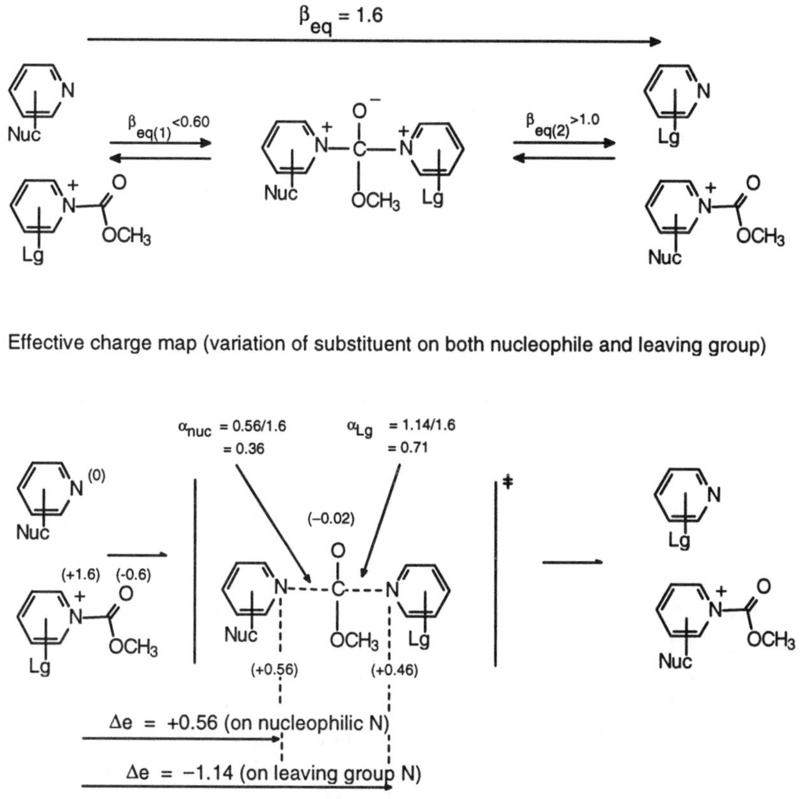

Scheme 4.9. Effective charge and Leffler parameter map for the pyridinolysis of acyl-pyridinium ions. Values in parentheses refer to effective charge derived from Brønsted β values.

Reaction of substituted phenoxide ions with 4-nitrophenyl acetate has been shown to involve a concerted pathway using similar arguments.[46] The Brønsted plot is illustrated in Figure 4.3 and the reaction map, Scheme 4.10, derived from structure reactivity studies indicates that there is substantial coupling between bond fission and bond formation. The data also predict that the second step of the stepwise process is more sensitive to change in the attacking phenolate ion than is the first step; such a conclusion is not consistent with the stepwise mechanism for this reaction.

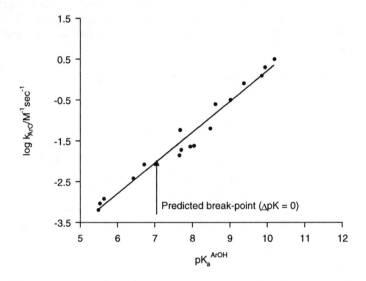

Figure 4.3. The quasi-symmetrical reaction of phenoxide ions with 4-nitrophenyl acetate (Data are from *Ref. 46*); line is the best fit linear regression to the data.

4.1.4 Estimation of rate constants

Arguments based on estimated rate constants for decay of tetrahedral intermediates have been advanced to indicate that the adduct can be too unstable to exist for weakly basic nucleophile donors and acceptors. Thus Ritchie, Van Verth and Virtanen (1982)[47] estimated that the decomposition of the species **4.1** has a rate constant of about $10^{13} sec^{-1}$ and $10^{19} sec^{-1}$ in water and DMSO respectively. The cyanide adduct, **4.2,** was shown to be "stable" in water but not in DMSO. These estimates have important consequences for the operation of mechanisms of acyl transfers involving these putative intermediates.

Guthrie[48,49] demonstrated that the adduct from methyl acetate and hydroxide ion has a free energy of activation for decomposition of 7.4 kcal/mole which corresponds to a half life of about 10^{-7} seconds. The adduct between phenoxide ions has a lifetime estimated to be too short for it to exist and the reaction is thus considered to be concerted in agreement with the results from the quasi-symmetrical reaction method. Relatively accurate estimates have also been made of the lifetimes of putative tetrahedral intermediates which might be expected in various acyl group transfer reactions[48-54] and structures such as **4.3** and **4.4** have rate constants for decomposition in the range 10^7 to $10^8 sec^{-1}$.

Putative stepwise mechanism (variation of substituent on nucleophile)

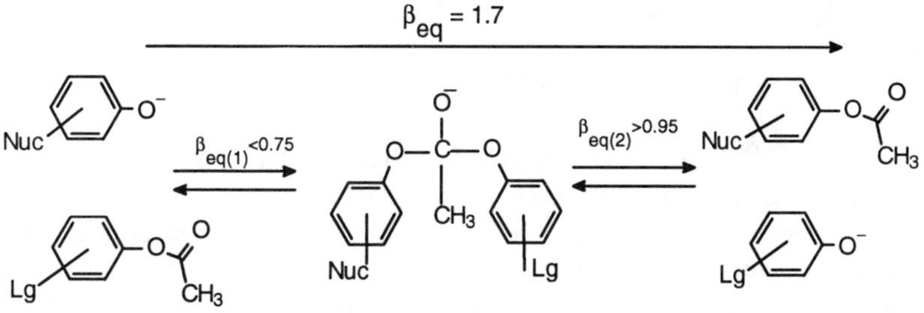

Effective charge map (variation of substituent on both nucleophile and leaving group)

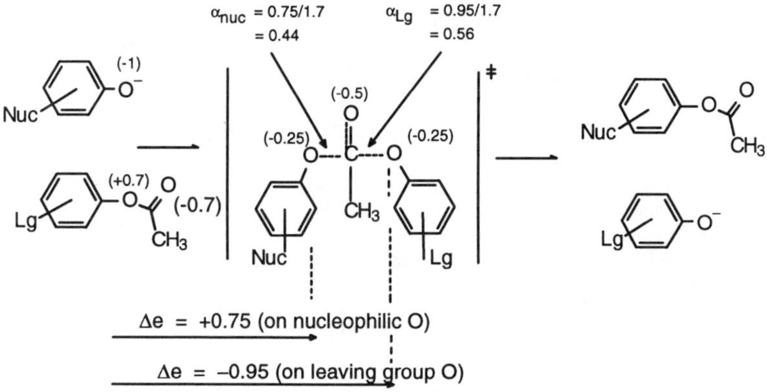

$$\Delta e = +0.75 \text{ (on nucleophilic O)}$$

$$\Delta e = -0.95 \text{ (on leaving group O)}$$

Scheme 4.10. Effective charge and Leffler parameter map for the phenolysis of phenyl acetates.

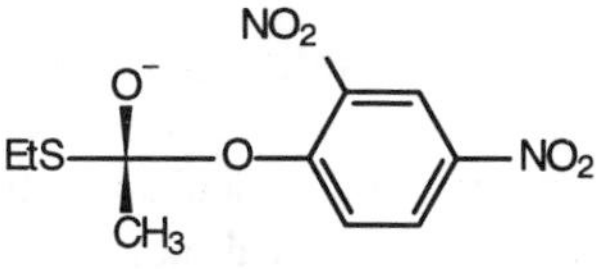

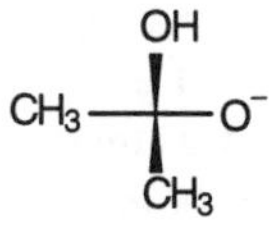

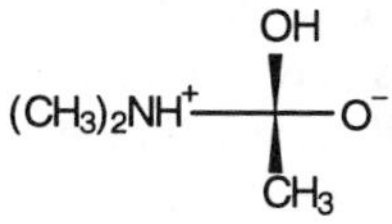

4.1.5 Energy considerations

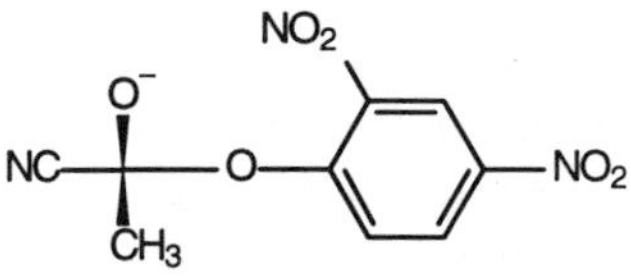

Free energies of activation for the reaction of substituted phenolate anions with substituted phenyl acetates may be estimated from linear free energy relationships of the type for reaction of phenolate ions with 4-nitrophenyl acetate (Equation (4.1) and Figure (4.3)).[46,55-57]

$$\log k_{ArO} = 0.75 \, pK_a^{ArOH} - 7.28 \tag{4.1}$$

Guthrie[48,49,58,59] described methods for obtaining the free energies of the putative acylium ion and tetrahedral adducts which have energies exceeding that of the observed transition state (Chapter 2); both intermediates are thus excluded as stations on the reaction coordinate for the displacement reaction. Table 4.1 collects free energies for the intermediates and transition states for various nucleophilic displacement reactions at acyl centres. Free energy surfaces

Table 4.1.
Free energies (kcal/mole at 298°) for intermediates and transition states for identity displacements at the carbonyl centre.[a]

Nucleophile	$\Delta G_{acylium}$	$\Delta G_{tet.int.}$	$\Delta G^{\ddagger}_{calc.}$	$\Delta G^{\ddagger}_{obs.}$
4-nitrophenolate ion + 4-nitrophenyl acetate[a]	24.66	22.03	15.68[f]	17.8[b]
phenolate ion + phenyl acetate[a]	32.59	16.44	12.49[f]	16.26[b]
4-cyanophenolate ion + 4-cyanophenyl acetate[a]	27.16	19.33	15.24[f]	18.00[b]
4-chloro-2-nitrophenolate ion + 4-chloro-2-nitro-phenyl acetate[a]	23.04	25.61	15.81[f]	17.84[b]
imidazole + 4-nitrophenyl acetate[c]	26.42	19.46	17.44[d]	15.46[e]
imidazole + phenyl acetate[c]	32.59	19.46	18.78[d]	17.97[e]

a) Data from ref. 58; b) Calculated from data in Ref. 56; c) Data from Ref. 59; d) Calculated using the quadratic equation for the free energy surface(Ref. 59); e) calculated from Ref. 5; f) Calculated with the quartic equation for the free energy surface (Ref. 58).

may be calculated by Guthrie's techniques using the quartic approximation (Section 2.2.5) for the identity reaction of phenolate ions with the corresponding acetate esters employing energies for the corners calculated from thermochemical data and intrinsic barrier energies of 1.41 kcal/mole (Figure 4.4a). Similar free energy surfaces may be constructed for the reaction of imidazole with phenyl acetates using intrinsic barrier energies of 1.3 and 5 kcal/mole for the N-C and C-O coordinates respectively (Figure 4.4b). Free energy surfaces may also be calculated for the reactions of phenolate ions with hypothetical pK_a values for identity processes and processes thermodynamically favourable (*downhill,* Figure 4.4c) and unfavourable (*uphill,* Figure 4.4d). Figures 4.4c and 4.4d illustrate how the transition state structure varies as a function of the energies of the encounter complexes represented by the corners of the More O'Ferrall-Jencks diagrams.

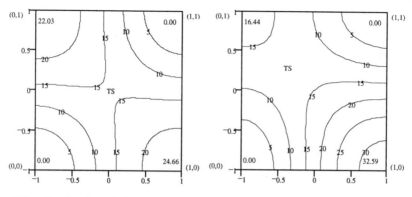

4-Nitrophenolate ion + 4-nitrophenyl acetate   Phenolate ion + phenyl acetate

Figure 4.4a. Free energy surfaces calculated via the quadratic equation for reactions of phenolate ions with phenyl acetates. $(0,0)$ = phenolate ion + phenyl ester; $(1,1)$ = product + phenolate ion; $(0,1)$ = tetrahedral adduct; $(1,0)$ = acylium ion + phenolate ions. The energies (kcal/mole at 25°) of the encounter complexes represented by the four corners of the More O'Ferrall-Jencks diagrams are obtained from Table 4.1.

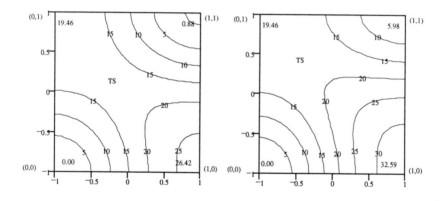

Imidazole + 4-nitrophenyl acetate   Imidazole + phenyl acetate

Figure 4.4b. Free energy surfaces calculated via the quadratic equation for reactions of imidazole with phenyl acetates. $(0,0)$ = imidazole + phenyl ester; $(1,1)$ = product + phenolate ion; $(0,1)$ = tetrahedral adduct; $(1,0)$ = acylium ion + phenolate ion + imidazole. The energies are represented as in Figure 4.4a.

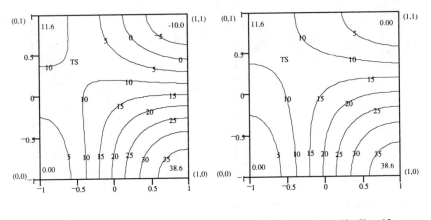

Downhill reaction with phenol pKₐ = 12   Identity reaction with pKₐ = 12

Figure 4.4c. Free energy surfaces calculated via the quadratic equation for the downhill and identity reactions of hypothetical phenolate ions with phenyl acetates. Coordinates and energies are as represented in Figure 4.4a.

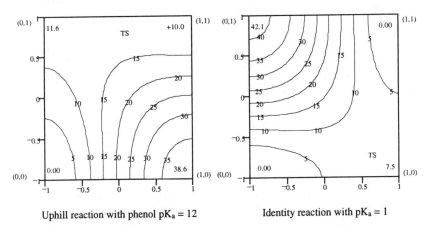

Uphill reaction with phenol pKₐ = 12   Identity reaction with pKₐ = 1

Figure 4.4d. Free energy surfaces calculated via the quadratic equation for uphill reactions of hypothetical phenolate ions with phenyl acetates. Coordinates and energies are as represented in Figure 4.4a.

4.1.6 Isotope effects

The application of isotope effects to questions of concertedness has mostly been to the bond undergoing fission, and measurements of coupling by comparing formation and fission in displacements at carbonyl centres are not available. Probably the most thorough isotope study is that of Hengge and Hess (1994),[60] who determined the heavy atom isotope effects for the leaving oxygen, the nitrogen of the nitro group, the carbonyl oxygen and the β-deuterium on the acetyl group in the reaction of nucleophiles with 4-nitrophenyl acetate. The isotope effects for the nitrogen and the leaving oxygen can be utilised to estimate a Leffler α_{lg} value[61] and these are substantial even though the formation of the putative tetrahedral intermediate must be rate-limiting. A decrease in the carbonyl π-bond caused by the attack of the nucleophile at the carbonyl centre is reflected in the secondary isotope effect on the carbonyl oxygen and in a decrease in hyperconjugation with the C-H bonds resulting in an inverse β-deuterium isotope effect.

Studies have been carried out on systems which have a change in rate-limiting step which can be monitored by the heavy atom isotope effect. There is good evidence that aminolysis of esters involves a stepwise process[47] and a study of the $^{18}k_{lg}/^{16}k_{lg}$ as a function of increasing pH[62] confirms this by indicating a sharp break at pH 9-10 from rate-limiting bond fission to rate-limiting nucleophilic addition. This experiment provides convincing evidence of the veracity of the heavy atom isotope effect studies and moreover gives a lower limit of $^{18}k_{lg}/^{16}k_{lg}$ when the bond fission is not rate-limiting.

4.1.7 Cross-coupling studies

Interaction between bond formation and bond fission can be quantified in the form of cross-coupling coefficients (Section 2.3.2). An interaction between forming and breaking bonds is consistent with a concerted mechanism.[57,63] In the case of phenolysis of phenyl esters β_{nuc} and β_{lg} show marked dependencies on pK_{lg} and pK_{nuc} respectively and this can be quantified by use of *identity rate constant* methodology to yield a Lewis-Kreevoy correlation which fits the theoretical paraboloid correlation (Figure 4.5). The slope of the Lewis-Kreevoy correlation (β_{ii}) is related to the β values for forward (β_{nuc}) and reverse (β_{lg}) by Equation (4.2) which can be derived from the Marcus equation neglecting its quadratic terms.[64] The values of β_{lg} and β_{nuc} can vary linearly with pK_{nuc} and pK_{lg} respectively (Equations (4.3) and (4.4))[57] and combining with Equation (4.2) and integration yields Equations (4.5) and leads to the quadratic Equation (4.6) which forms the basis of the Lewis-Kreevoy correlation.

$$\partial \log k_{ii}/\partial pK_{ii} = \beta_{ii} = \beta_{nuc} + \beta_{lg} \tag{4.2}$$

$$\beta_{lg} = p_{xy} \, pK_{nuc} + L \tag{4.3}$$

$$\beta_{nuc} = p_{xy} \, pK_{lg} + N \tag{4.4}$$

$$\beta_{ii} = 2p_{xy} \, pK_{ii} + (L + N) = d\log k_{ii}/dpK_{ii} \tag{4.5}$$

$$\log k_{ii} = p_{xy} \, pK_{ii}^{2} + (L + N)pK_{ii} + C \tag{4.6}$$

In the case of phenyl acetates Equations (4.3) and (4.4) have parameters $p_{xy} = 0.16$, $N = -0.21$ and $L = -1.91$ and substituting in Equation (4.6) gives a parabola (Figure 4.5) which only requires vertical adjustment (C) to fit the experimental data for $\log k_{ii}$. The More O'Ferrall-Jencks diagram (Figure 4.6) indicates the structure of the identity transition state as a function of increasing basicity of the nucleophile (and leaving group); it parallels that of the surface contours calculated by Guthrie's technique (Figure 4.4) and indicates the transition structures for the identity reactions of 2,4-dinitrophenyl acetate, phenyl acetate and 2,4-dinitrophenyl 4-methoxy-2,6-dimethylbenzoate. The transition structures of concerted identity reactions alway falls on the tightness diagonal (τ) of the More O'Ferrall-Jencks diagram.

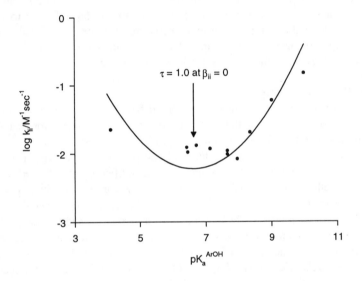

Figure 4.5. Lewis-Kreevoy correlation for k_{ii} (the identity rate constant) for the phenolysis of substituted phenyl acetates; the line is a parabola drawn using the Equation (4.6) and *independent* parameters given in the text. Data are from *Refs. 55,56*. See Equation (6.2) p. 162 for $\tau = 1$.

The gas phase work of Brauman[42,43] and the theoretical studies of Jorgensen[65] show how the stepwise and concerted processes in the gas phase are connected. High level theoretical studies[65] on the reactions of chloride ion with formyl and acetyl chloride in the gas phase indicate that two intermediates are formed which are not symmetrical and are loose ion-neutral clusters. The reactant and product clusters interconvert *via* a transition state with C_s symmetry and the calculations exclude a planar stucture with C_{2v} symmetry. In the limiting transition state where leaving group and nucleophile have zero bonding to the central carbon (1,0 in Figure 4.6) the stereochemistry should be largely controlled by electrostatics; stereochemistry would then be planar in order to minimise the repulsion between nucleophile and leaving group. The theoretical results imply that non-planarity sets in early as the tightness parameter (τ, see Equations 6.1 - 6.3) increases from zero. However, these results refer to the gas phase and in polar solvents the electrostatic effect could be minor.

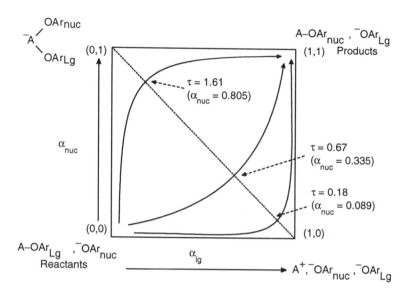

Figure 4.6. More O'Ferrall-Jencks diagram for the identity reactions of substituted phenolate ions with substituted phenyl esters. Transition states for increasing nucleophile basicity are at the intersection of reaction coordinates and the tightness diagonal: $\tau = 0.18$, 2,4-dinitrophenyl 4-methoxy-2,4-dimethylbenzoate; $\tau = 0.67$, 2,4-dinitrophenyl acetate; $\tau = 1.61$, phenyl acetate (data from *Ref.* 55).

4.1.8 Molecular orbital theory

The stereochemistry in substitution at the carbonyl centre (Scheme 4.8) involves a tetrahedral configuration at $\tau = 2$ to planar when $\tau = 0$ with intermediate stereochemistry for intervening values of τ on the More O'Ferrall-Jencks diagram (Figure 4.6) (see Page, and Williams, 1997).[61] The stereochemistry of nucleophilic attack on the carbonyl group has been addressed by x-ray crystallographic studies of stable species containing both nucleophile and carbonyl which show that the nucleophilic atoms are on lines with a mean angle of between 105° and 110° with the carbonyl bond in the vertical plane (**4.5,** Scheme 4.11)[66]. The orbital description of the addition step (**4.6**) involves the lone pair forming on the carbonyl oxygen (i) with an orbital almost antiperiplanar to the entering pair of electrons and that forming on the leaving atom (ii) also antiperiplanar to the entering pair of electrons (the antiperiplanar lone pair hypothesis - "ALPH"). The decomposition conformation of the tetrahedral adduct therefore possesses lone pairs on the C-O oxygen antiperiplanar to the leaving bond (**4.7**).[67,68] Most of the stereochemical results for nucleophilic attack at a carbonyl centre can also be explained by the principle of least nuclear motion,[69] but both least motion and ALPH refer with some exceptions to relatively small energy differences.

The introduction of 2,6-dimethyl substituents into benzoic acid derivatives hinders nucleophilic attack on the carbonyl group and consequently enables 2,6-dimethylbenzoate esters to undergo acid catalysed hydrolysis by a dissociative pathway. The dissociative mechanism can also be accelerated and the dramatic difference in reactivity between the conjugate base of 2,6-dimethyl-4-hydroxybenzoate ester and that of the ester without the labile proton of the hydroxy group is ascribed to the carbonyl function being forced into the conformation where the leaving group is "antiperiplanar" to the lone pairs of the oxyanion in the substituted ester (Scheme 4.12); this enables the elimination reaction to compete successfully with the BAc2 type attack normally associated with ester hydrolysis.[70,71]

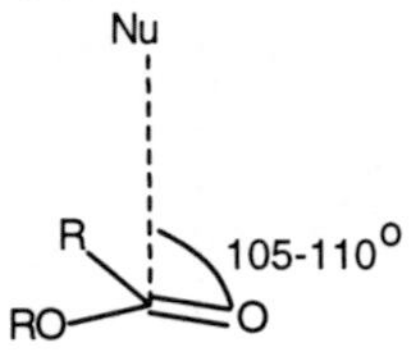

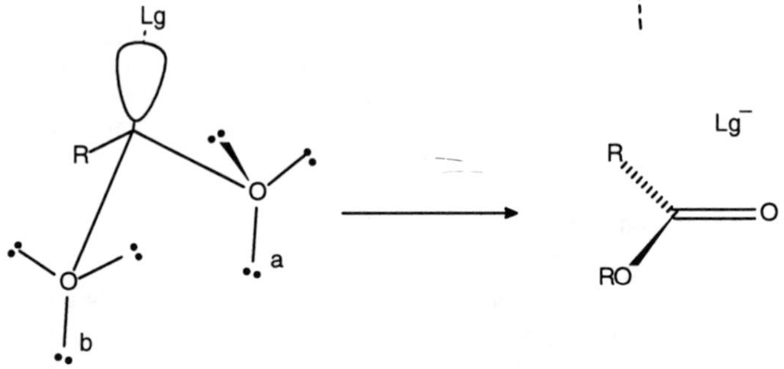

Scheme 4.11. Stereochemistry of nucleophilic attack on carbonyl carbon.

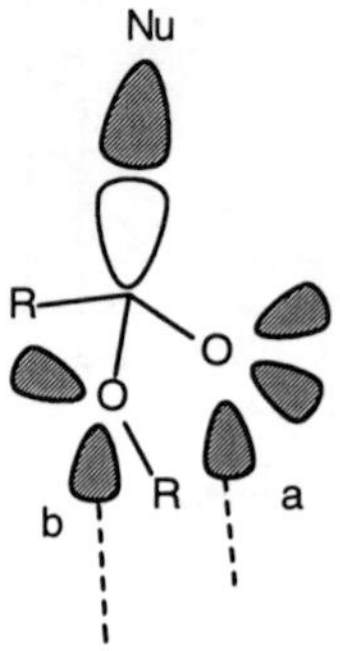

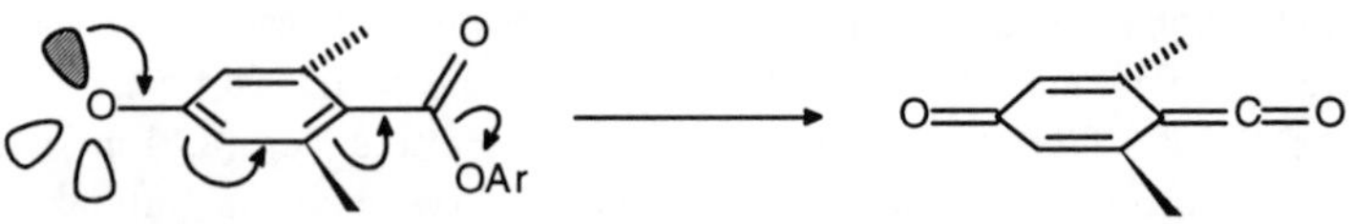

Scheme 4.12. Molecular orbital configuration in the dissociative hydrolysis of phenyl 2,6-dimethyl-4-hydroxybenzoate esters.

Molecular orbital calculations indicate that the direct displacement of phenol by amine is the preferred route in the gas phase.[72,73] These calculations refer to displacement at a carbonyl centre by a nucleophile bearing a proton which is required to transfer to the leaving group in the product. These reactions can be catalysed by bifunctional reagents such as α-pyridone or polyether; transition state structures such as those in **4.8** and **4.9** were shown to take the major part of the reaction flux.

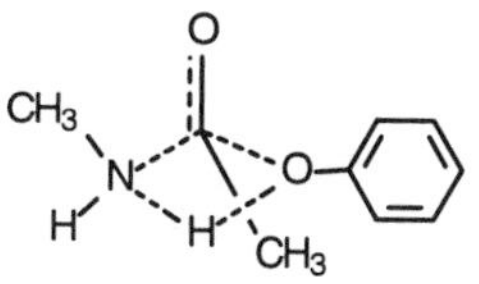

4.8 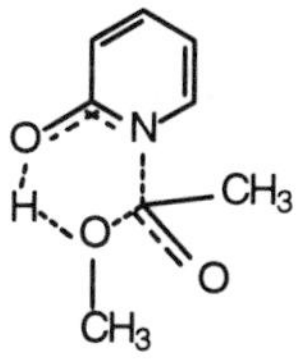   4.9

Semi-empirical molecular orbital studies (PM3) indicate that the gas phase reaction of phenolate ions with formate esters involve tetrahedral intermediates. In the case of the reaction of 4-nitrophenyl formate with phenolate ions the energy level of the transition state for fission of the ArO-C bond in the intermediate is almost indistinguishable from that of the intermediate consistent with an enforced concerted process.[74] Solvation or complexation appears to depress the transition state energy for bond fission more than that for bond formation and hence enhances the concerted mechanism.[72-74]

4.2 VINYL GROUP TRANSFER

4.2.1 Mechanisms

Nucleophilic displacements at the vinyl centre are analogous to those at the carbonyl centre and have been the subject of considerable scrutiny. The mechanisms available for this process follow those of displacements at the carbonyl centre. Study of the reaction has an advantage over that of carbonyl group transfer because stereochemistry can be employed as a tool and a substantial body of evidence has been accumulated on this topic. However this advantage is offset by the systems being experimentally more difficult to study and their smaller impact on mainstream bio-organic chemistry.

The displacement of a nucleofuge from a vinyl centre only occurs readily when the adjacent sp^2 carbon is activated by electron withdrawing groups. Nucleophilic substitution at vinyl chlorides, of the type shown in Scheme 4.13, is very sluggish due to the difficulty of placing charge onto the adjacent carbon and of attacking an electron-rich centre. The concerted fission of the leaving group bond could assist such a displacement reaction but elimination to give the acetylene would occur readily if there were a β-hydrogen available (X or Y = H).

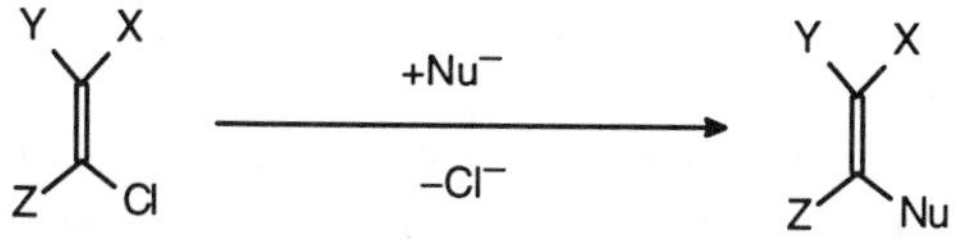

Scheme 4.13. Nucleophilic displacement at vinylic carbon.

4.2.2 In-line attack

The vinyl analogue of the aliphatic S_N2 displacement mechanism involves attack in-line with leaving group departure through a planar transition state structure (Scheme 4.14). The logical outcome of this mechanism is that the stereochemistry of the product would be inverted.

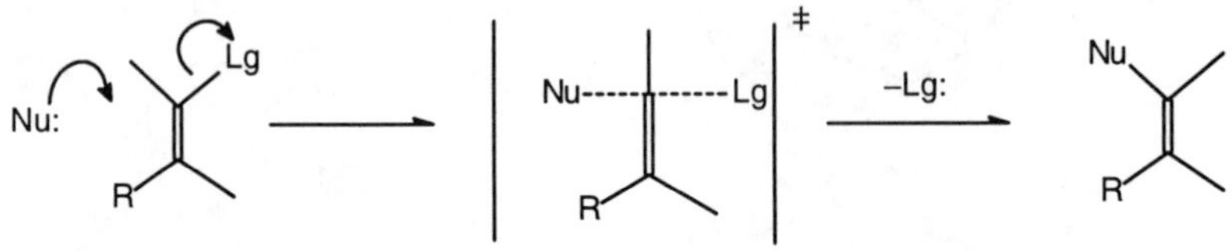

Scheme 4.14. Inversion in vinylic displacement *via* an in-line concerted mechanism.

Although early theoretical considerations have a preference for orthogonal attack of the nucleophile at the planar vinyl centre[75,76] a recent study using a sophisticated computational level[13] suggests that in-line concerted attack is still possible. The vinyl system can be constrained experimentally in such a way as to force the nucleophilic displacement into an in-line mechanism. Modena and his co-workers[14,77-80] have studied the rearrangement of thiirenium ions into thietium ions (Scheme 4.15).

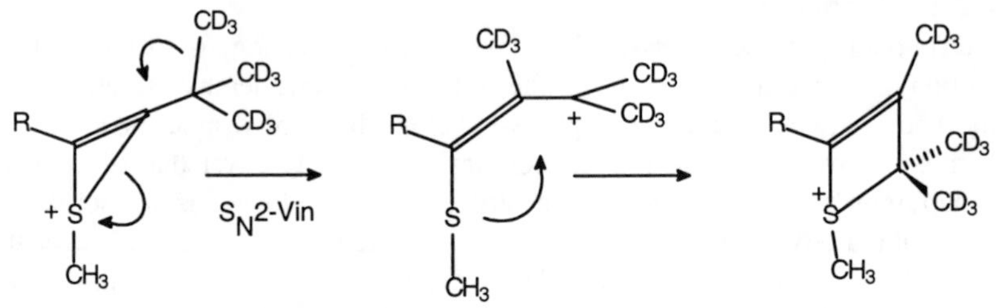

4.10   4.11

Scheme 4.15. Labelling evidence for an in-line concerted mechanism (S_N2-Vin) for the rearrangement of thiirenium ions into thietium ions; $R = C(CH_3)_3$..

The rearrangement (Scheme 4.15) was established as involving a concerted, in-line, displacement mechanism at the vinyl centre by the following arguments: (i) the product **4.11** is formed quantitatively from **4.10** and the partially deuteriated thietium ions are the *sole* products; (ii) the existence of secondary kinetic isotope effects.

Okayama[81] found that the displacement reactions of vinyl chloroiodanes involved inversion of configuration (Scheme 4.16); while this result is consistent with a concerted in-line displacement mechanism it does not *exclude* an addition-elimination process.

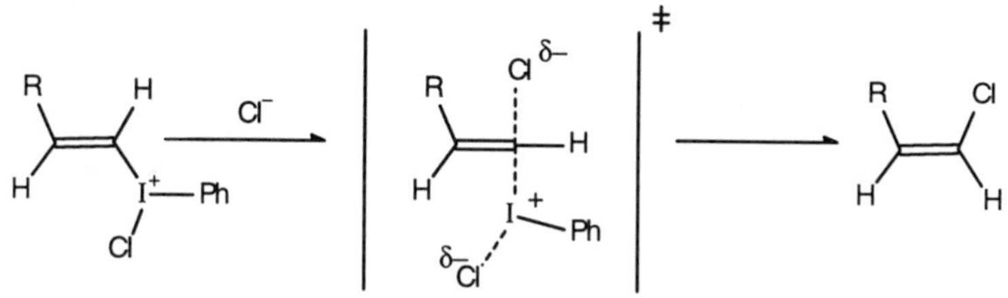

Scheme 4.16. In-line displacement at a vinyl centre with an iodane leaving group.

4.2.3 Front-side attack

A substantial body of stereochemical data for substitution at the vinyl centre has accumulated and it was originally thought that the highly stereoselective reactions were good evidence for concerted displacements. Retention of configuration has been observed in displacements at the vinyl carbon in many reactions, and this is consistent with a concerted displacement mechanism (Scheme 4.17) involving a non-planar transition state structure as shown. This has been subsequently discounted because of the observation that some *cis* and *trans* isomers give a product with the same stereochemistry; partial retention and inversion is also observed with some reactions. The course of the stepwise substitution process involves initial formation of the bond to the incoming nucleophile followed by a rotation in the adduct and then an elimination as illustrated in Scheme 4.18. This mechanism raises interesting questions (not, however, relevant to this text) regarding the direction and magnitude of the rotation and the stability of the conformers leading to elimination.

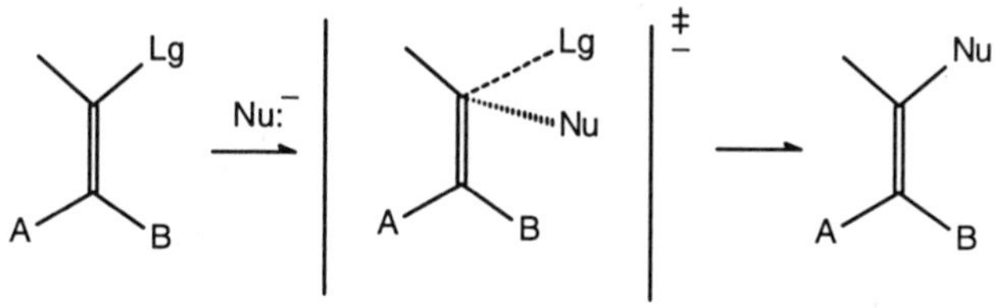

Scheme 4.17. Retention of configuration in vinyl displacement reactions *via* concerted front-side displacement.

The direct observation of intermediates excludes the concerted path for the reactions under examination and leaves the question of concertedness still open. The first definitive evidence for an intermediate was the demonstration of an acetylene in the attack of alkoxide ion on β-sulphonylvinyl chlorides (Scheme 4.19).[80] The observation of an addition intermediate was made by van der Sluijs and Stirling[82] in the reaction of alkoxides with β-sulphonylvinyl phenyl ethers (Scheme 4.20). The displacement of phenoxide ion leaving groups from β-nitrostilbenes was recently demonstrated to involve an addition intermediate (Scheme 4.20).[83-87] However, the question of concertedness in other systems must still remain open judging from the experience in carbonyl chemistry.

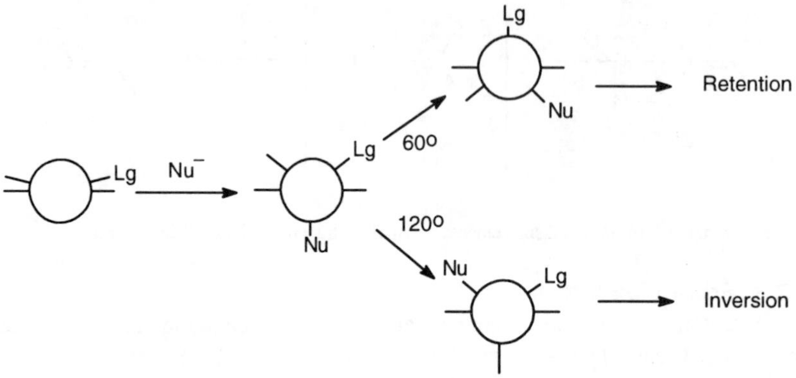

Scheme 4.18. Newman projections illustrate how a stepwise addition-elimination mechanism can give retention *or* inversion in vinyl substitution.

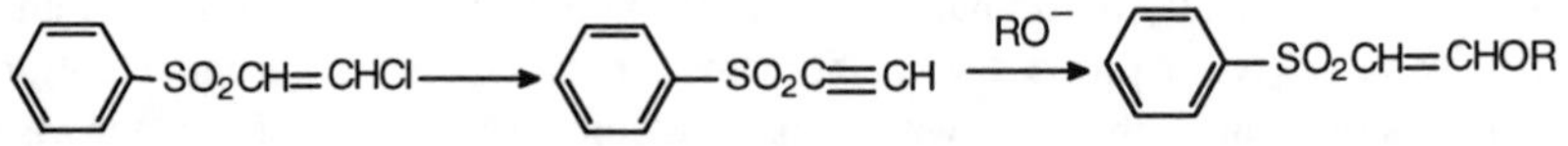

Scheme 4.19. An elimination-addition mechanism of substitution at activated vinyl centres.[80]

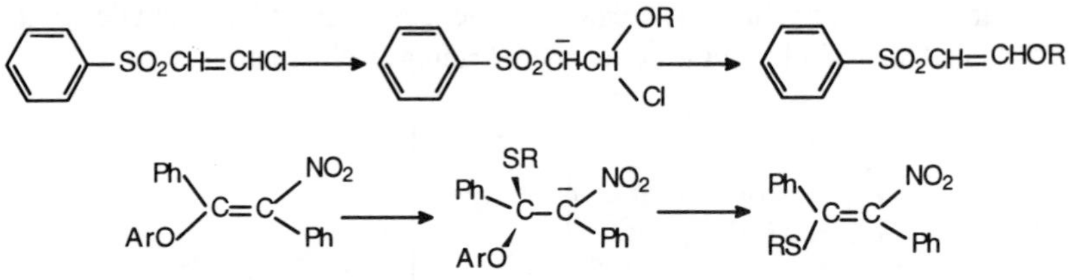

Scheme 4.20. Addition-elimination mechanisms in substitution at activated vinyl centres.[82-87]

4.3   NUCLEOPHILIC AROMATIC SUBSTITUTION

Until recently there were no definitive reports of concerted mechanisms of nucleophilic aromatic substitutions although Illuminati (1964)[88] had mooted its possibility in heterocyclic systems. Many stable Meisenheimer complexes have been characterised, but the formation of intermediate adducts on the reaction path has not often been demonstrated.

The existence of stable Meisenheimer addition complexes[89] dominates the field of mechanism in nucleophilic aromatic substitution. Many aromatic substitution reactions involve groups which activate the aromatic centre by electron withdrawal and the nitro group, a favourite substituent, also has the potential of stabilising an adduct by strongly localising the negative charge on its oxygens. Substitution at a heterocyclic centre, as in the hydrolysis of cyanuric chlorides, has provided a system where the stabilisation is similar to that for nitro groups and yet the nucleus is activated to nucleophilic attack.

The displacement of 4-nitrophenolate ion by substituted phenolate ions from 2-(4-nitrophenyl)-4,6-dimethoxytriazine (Scheme 4.21)[90,91] was studied by use of the *quasi-symmetrical technique* and the kinetics of the reaction obey a linear Brønsted law (Figure 4.7).* Linearity over the range of pK_a values above and below the predicted break-point is prima-facie evidence for a mechanism with only one transition state.

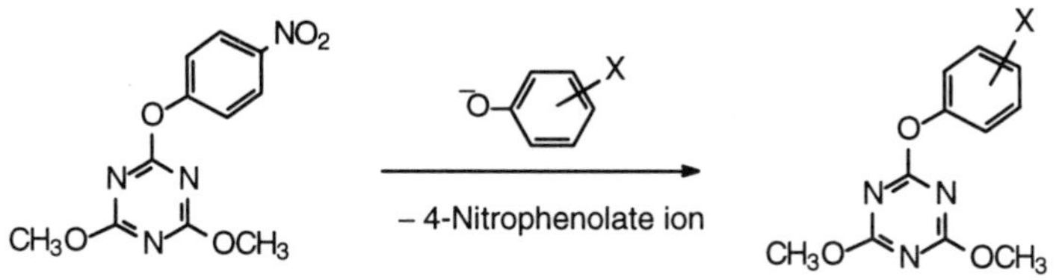

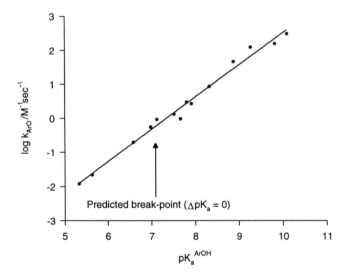

Scheme 4.21. Displacement of phenoxide ions from 1,3,5-triazine centres.[90,91]

Figure 4.7. Phenolysis of 4-nitrophenyl triazine; data are from *Ref. 90.*

---

*The 1,3,5-triazine system provides an excellent test-bed for studying nucleophilic aromatic substitution because the endocyclic nitrogen atoms confer a suitable reactivity to displacement at the 2,4 or 6 centres. Unlike aromatic systems activated with the nitro group (of enhancing power similar to that of the endocyclic nitrogen) the system possesses relatively weak electronic absorption in the UV region making it suitable for the study of nucleophiles of leaving groups in this region of the spectrum.

If the putative two-step mechanism (Scheme 4.22) is true then the change in effective charge from reactants to intermediate and from intermediate to products may be estimated from the β_{nuc} for the case where the second step is rate-limiting and the β_{eq} for the overall reaction (1.48). The value of β_{nuc} for the second step rate-limiting is 0.95 so that formation of intermediate has $\beta_{eq(1)} <$ 0.95 (see page 102). Thus the second step has $\beta_{eq(2)} > 1.48 - 0.95 = 0.53$. This value is large compared with that for $\beta_{eq(1)}$ and would not be expected for a step where *no substantial* bonding change is present.

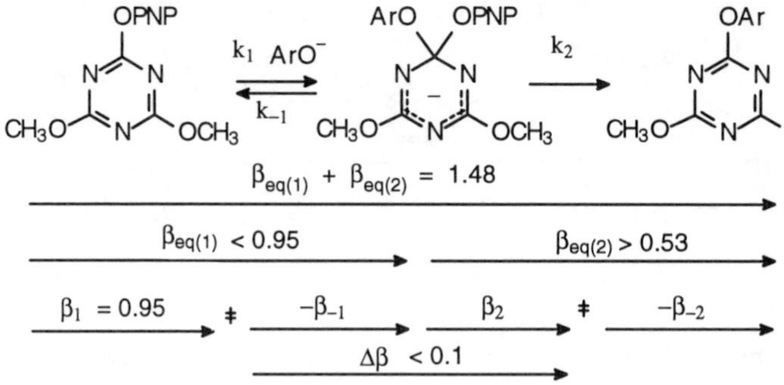

Scheme 4.22. Effective charge map for the putative stepwise displacement by phenoxide ions of 4-nitrophenoxide ion ($^-$OPNP) from the 2-(4'-nitrophenoxy)-1,3,5-triazine ether.

The value of $\Delta\beta < 0.1$ is 7% of the total possible change of 1.48 units of effective charge on the nucleophile. The change in charge between the intermediate and either of the transition structures must be less than *half* of $\Delta\beta$. If the intermediate is to have any existence it would have to support a change in effective charge on the attacking oxygen from the intermediate to one of the transition states (forward or reverse) which must be less than 3.5% of the total change in effective charge in the overall reaction.

The dependencies of β_{lg} and β_{nuc} on the pK_a of nucleophile and nucleofuge respectively are illustrated in Figure 4.8. The values of β fit Equations (4.3) and (4.4) with $p_{xy} = 0.0561$, L = 0.439 and N = -0.976 and identity rate constants determined from these equations fit a quadratic equation which can be predicted from Equation (4.6). The data are illustrated in Figure (4.9) with the parameter C of Equation (4.6) adjusted to give the best vertical fit.

The variation of β_{nuc} and β_{lg} as a function of pK_{lg} and pK_{nuc} is good evidence for a change in structure of the transition state resulting from a change in the entering or departing ligand. In the present case the effect of changing the basicity on the Brønsted slope occurs by a compensation effect whereby the structural change of the transition state in the diagram (Figure 4.10) is along the line of either constant β_{nuc} or constant β_{lg}. As the pK_a of the phenolate ion is decreased both 0,0 and 0,1 corners of the diagram become more stable. Thus the increasing stability of the phenolate ion effectively moves the structure of the transition state towards the 0,1 corner (perpendicular to the reaction coordinate) and towards the 1,1 corner (along the reaction coordinate). The constant slope of the Brønsted plot would result from movement of the transition state structure along the horizontal axis due to cancellation of vertical motions.

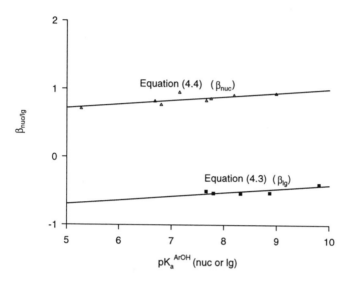

Figure 4.8. Cross-correlation effects in phenolysis at the triazine nucleus; the data are from *Ref. 90* and the slopes of the lines ($p_{xy} = 0.0561$) are identical.

Equation (4.6) for the triazinyl displacement reaction indicates that a reaction with $\tau = 1$ ($\beta_{ii} = 0$) occurs for an identity ligand with pK_{ii} of 4.79. Mechanisms with anionic and cationic triazinyl intermediates ($\tau = 2$ and 0 respectively) occur at $pK_{ii} = 18.00$ and -8.42 (see Figure 4.10). It is interesting to speculate on the geometry of the transition state structure for the concerted mechanism which is likely to depend on the nature of the ligands involved. For identity reactions with high τ the geometry is expected to possess an almost tetrahedral arrangement at the central carbon (**4.12**). An identity reaction with low τ will possess a geometry with a much wider angle between leaving and entering atoms than obtained with the tetrahedral adduct (**4.13**).

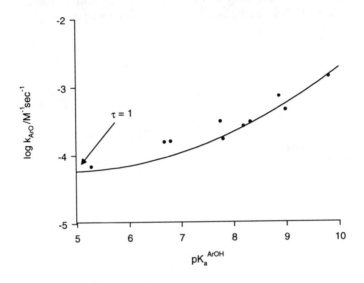

Figure 4.9. Lewis-Kreevoy plot for the identity reaction of phenoxide ions with 2-phenoxy-1,3,5-triazines; the data are from *Ref. 90* and fit the predicted line drawn using Equation (4.6) and the parameters given in the text.

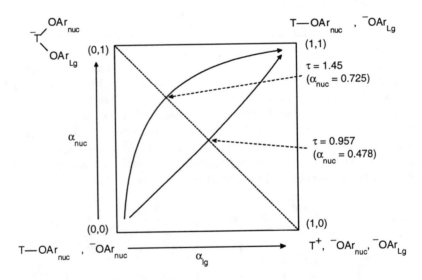

Figure 4.10. More O'Ferrall-Jencks diagram for the identity reactions of phenoxide ions with 2-phenoxy-1,3,5-triazines; $\tau = 1.45$, 2-phenoxy-4,6-dimethoxy-1,3,5-triazine;. $\tau = 0.957$, 2-(3,4-dinitrophenoxy)-4,6-dimethoxy-1,3,5-triazine (*Ref. 90*). T is the 1,3,5-triazinyl group.

The *quasi-symmetrical* displacement of pyridine groups from the triazine nucleus (Scheme 4.23) is an interesting reaction because it provides the first case where a break occurs at a *predicted* pK_a in a stepwise process.[92,93] The second-order rate constants for pyridinolysis fit Equation (2.8) and the non-linear Brønsted plots are illustrated in Figure 4.11. The reaction can be followed in both forward and reverse directions and the global fit exhibits break-points at the predicted pK_a^{pyr} (Figure 4.11). The identity of the break-points is excellent evidence for a stepwise process. The results are conclusive evidence against a concerted mechanism in this case. The detailed structures of the transition states are illustrated in Scheme 4.24 and the system provides the first example of a full effective charge map for an A_N+D_N process.

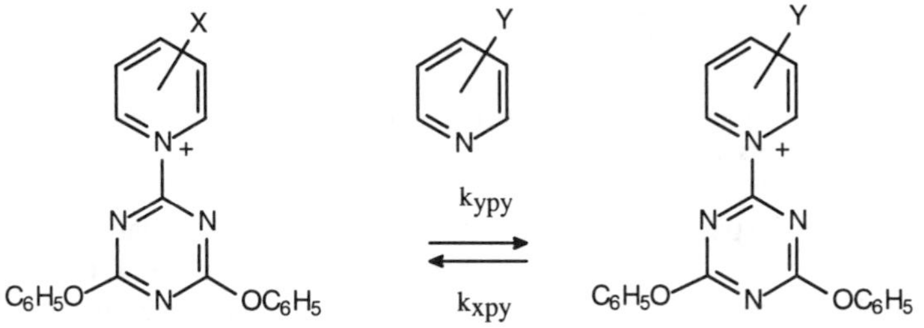

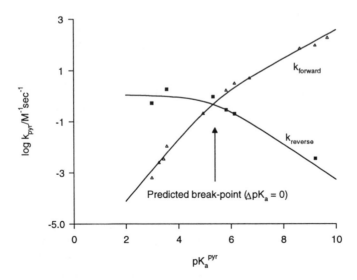

Scheme 4.23. Displacement of pyridines from 1,3,5-triazin-2-yl pyridinium ions.

Figure 4.11. Displacement of pyridines by pyridine nucleophiles from the triazine nucleus (Scheme 4.23). Lines are calculated from Equations (2.9) and (2.10) with data and parameters from *Ref. 92*.

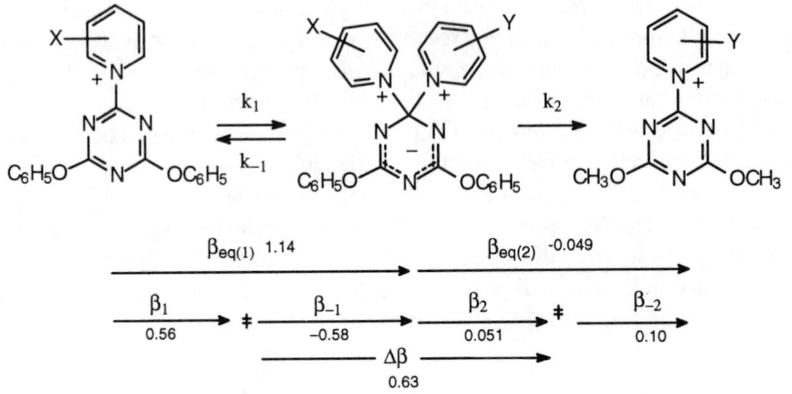

Scheme 4.24. Effective charge map for the identity displacement of pyridines from 1,3,5-triazin-2-yl pyridinium ions; Data from *Ref. 92.*

Femtosecond spectroscopy has been applied to the displacement of an iodide atom by a chlorine atom at a phenyl centre[94] and the observation of a short-lived addition intermediate indicates that in this case the substitution is stepwise.

4.4 NUCLEOPHILIC SUBSTITUTION AT DIGONAL CARBON

Early studies of nucleophilic substitution at an acetylenic carbon (Equation (4.7)) showed that the reaction was very sluggish compared with other substitution reactions. However, since 1962 there have been many facile syntheses effected according to this route (Miller, 1976)[95] and this has raised the possibility of a concerted mechanism. The reaction is complicated by the existence of three known stoichiometric routes (Scheme 4.25)[95-98] which serves to illustrate that it is sometimes very difficult to isolate a single reaction path such as substitution at C1 and study it by altering the substituents; it is possible and sometimes likely that altering the conditions of a reaction by, for example, changing substituents will favour routes other than that under examination.

$$R-C\equiv C-Lg \xrightarrow{Nu^-} R-C\equiv C-Nu + Lg^- \qquad (4.7)$$

At this time there is no evidence for an S_N1 mechanism and gas phase heats of reaction indicate that the acetylenic cation C_2H^+ is by far the most difficult to form from the parent acetylene.[99-101] The concerted displacement mechanism has not been considered seriously but it would involve *front-side* attack and possess a planar transition state structure (**4.14**).

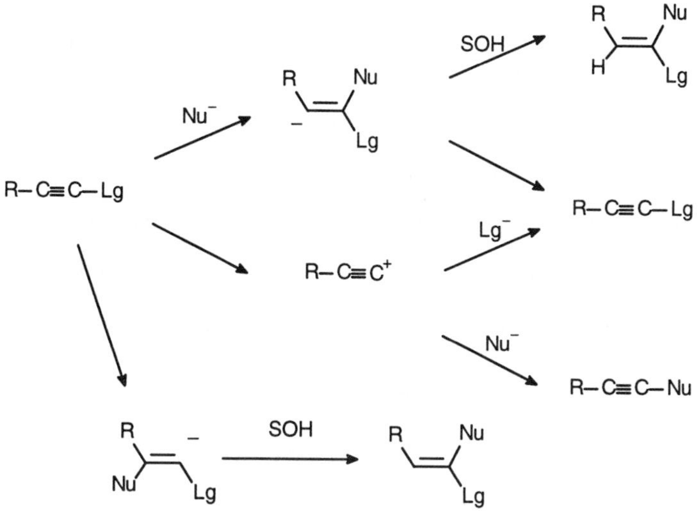

Scheme 4.25. Mechanisms for reactions of nucleophiles with acetylenic carbon.

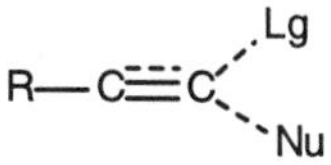

$$R—C{\equiv}C{\underset{Nu}{\overset{Lg}{<}}} \qquad 4.14$$

FURTHER READING

Bender, M. L., Mechanisms of catalysis of nucleophilic reactions of carboxylic acid derivatives, *Chem. Revs.*, *60*, 53, 1960.

Bentley, T. W., Llewellyn, G., and McAlister, J. A., S_N2 Mechanism for alcoholysis, aminolysis, and hydrolysis of acetyl chloride, *J. Org. Chem.*, *61*, 7927, 1996.

Capon, B., Ghosh, A. K., and Grieve, D. McL. A., Direct observation of simple tetrahedral intermediates, *Accs. Chem. Res.*, *14*, 306, 1981.

Crampton, M. R., and Willison, M. J., The stabilities of Meisenheimer complexes. Part VIII. Equilibrium and kinetic data for spiro-complex formation in water, *J. Chem. Soc. Perkin Trans. 2*, 1681, 1974.

Illuminati, G., Nucleophilic heteroaromatic substitution, *Adv. Het. Chem.*, *3*, 285, 1964.

Jencks, W. P., When is an intermediate not an intermediate? Enforced mechanisms of general acid-base catalysed, carbocation, carbanion, and ligand exchange reactions, *Accs. Chem. Res.*, *13*, 161, 1980.

Johnson, S. L., Nucleophilic catalysis of ester hydrolysis and related reactions, *Adv. Phys. Org. Chem.*, 5, 237, 1967.

Lee, I., Cross-interaction constants and transition state structure in solution, *Adv. Phys. Org. Chem.*, *27*, 57, 1992.

McClelland, R. A., and Santry, L. J., Reactivity of tetrahedral intermediates, *Accs. Chem. Res.*, *16*, 394, 1983.

Miller, S. I., Stereoselection in nucleophilic substitution at an sp^2 carbon, *Tetrahedron*, *33*, 1211, 1977.

Modena, G., Reactions of nucleophiles with ethylenic substrates, *Accs. Chem. Res.*, *4*, 73, 1971.

Page, M. I., and Williams, A., *Organic and Bio-Organic Mechanisms*, Addison Wesley Longman, Harlow, Essex, 1997.

Pietra, F., Mechanisms for nucleophilic and photonucleophilic aromatic substitution reactions, *Quart. Rev. Chem. Soc., 23,* 504, 1969.

Rappoport, Z., Solvolysis of α-arylvinyl derivatives, *Accs. Chem. Res., 9,* 265, 1976.

Rappoport, Z., Nucleophilic vinylic substitution. A single- or a multi-step process? *Accs. Chem. Res., 14,* 7, 1981.

Rappoport, Z., Nucleophilic vinylic substitution, *Adv. Phys. Org. Chem., 7,* 1, 1969.

Rappoport, Z., The rapid steps in nucleophilic vinylic "addition-elimination" substitution. Recent developments, *Accs. Chem. Res., 25,* 474, 1992.

Terrier, F., *Nucleophilic Aromatic Displacement,* VCH, New York, 1991.

Williams, A., Effective charge and Leffler's index as mechanistic tools for reactions in solution, *Accs. Chem. Res., 17,* 425, 1984.

Williams, A., and Douglas, K. T., Elimination-addition mechanisms of acyl transfer reactions, *Chem. Revs., 75,* 627, 1975.

Zoltewicz, J. A., New directions in aromatic nucleophilic substitution, *Topics in Current Chem., 59,* 33, 1975.

REFERENCES

1. **Dewar, M. J. S.,** *The Electronic Theory of Organic Chemistry,* Oxford University Press, Oxford, 1948, p. 117.

2. **Ingold, C. K.,** *Structure and Mechanism in Organic Chemistry,* 1st edn., G. Bell and Sons Ltd. (see also 2nd edn., Cornell University Press, Ithaca, 1969), London, 1953.

3. **Ingold, C. K.,** Principles of an electronic theory of organic reactions, *Chem. Revs., 15,* 225, 1934.

4. **Bender, M. L.,** Oxygen exchange as evidence for the existence of an intermediate in ester hydrolysis, *J. Amer. Chem. Soc., 73,* 1626, 1951.

5. **Kivinen, A.,** Mechanisms of substitution at the COX group, in *The Chemistry of Acyl Halides,* Patai, S., ed., Interscience, New York, 1972, p. 177.

6. **Kivinen, A.,** The kinetics of the solvolysis of acid clorides. III. The influence of solvent composition on reaction rate, with special reference to the solvolysis of ethyl chloroformate, *Acta Chem. Scand., 19,* 845, 1965.

7. **Jencks, W. P., and Gilchrist, M.,** Nonlinear structure-reactivity correlations. The reactivity of nucleophilic reagents towards esters, *J. Amer. Chem. Soc., 90,* 2622, 1968.

8. **Melander, L.,** Mechanism of nitration of the aromatic nucleus, *Nature, 163,* 599, 1949.

9. **Bonner, T. G., Bowyer, F., and Williams, G.,** Nitration in sulphuric acid. Part IX. The rates of nitration of nitrobenzene and pentadeuterionitrobenzene, *J. Chem. Soc.,* 2650, 1953.

10. **Melander, L.,** The mechanism of electrophilic aromatic substitution. An investigation by means of the effect of isotopic mass on reaction velocity, *Arkiv Kemi, 2,* 213, 1950.

11. **Gold, V.,** Orbital hybridisation and some other considerations concerning the transition state of bimolecular organic substitution reactions, *J. Chem. Soc.,* 1430, 1951.

12. **Salem, L.,** Intermolecular orbital theory of the interaction between conjugated systems. II. Thermal and photochemical cycloadditions one step reaction, concerted is not two step, *J. Amer. Chem. Soc., 90,* 553, 1968.

13. **Glukhovtsev, M. N., Pross, A., and Radom, L.,** Is S_N2 substitution with inversion of configuration at vinylic carbon feasible? *J. Amer. Chem. Soc., 116,* 5961, 1994.

14. **Lucchini, V., Modena, G., and Pasquato, L.,** An authentic case of in-plane nucleophilic vinylic substitution: the anionotropic rearrangement of di-*tert*-butyl thiirenium ions into thietium ions, *J. Amer. Chem. Soc., 115,* 4527, 1993.

15. **Applequist, D. E., and Kaplan, L.,** The decarbonylation of aliphatic and bridge head aldehydes, *J. Amer. Chem. Soc., 87,* 2194, 1965.

16. **Applequist, D. E., and Klug, J. H.,** Energies of the cycloalkyl and 1-methyl-cycloalkyl free radicals by the decarbonylation method, *J. Org. Chem., 43,* 1729, 1978.

17. **Bacaloglu, R., Blasko, A., Bunton, C. A., and Ortega, F.,** Single electron transfer in deacylation of ethyl dinitrobenzoates, *J. Amer. Chem. Soc., 112,* 9336, 1990.

18. **Giese, B.,** *Radicals in Organic Synthesis: Formation of Carbon Carbon Bonds,* Pergamon Press, Oxford, 1986.

19. **Boger, D. L., and Mathvink, R. J.,** Acyl radicals: intermolecular and intramolecular alkene addition reactions, *J. Org. Chem., 57,* 1429, 1992.

20. **Crich, D., and Fortt, S. M.,** Acyl radical cyclisations in synthesis Part I Substituent effects on the mode and efficiency of cyclisation of 7-heptenoyl radicals, *Tetrahedron, 45,* 6581, 1989.

21. **Pfenninger, J. P., Henberger, C., and Graf, W.,** The radical induced stannane reduction of seleno esters and seleno carbonates: a new method for the degradation of carboxylic acids to nor-alkanes and for decarbonylation of alcohols to alkanes, *Helv. Chim. Acta, 63,* 2328, 1980.

22. **Capon, B., Ghosh, A. K., and Grieve, D. McL. A.,** Direct observation of simple tetrahedral intermediates, *Accs. Chem. Res., 14,* 306, 1981.

23. **McClelland, R. A., and Santry, L. J.,** Reactivity of tetrahedral intermediates, *Accs. Chem. Res., 16,* 394, 1983.

24. **Castro, E. A., Aranada, C., A., and Santos, J. G.,** Kinetics and mechanism of the pyridinolysis of 2,4-dinitrophenyl and 2,4,6-trinitrophenyl O-ethyl dithiocarbonates, *J. Org. Chem., 62,* 126, 1997.

25. **Castro, E. A., Cubillos, M., Santos, J. G., and Téllez, J.,** Kinetics and mechanism of the pyridinolysis of alkyl aryl thionocarbonates, *J. Org. Chem., 62,* 2512, 1997.

26. **Castro, E. A., Pizarro, M. I., and Santos, J. G.,** Kinetics and mechanism of the pyridinolysis of O-ethyl S-aryl thiocarbonates in aqueous solution, *J. Org. Chem., 61,* 5982, 1996.

27. **Castro, E. A., Ibáñez, F., Lagos, S., Schick, M., and Santos, J. G.,** Kinetics and mechanism of the pyridinolysis of 2,4,6-trinitrophenyl acetate and 2,4,6-trinitrophenyl methyl carbonate, *J. Org. Chem., 57,* 2691, 1992.

28. **Castro, E. A., Santos, J. G., Téllez, J., and Umaña, M. I.,** Structure-reactivity correlations in the aminolysis and pyridinolysis of bis(phenyl) and bis(4-nitrophenyl) thionocarbonates, *J. Org. Chem., 62,* 6568, 1997.

29. **Castro, E. A., Ibáñez, F., Salas, M., and Santos, J. G.,** Concerted mechanism of the aminolysis of O-ethyl S-(2,4-dinitrophenyl) thiocarbonate, *J. Org. Chem., 56,* 4819, 1991.

30. **Castro, E. A., and Moodie, R. B.,** Pyridinolysis of methyl chloroformate and the interpretation of curved Brønsted plots in nucleophilic reactions, *J. Chem. Soc. Chem. Commun.,* 828, 1973.

31. **Williams, A., and Douglas, K. T.,** Elimination-addition mechanisms of acyl transfer reactions, *Chem. Revs., 75,* 627, 1975.

32. **Kim, J. K., and Caserio, M. C.,** Acyl-transfer reactions in the gas phase. The question of tetrahedral intermediates, *J. Amer. Chem. Soc., 103,* 2124, 1981.

33. **Kim, J. K., and Caserio, M. C.,** Acyl transfer reactions in the gas phase. Ion-molecule chemistry of vinyl acetate, *J. Amer. Chem. Soc., 104,* 4624, 1982.

34. **Pau, J. K., Kim, J. K., and Caserio, M. C.,** Mechanisms of ionic reactions in the gas phase. Displacement reactions at carbonyl carbon, *J. Amer. Chem. Soc., 100,* 3831, 1978.

35. **Katritzky, A. R., Burton, R. D., Shipkova, P. A., Qi, M., Watson, C. H., and Eyler, J. R.,** Collisionally activated dissociation of N-acylpyridinium cations, *J. Chem. Soc. Perkin Trans. 2,* 835, 1998.

36. **Bender, M. L., and Chen, M. C.,** Acylium ion formation in the reactions of carboxylic acid derivatives, *J. Amer. Chem. Soc., 85,* 37, 1963.

37. **Olah, G. A., and White, A. M.,** Stable carbonium ions. XLIV. Cleavage of protonated aliphatic carboxylic acids to alkyl oxocarbonium ions, *J. Amer. Chem. Soc., 89,* 3591, 1967.

38. **Olah, G. A., Kuhn, S. J., Flood, S. H., and Hardie, B. A.,** Aromatic substitution. XXII. Acetylation of benzene, alkylbenzenes, and halobenzenes with methyloxocarbonium (acetylium) hexafluoro- and hexachloroantimonate, *J. Amer. Chem. Soc., 86,* 2203, 1964.

39. **Olah, G. A., Kuhn, S. J., Tolgyesi, W. S., and Baker, E. B.,** Stable carbonium ions. II. Oxocarbonium (acylium) tetrafluoroborates, hexafluorophosphates, hexafluoroantimonates

and hexafluoroarsenates. Structure and chemical reactivity of acyl fluoride. Lewis acid fluoride complexes, *J. Amer. Chem. Soc., 84*, 2733, 1962.

40. **Cevasco, G., Guanti, G., Hopkins, A. R., Thea, S., and Williams, A.,** A novel dissociative mechanism in acyl transfer from aryl 4-hydroxybenzoate esters in aqueous solution, *J. Org. Chem., 50*, 479, 1985.

41. **Olah, G. A., O'Brien, D. H., and White, A. M.,** Stable carbonium ions. LII. Protonated esters and their cleavage in fluorosulfonic acid-antimony pentafluoride solution, *J. Amer. Chem. Soc., 89*, 5694, 1967.

42. **Asubiojo, O. I., and Brauman, J. I.,** Gas phase nucleophilic displacement reactions of negative ions with carbonyl compounds, *J. Amer. Chem. Soc., 101*, 3715, 1979.

43. **Han, C-C., and Brauman, J. I.,** Intermediates and transition states in chloride ion/acyl chloride displacement reactions, *J. Amer. Chem. Soc., 109*, 589, 1987.

44. **Curran, T. P., Farrar, C. R., Niazy, O., and Williams, A.,** Structure activity studies on the equilibrium reaction between phenolate ions and 2-aryloxazolin-5-ones - data consistent with a concerted acyl group transfer mechanism, *J. Amer. Chem. Soc., 102*, 6828, 1980.

45. **Chrystiuk, E., and Williams, A.,** A single transition state in the transfer of methoxycarbonyl group between isoquinoline and substitutes pyridines, *J. Amer. Chem. Soc., 109*, 3040, 1987.

46. **Ba-Saif, S. A., Luthra, A. K., and Williams, A.,** Concertedness in acyl group transfer: a single transition state in acetyl transfer between phenolate ion nucleophiles, *J. Amer. Chem. Soc., 109*, 6362, 1987.

47. **Ritchie, C. D., Van-Verth, J. E., and Virtanen, P. O. L.,** Cation-anion combination reactions. 21. Reactions of thiolate ions, cyanide ion, and amines with cations and carbonyl compounds in Me$_2$SO solution, *J. Amer. Chem. Soc., 104*, 3491, 1982.

48. **Guthrie, J. P.,** Hydration of carboxylic acids and esters. Evaluation of the free energy change for addition of water to acetic and formic acids and their methyl esters, *J. Amer. Chem. Soc., 95*, 6999, 1973.

49. **Guthrie, J. P.,** Hydration of carboxamides. Evaluation of the free energy change for addition of water to acetamide and formamide derivatives, *J. Amer. Chem. Soc., 96*, 3608, 1974.

50. **Satterthwait, A. C., and Jencks, W. P.,** The mechanism of aminolysis of acetate esters, *J. Amer. Chem. Soc., 96*, 7018, 1974.

51. **Johnson, S. L.,** 1967 Nucleophilic catalysis of ester hydrolysis and related reactions, *Adv. Phys. Org. Chem.*, 5, 237, 1967.

52. **Funderburk, L. H., Aldwin, L., and Jencks, W. P.,** Mechanisms of general acid and base catalysis of the reactions of water and alcohols with formaldehyde, *J. Amer. Chem. Soc., 100*, 5444, 1978.

53. **Funderburk, L. H., and Jencks, W. P.,** Structure-reactivity coefficients for general acid catalysi of semicarbazone formation, *J. Amer. Chem. Soc., 100*, 6708, 1978.

54. **Capon, B., and Grieve, D. McL. A.,** Tetrahedral intermediates. Part 1. The generation and characterisation of some hemiortho esters, *J. Chem. Soc. Perkin Trans. 2*, 300, 1980.

55. **Ba-Saif, S. A., Colthurst, M., Waring, M. A., and Williams, A.,** An open transition state in carbonyl acyl group transfer in aqueous solution, *J. Chem. Soc. Perkin Trans. 2*, 1901, 1991.

56. **Ba-Saif, S., Luthra, A. K., and Williams, A.,** Concerted acetyl group transfer between substituted phenolate ion nucleophiles: variation of transition state structure as a function of substituent, *J. Amer. Chem. Soc., 111*, 2647, 1989.

57. **Jencks, D. A., and Jencks, W. P.,** On the characterization of transition states by structure-reactivity coefficients, *J. Amer. Chem. Soc., 99*, 7948, 1977.

58. **Guthrie, J. P.,** Concerted mechanism for alcoholysis of esters: an examination of the requirements, *J. Amer. Chem. Soc., 113*, 3941, 1991.

59. **Guthrie, J. P., and Pike, D. C.,** Hydration of acylimidazoles: tetrahedral intermediates in acylimidazole hydrolysis and nucleophilic attack by imidazoles on esters. The question of concerted mechanisms for acyl transfer, *Can. J. Chem.*, 65, 1951, 1987.

60. **Hengge, A. C., and Hess, R. A.,** Concerted or stepwise mechanisms for acyl transfer reactions pof *p*-nitrophenyl acetate? Transition state structures from isotope effects, *J. Amer. Chem. Soc., 116,* 11256, 1994.

61. **Page, M. I., and Williams, A.,** *Organic and Bio-Organic Mechanisms,* Addison Wesley Longman, Harlow, Essex, 1997, p. 73.

62. **Hess, R. A., Hengge, A. C., and Cleland, W. W.,** Kinetic isotope effects for acyl transfer from *p*-nitrophenyl acetate to hydroxylamine show a pH-dependent change in mechanism, *J. Amer. Chem. Soc., 119,* 6980, 1997.

63. **Lee, I.,** Cross-interaction constants and transition state structure in solution, *Adv. Phys. Org. Chem., 27,* 57, 1992.

64. **Lewis, E. S., and Hu, D. D.,** Methyl transfers. 8. The Marcus equation and transfer between arenesulfonates, *J. Amer. Chem. Soc., 106,* 3292, 1984.

65. **Blake, J. F., and Jorgensen, W. L.,** *Ab initio* study of the displacement reactions of chloride ion with formyl and acetyl chloride, *J. Amer. Chem. Soc., 109,* 3856, 1987.

66. **Burgi, H. B., and Dunitz, J. D.,** From crystal statics to chemical-dynamics, *Accs. Chem. Res., 16,* 153, 1983.

67. **Deslongchamps, P.,** *Stereoelectronic Effects in Organic Chemistry,* Pergamon Press, Oxford, 1983.

68. **Kirby, A. J.,** *The Anomeric Effect and Related Stereoelectronic Effects at Oxygen,* Springer-Verlag, Berlin, 1983.

69. **Sinnott, M. L.,** The principle of least nuclear motion and the theory of stereoelectronic control, *Adv. Phys. Org. Chem., 24,* 113, 1988.

70. **Thea, S., Guanti, G., Kashefi-Naini, N., and Williams, A.,** Steric and electronic control of the dissociative hydrolysis of 4-hydroxybenzoate esters, *J. Chem. Soc. Chem. Commun.,* 529, 1983.

71. **Thea, S., Cevasco, G., Guanti, G., Kashefi-Naini, N., and Williams, A.,** Reactivity in the *para*-oxoketene route of ester hydrolysis: the effect of internal nucleophilicity and the irrelevance of "B" strain, *J. Org. Chem., 50,* 1867, 1985.

72. **Wang, L-h., and Zipse, H.,** Bifunctional catalysis of ester aminolysis - a computational and experimental study, *Liebig's Ann.,* 1501, 1996.

73. **Zipse, H., Wang, L-h., and Houk, K. N.,** Polyether catalysis of ester aminolysis - a computational and experimental study, *Liebig's Ann.,* 1511, 1996.

74. **Park, Y. S., Kim, C. K., Lee, B-s., Lee, I., Lim, W. M., and Kim, W-k.,** Theoretical studies on the reactions of substituted phenolate anions with formate esters, *J. Phys. Org. Chem., 8,* 325, 1995.

75. **Stohrer, W-D.,** Ein Argument für den konzertierten Verlauf der nucleophilen vinylischen Substitution mit Konfigurationserhalt, *Tetrahedron Letters,* 207, 1975.

76. **Stohrer, W-D.,** An MO model for the S_N2 - Reaction with retention, *Chem. Ber., 109,* 285, 1976.

77. **Lucchini, V., Modena, G., and Pasquato, L.,** A novel type of selectivity in anionotropic rearrangements, *J. Amer. Chem. Soc., 110,* 6900, 1988.

78. **Lucchini, V., Modena, G., and Pasquato, L.,** Anionotropic rearrangements of *tert*-butyl- and adamantyl-thiiranium ions into thietanium ions. A novel case of selectivity, *J. Amer. Chem. Soc., 113,* 6600, 1991.

79. **Lucchini, V., Modena, G., and Pasquato, L.,** S_N2 and Ad_N-E Mechanisms in bimolecular nucleophilic substitutions at vinyl carbon. The relevance of the LUMO symmetry of the electrophile, *J. Amer. Chem. Soc., 117,* 2297, 1995.

80. **Di Nunno, L., Modena, G., and Scorrano, G.,** Nucleophilic reactions in ethylenic derivatives Part IX. "Direct Substitution" and "Elimination-addition" mechanisms in the reaction of arylsulphonylchloroethylenes with alkoxide ions, *J. Chem. Soc. B,* 1186. 1966.

81. **Okayama, T., Takino, T., Sato, K., and Ochiai, M.,** In-plane vinylic S_N2 substitution and intramolecular β-elimination of β-alkyl vinyl(chloro)-l^3-iodanes, *J. Amer. Chem. Soc., 120,* 2275, 1998.

82. **Sluijs, M. J. van der., and Stirling, C. J. M.,** Elimination and addition reactions Part XXV. Addition-elimination reactions of phenoxy vinyl sulphones, *J. Chem. Soc. Perkin Trans. 2,* 1268, 1974.

83. **Bernasconi, C. F., Fassberg, J., Killion, R. B., and Rappoport, Z.,** Kinetics of reaction of thiolate ions with α-nitro-β-substituted stilbenes in 50% Me_2SO - 50% water. Observations of the intermediate in nucleophilic vinylic substitution reactions, *J. Amer. Chem. Soc., 112,* 3169, 1990.

84. **Bernasconi, C. F., Fassberg, J., Killion, R. B., and Rappoport, Z.,** Kinetics of the reaction of β-methoxy-α-nitrostilbene with thiolate ions - 1st direct observation of the intermediate in a nucleophilic vinylic substitution, *J. Amer. Chem. Soc., 111,* 6862, 1989.

85. **Bernasconi, C. F., Schuck, D. F., Ketner, R. F., Weiss, M., and Rappoport, Z.,** Kinetic analysis of elementary steps in nucleophilic vinylic substitution reactions of α-nitro-β-X-stilbenes (X=OCH2CF3, OCH3, NO2) with various nucleophiles - detection of the intermediate in the reaction of α-nitro-β-(2,2,2-trifluoroethoxy)stilbene withHOCH2CH2S$^-$ and of β-methoxy-nitrostilbene with CF3CH2CH2O$^-$, *J. Amer. Chem. Soc., 116,* 11764, 1994.

86. **Bernasconi, C. F., Leyes, A. E., Rappoport, Z, Eventova, I.,** Reaction of N-methylmethoxyamine with β-methoxy-nitrostilbene - 1st direct observation of the intermediate in a nucleophilic vinylic substitution with an amine nucleophile, *J. Amer. Chem. Soc., 115,* 7513, 1993.

87. **Bernasconi, C. F., Leyes, A. E., Eventova, I., and Rappoport, Z.,** Kinetics of the reactions of β-methoxy-α-nitro stilbene with methoxyamine and N-methylmethoxyamine - direct observation of the intermediate in nucleophilic vinylic substitution, *J. Amer. Chem. Soc., 117,* 1703, 1995.

88. **Illuminati, G.,** Nucleophilic heteroaromatic substitution, *Adv. Het. Chem., 3,* 285, 1964.

89. **Zoltewicz, J. A.,** New directions in aromatic nucleophilic substitution, *Topics in current Chem., 59,* 33, 1975.

90. **Renfrew, A. H. M., Rettura, D., Taylor, J. A., Whitmore, J. M. J., and Williams, A.,** Stepwise versus concerted mechanisms at trigonal carbon: transfer of the 1,3,5-triazinyl group between aryl oxide ions in aqueous solution, *J. Amer. Chem. Soc., 117,* 5484, 1995.

91. **Renfrew, A. H. M., Taylor, J. A., Whitmore, J. M. J., and Williams, A.,** A single transition state in nucleophilic aromatic substitution - reaction of phenolate ions with 2-(4-nitrophenoxy)-4,6-dimethoxy-1,3,5-triazine in aqueous solution, *J. Chem. Soc. Perkin Trans. 2,* 1703, 1993.

92. **Cullum, N. R., Renfrew, A. H. M., Rettura, D., Taylor, J. A., Whitmore, J. M. J., and Williams, A.,** Effective charge on the nucleophile and leaving group during the stepwise transfer of the triazinyl group between pyridines in aqueous solution, *J. Amer. Chem. Soc., 117,* 9200, 1995.

93. **Renfrew, A. H. M., Taylor, J. A., Whitmore, J. M. J., and Williams, A.,** Timing of bonding changes in fundamental reactions in solutions - pyridinolysis of a triazinylpyridinium salt, *J. Chem. Soc. Perkin Trans. 2,* 2383, 1994.

94. **Zhong, D., Ahmad, S., Cheng, P. Y., and Zewail, A. H.,** Femtosecond nucleophilic substitution reaction dynamics, *J. Amer. Chem. Soc., 119,* 2305, 1997.

95. **Miller, S. I., and Dickstein, J. I.,** Nucleophilic substitution at acetylenic carbon. The last holdout, *Accs. Chem. Res., 9,* 358, 1976.

96. **Brandsma, L., Bos, H. J. T., and Arens, F.,** Ethynyl ethers and thioethers, in *The Chemistry of Acetylenes,* Viehe, H. G., ed., Marcel Dekker, New York, 1969, Chapter 11.

97. **Viehe, H. G.,** Ynamines, in *The Chemistry of Acetylenes,* Viehe, H. G., ed., Marcel Dekker, New York, 1969, Chapter 12.

98. **Fujii, A., Dickstein, J. L., and Miller, S. I.,** Stereoselection in nucleophilic substitution at an sp^2 carbon, *Tetrahedron, 33,* 1211, 1977.

99. **Dill, J. D., Schleyer, P. von R., and Pople, J. A.,** Ab initio studies of aminophenyl cations, *Tetrahedron, 31,* 2857, 1975.

100. **Lossing, F. P., and Semeluk, G. P.,** Free radicals by mass spectrometry. XLII. Ionisation potentials and ionic heats of formation for C_1-C_4 alkyl radicals, *Can. J. Chem., 48,* 955, 1970

101. **Lossing, F. P.,** Free radicals by mass spectrometry. XLIII. Ionisation potentials and ionic heats of formation for vinyl, allyl and benzyl radicals, *Can. J. Chem., 49,* 357, 1971.

Chapter 5

Nucleophilic Displacements at Saturated Carbon

The classical reaction involving nucleophilic displacement at aliphatic carbon has been a test-bed of mechanistic theories since they were first studied almost a century ago. The perceived mechanism for this reaction developed in step with the development of mechanistic ideas. Ingold and Hughes described two main mechanisms, namely the concerted S_N2 (A_ND_N) process and the stepwise S_N1 (D_N+A_N) process, but both pathways have now been subjected to considerable study to the extent that at one time the validity of the concerted S_N2 path was in serious doubt.[1]

The points which have arisen for nucleophilic substitution at saturated carbon are mainly due to experimental data from the following types of study: (i) stereochemistry which gives data consistent with but does not *require* a concerted mechanism; (ii) determination of the lifetimes of carbenium ion intermediates has enabled predictions to be made regarding enforced concertedness; (iii) isotope effects have enabled the strengths of bonds to be determined in the transition state structures; (iv) positional isotope exchange has given information regarding the stability of ion-pair intermediates; (v) gas phase and theoretical studies have yielded information about the potential energy surfaces to be expected in nucleophilic substitution at saturated carbon.

5.1 MECHANISMS

The More O'Ferrall-Jencks diagram (Figure 5.1) provides a means for discussing the relationships between the mechanisms for nucleophilic aliphatic substitution[1-3] already delineated by Ingold. The diagram indicates that, for the purposes of providing a surface, the 0,1 position corresponds to a hypothetical pentacoordinate intermediate;[2,4] it is feasible to conceive of such a structure as a station on the surface although the corresponding molecule has only been isolated in exceptional circumstances. There is no inconsistency in postulating such a hypothetical structure for a corner when it is considered that all non-corner structures on the surface are themselves hypothetical. There has been discussion of the existence of such a pentacoordinate intermediate; hypervalent anionic carbon structures are well characterised in organometallic chemistry[5-16] and the intermediate constitutes part of the E2C mechanism of elimination.[9] A hypervalent molecule directly analogous to the structure of the (0,1) corner has been stabilised by the use of special structural constraints and isolated (**5.1**, Scheme 5.1). The structure was shown to be neither **5.2** nor a rapidly interconverting equilibrium mixture (**5.3**).[17-21] The species represented at the 1,0 corner of the diagram, a carbenium ion, requires no argument to justify it as a putative station on the surface.

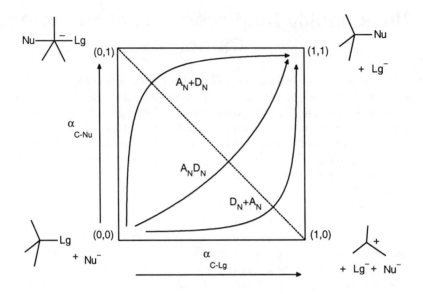

Figure 5.1. More O'Ferrall-Jencks diagram for displacement mechanisms at saturated carbon. The corners represent the energies of the associated *encounter* complexes.

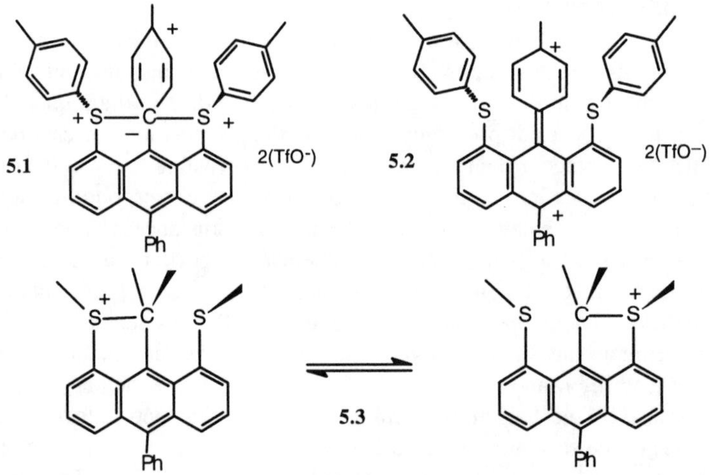

Scheme 5.1 Hypervalent carbon in an intermediate as a putative alternative to the A_ND_N transition structure (TfO⁻ = triflate ion).

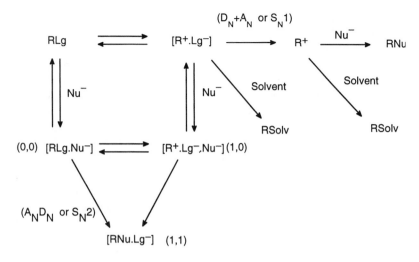

Scheme 5.2. Nucleophilic substitution at saturated carbon. In this scheme the rearrangements within the encounter complexes (structures in square brackets), other than those involving bonding changes, are not delineated. Figures in parenthesis refer to the corner coordinates of the More O'Ferrall-Jencks diagram (Figure 5.1).

Scheme 5.2 illustrates the changes in reaction mechanism that occur with decreasing lifetime of the carbenium ion intermediate in aqueous solution.[22,23] Substantial theoretical and experimental work has been carried out on the question of concertedness in single electron transfer mechanisms of nucleophilic substitution and displacement reactions by radicals.[24-29]

5.2   THE CARBENIUM ION

The carbenium ion is often a very reactive species in the presence of nucleophiles and consequently few nucleophilic substitution reactions have revealed stable carbenium ions. The expression of a carbenium ion intermediate is strongly affected by the solvent which itself is often nucleophilic in nature.

Many carbenium ions have lifetimes which put displacement reactions in a borderline region of mechanism between stepwise and concerted processes.[30-32] The existence of reactive carbenium ion intermediates in aliphatic substitution reactions has been well documented and is best demonstrated by racemisation and the observation of a change in rate-limiting step caused by the presence of a common ion.

The fission of a series of substituted acetophenones proceeds via oxocarbenium ion intermediates. The rate of ketal cleavage to yield bisulphite adducts is independent of sulphite ion concentration (Young and Jencks,1977).[33] The lifetimes of the carbenium ion intermediates range from 10^{-7} to 10^{-9} seconds as determined from the rate constant of the reaction with sulphite ion and water assuming combination of the carbenium ion and sulphite ion to be diffusion controlled (5×10^9 M^{-1}sec^{-1}).

5.3 ION-PAIR INTERMEDIATES

Hammett (1940)[34] and Winstein (1965)[35] first raised the possibility that the ion-pair mechanism could be involved in aliphatic substitution reactions. The D_N+A_N (S_N1) mechanistic extremum is *required* to involve a contact ion-pair intermediate[*] (for a neutral substrate) on its way to expression of the free carbenium ion which then reacts with solvent or a nucleophile (Scheme 5.2). This mechanism cannot give rise to the special effects such as partial inversion of configuration because the carbenium ion must be stable enough to survive the diffusion process (and also the "tumbling" period). The shorter-lived carbenium ion will react before it has the chance to escape from the contact ion-pair ($[R^+.Lg^-]$) and, in the extreme case, will react before it can "tumble" relative to the nucleophile. These conditions give rise to the stereochemical effects of partial inversion or retention during solvolysis and also to dependence of the rate constant on the concentration of the added nucleophile.

The ion-pair intermediate was of considerable interest in the 1970's[36-51] when it became part of a unified mechanism of nucleophilic aliphatic substitution (Sneen, 1973).[40] In the case of the solvolysis and azidolysis of 2-octyl sulphonate ester the product 2-octyl azide is inverted and it may be concluded that an intermediate (an ion-pair) is formed in the rate-determining step which does not include azide ion but which must be asymmetrical. In this mechanism the ion-pair reacts with nucleophile (Scheme 5.3) to give product and if there is a competition between two nucleophiles such as solvent (water) and the azide ion the partitioning ratio of the two products is predicted to be proportional to the concentration of the trapping nucleophile. The rate constant should increase to a maximum as the rate-limiting step changes from attack of nucleophile on carbenium ion (k_N) to formation of the carbenium ion (k_a) (Equations (5.1) and (5.2)).

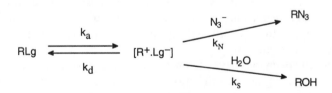

Scheme 5.3. Sneen's unified mechanism for nucleophilic substitution.

$$k_{obs} = k_a(k_N[N_3^-] + k_s)/(k_d + k_N[N_3^-] + k_s) \qquad (5.1)$$

$$[RN_3]_{final}/[ROH]_{final} = k_N[N_3^-]/k_s \qquad (5.2)$$

---

[*] A contact ion-pair is that formed immediately upon fission of the alkyl-leaving group bond. A solvent separated ion-pair results from exchange of solvent for the leaving group which remains within the complex; this is essentially an "encounter complex" formed by reaction rather than by collision.[52]

The effect of azide ion concentration on the value of k_{obs} is illustrated in Figure 5.2. The mechanism of Scheme 5.3 is a special case of Sneen's unifying mechanism; while this is undoubtedly correct, subsequent studies have indicated that it is not universal.

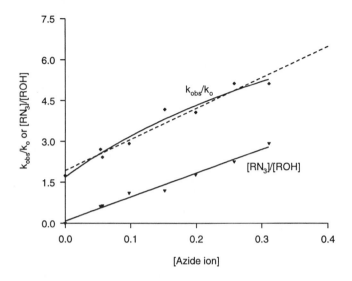

Figure 5.2. Inhibition of solvolysis of an alkyl azide by added azide ion. Data are from *Refs. 37* and *38*. The dotted line refers to a linear correlation whereas the curve is from Equation 5.1.

It is salutary to consider the results illustrated in Figure 5.2. The rate constant ratio has been corrected for *salt effect* yet only exhibits a slight curvature; this could be due to factors other than the change in rate-limiting step caused by increasing the $k_N[N_3^-]$ term such as a varying ionic strength and a *specific* salt effect. The argument for Scheme 5.3 rests on the product ratio being proportional to $[N_3^-]$ and inversion over the range of $[N_3^-]$ where the rate-limiting step changes. A larger curvature would provide more compelling evidence for Scheme 5.3 and as it is the data can also fit a linear correlation as shown in Figure 5.2. The best evidence for the existence of encounter complexes with ion-pairs of significant lifetime is from positional isotope exchange experiments[53-59] which demonstrate that internal return occurred in many cases (Scheme 5.4). Another technique was applied by Streitwieser[60,61] who showed that increased racemisation occurred in the tosyl ester during the acetolysis of 2-octyl toluenesulphonate in the presence of added lithium tosylate common ion (Scheme 5.4).

It was suggested that the ion-pair mechanism explains all nucleophilic displacement reactions and that the S_N2 mechanism is simply explained by rate-limiting backside attack of the nucleophile on the carbenium ion of a contact ion-pair (Sneen).[37,38,40] This mechanism undoubtedly occurs[62] but not to the exclusion of the classical concerted model.[63-65] For the ion-pair mechanism to operate in simple aliphatic displacements at primary alkyl halides an activation barrier would be required of between 15 and 40 kcal/mole higher than the observed values.[65-67] For example the $\Delta G°$ for formation of the carbenium ion-pair from methyl bromide is 66 kcal/mole whereas the $\Delta G^{\ddagger}$ for hydrolysis is 26 kcal/mole at normal temperature.

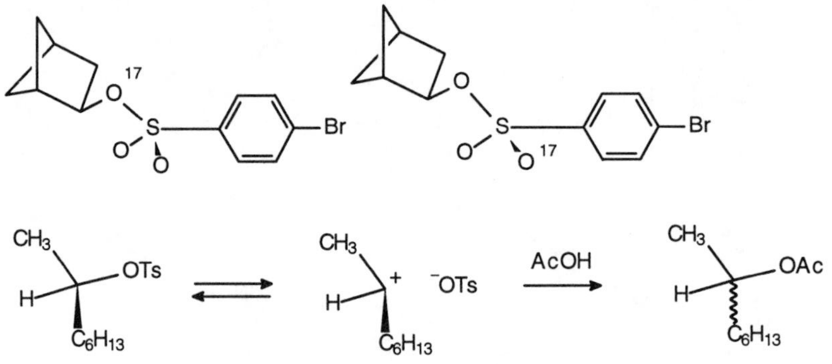

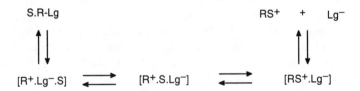

Scheme 5.4. Positional isotope exchange demonstrates ion-pairs in encounter complexes.

Solvent separated and contact ion-pairs will occur in sequence during reaction as a result of the initial ionisation when the counterions are barely separated. The constituents of the ion-pairs "tumble" relative to each other (half-life of $ca.$ 10^{-12} sec.) and after further time solvent is able to intervene to yield solvent separated ion-pairs which then react to give the product encounter-complex. The simplest mechanism for an SN1 (D_N+A_N) solvolysis mechanism would thus be as in Scheme 5.5. Reaction with nucleophiles other than solvent, giving rise to the second-order rate law, is best delineated[68-71] as

S.R-Lg   RS+   +   Lg−

$\uparrow\downarrow$ $\uparrow\downarrow$

[R+.Lg−.S] $\rightleftarrows$ [R+.S.Lg−] $\rightleftarrows$ [RS+.Lg−]

Scheme 5.5. The simplest mechanism for a D_N+A_N solvolysis reaction; S represents solvent molecules.

a pre-association mechanism if ion-pairs are involved (Scheme 5.6). This mechanism differs from that originally described by Sneen (1973)[40] in that the nucleophile is required to be within the encounter complex before the bond-fission step can lead to products. Sneen's mechanism involves collision of the nucleophile with the encounter complex (Scheme 5.3).

The reaction of a strong nucleophile with an unstable carbenium ion is expected to be very fast and indeed if this reaction is diffusion controlled it can be used as a "clock"[30-32] to determine the rate constant for the decomposition of the intermediate in water.[30-33] This technique has produced results for the stability of carbenium ions in reactions with mechanisms of the

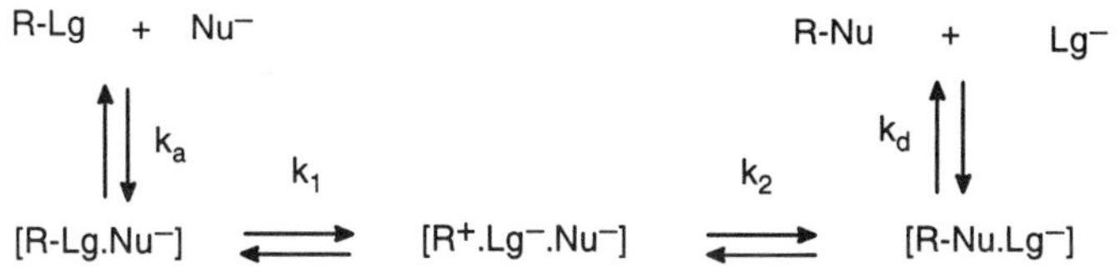

$$R\text{-}Lg \ + \ Nu^- \qquad\qquad\qquad R\text{-}Nu \ + \ Lg^-$$

$$\Big\uparrow\Big\downarrow k_a \qquad k_1 \qquad\qquad k_2 \qquad\qquad k_d \Big\uparrow\Big\downarrow$$

$$[R\text{-}Lg.Nu^-] \ \rightleftharpoons \ [R^+.Lg^-.Nu^-] \ \rightleftharpoons \ [R\text{-}Nu.Lg^-]$$

Scheme 5.6. Ion-pairs in bimolecular nucleophilic substitution.

D_N+A_N type enabling estimates to be made of the lifetimes of putative carbenium ion intermediates. The rate constants for hydration of the oxocarbenium ions from carbonyl compounds may be determined from inhibition of the solvolyses of α-azido ethers by added azide ion (Scheme 5.7); the rate constant follows Equation (5.3) and k_{az}/k_w may be deduced from

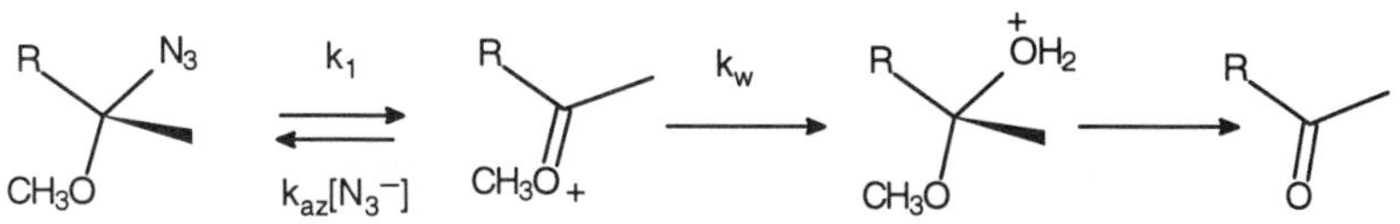

Scheme 5.7. Trapping the oxocarbenium ion with azide ion.

$$k_{obs} = k_1.k_w/(k_w + k_{az}[N_3^-]) \qquad\qquad (5.3)$$

experiments measuring k_{obs} as a function of azide ion concentration (Figure 5.3). The value k_{az} is taken to be $5.10^9 M^{-1}sec^{-1}$, the diffusion-controlled reaction of azide ion with the oxocarbenium ion is shown to be structure independent.[33,72-75] From this value the equilibrium constant $K_{eq} = k_1/k_{az}$ may be obtained. The value of k_w is related to k_1 by a linear free energy relationship from which k_w can be estimated given any k_1 value. The value of k_1 extrapolated for the $CH_3OCH_2^+$ cation is of the order of $10^{12}sec^{-1}$ indicating that concerted mechanisms of the methoxymethyl reagent should be enforced because stepwise mechanisms are not possible.[70,76] A similar conclusion can be made for the glycosyl oxocarbenium ion which has a lifetime too short to allow diffusional equilibrium with solute or solvent. The acid-catalysed anomerisation of methyl β-D-glucopyranoside (5.4) in CD_3OD gives >98% incorporation of CD_3OD into the α-product but with no detectable exchange of CD_3OD into the β-reactant (Capon and Thacker, 1967).[77] It is evident that the formation of the oxocarbenium ion (5.5) is not a universal feature of solvolysis mechanisms of oxyalkyl species and one of the criteria is the lifetime of this intermediate ion.[78-84]

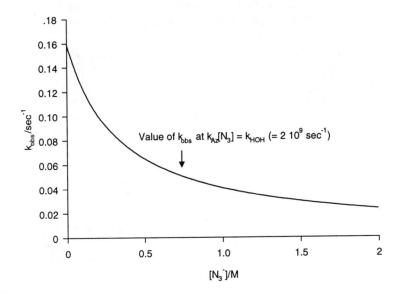

Figure 5.3. Solvolysis of 3-azido-1,3-dimethoxybutane ($CH_3CN_3(OCH_3)$ ·$CH_2CH_2OCH_3$). Line computed from Equation (5.3) from data in *Ref. 69*.

    5.4   **5.5**

5.4   THE CONCERTED MECHANISM

5.4.1 Application of "clocking" reactions

Even in 1973 Sneen concluded[40] that the traditional mechanism for the S_N2 reaction involving direct attack on covalent carbon had not been established. Stereochemistry is an unreliable diagnostic tool and the main evidence resides in estimations of the lifetimes of putative carbenium ion intermediates and in isotope effects. Amyes and Jencks[23,68,69] demonstrated that an acetal derivative of benzaldehyde (**5.6**) undergoes rapid concerted bimolecular substitution with reactive nucleophiles such as azide, thiocyanate and thiosulphate ions. A free energy relationship was determined (Young and Jencks, 1977)[33] between the rate constants for attack of water (k_w) on the oxocarbenium ions derived from acetophenone dimethyl acetals and for attack of the sulphite dianion on the corresponding ketones. Extrapolation indicates a lifetime of $\sim 10^{-15}$sec for the $(CH_3OCH_2)^+$ cation[33] so that concerted substitution mechanisms should be enforced.[85] Accordingly methoxymethyl derivatives were shown to undergo concerted bimolecular reactions with nucleophiles.[70,76] The methoxymethyl species **5.7** reacts with nucleophiles via a second-order rate law and the rate constants obey a Swain-Scott correlation with a value of $s = 0.28$ for the slope. The small value of s and the large negative value of β_{lg} (from -0.7 to -0.9) indicate a transition structure **5.8** that can be described as either an open transition structure for S_N2 displacement or as an oxocarbenium ion stabilised by weak interactions with nucleophile and electrofuge. The shape of the energy surface at the transition structure is relatively flat as judged from the zero value of p_{xy} (Section 2.3.2).

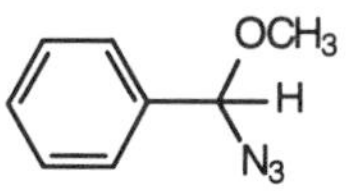

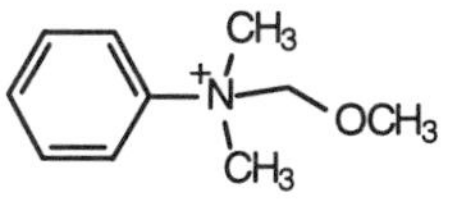

The question remains whether the reaction passes through an unstable intermediate or whether the intermediate is so unstable that there is no barrier for its formation or breakdown. The intermediate could be described by Scheme 5.8. If Nu were a weaker nucleophile than Lg then k_1 would be rate-limiting ($k_2 < k_{-1}$) and this could explain the different reactivities and different secondary isotope effects of different nucleophiles (Nu). However the attacking amines are up to 10^8-fold more basic than the leaving N,N-dinitro-3-aniline so that $k_2 > k_{-1}$ and the rate-limiting step should be k_1.[85]

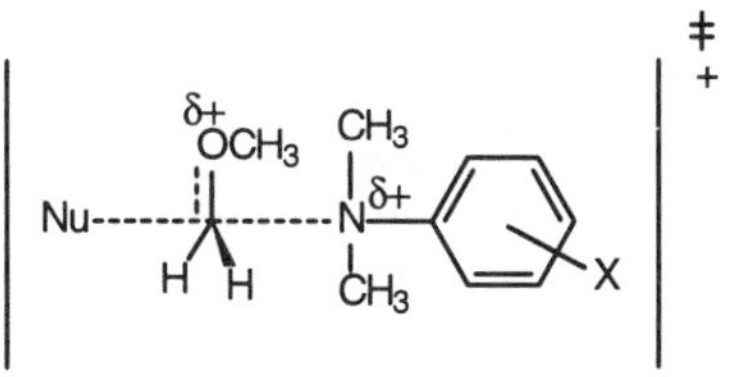

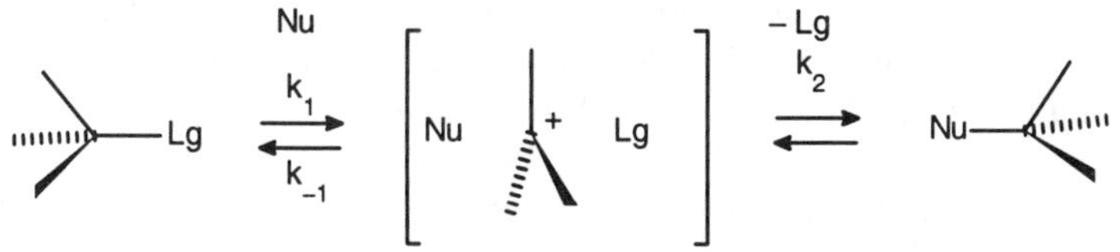

Scheme 5.8. Change in rate-limiting step for an S_N2 reaction with a carbenium ion intermediate.

All the nucleophilic reagents studied are stronger nucleophiles measured against carbenium ions than aniline as determined from Ritchie's N^+ values[86,87] and would therefore be expected to have k_1 rate-limiting as for the amines. If there were an intermediate the Swain-Scott correlation would be expected to exhibit a break at a value of n corresponding to the leaving aniline (Knier and Jencks, 1980).[70] The data (Figure 5.4) show no evidence of a break required for the change in rate-limiting step which would occur at an n value close to that for aniline (5.7). If the methoxymethyl carbenium ion were to exist as an intermediate (within the encounter complex with nucleophile and nucleofuge) in the reaction with water this would not be the case for stronger nucleophilic reagents such as thiolate anions which perforce must act via a concerted pathway.

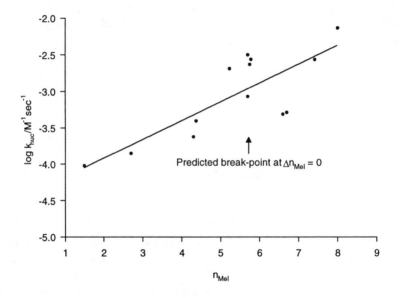

Figure 5.4. Reactions of nucleophiles with N-(methoxymethyl)-N,N-dimethyl-3-nitroanilinium ion in 30% acetone/water as a function of the Swain-Scott parameter n. Data is from *Ref. 70* and line is from a least squares linear fit.

The lifetimes of iminium ions in aqueous solution can be estimated[88,89] by assuming a rate constant of $5.10^9 M^{-1} sec^{-1}$ for diffusion-controlled trapping of the iminium ion by thiolate anions. The rate constants for reaction of water with the iminium ion (k_w, Scheme 5.9) were readily determined and the identity of k_{-1} as a diffusion rate constant was verified by the study of the ratio k_{-1}/k_1 as a function of viscosity (see Chapter 2). Further evidence for the diffusional character of k_{-1} is that the combination reaction of RS^- with the iminium ion has a rate constant insensitive to the pK_a of the thiol and that less reactive carbenium ions than the iminium ions have recombation rate constants with thiolate ions approaching that of the diffusion limit (Ritchie, 1986).[90]

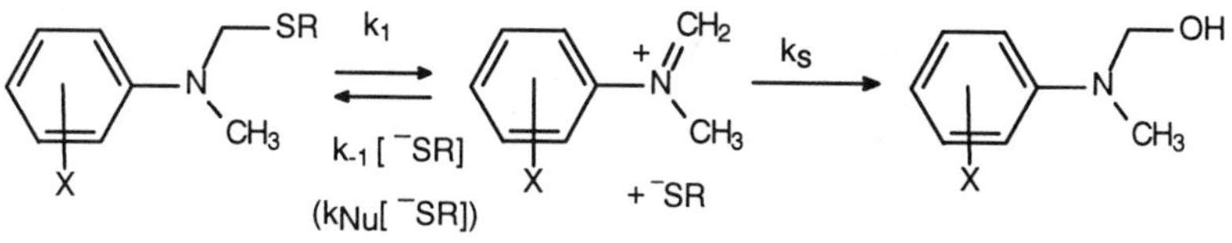

Scheme 5.9. Measuring the lifetime of the iminium ion.

The rate constant for decomposition of $CH_2=N^+(CH_3)C_6H_4$-4-NO_2 is $1.10^8 sec^{-1}$. The N^+ scale of nucleophilic reactivities[90] indicates that the thiophenoxide ion will be more reactive than water by a factor of ~10^8-fold thus giving a rate constant of $10^{16} M^{-1} sec^{-1}$. This requires that reaction of the parent thioether with a nucleophile must proceed via a concerted bimolecular route because there can be no significant barrier for the reaction of thiolate ion with the iminium cation.[91] However, the reaction with nucleophilic solvent proceeds by a stepwise mechanism through a liberated intermediate, which is also trapped by added thiolate anions. The lifetimes of aliphatic iminium ions in aqueous solution are likely to be larger that that of the iminium ions from anilines and the rate constant for reaction of water with $CH_2=N^+(CH_3)CH_2CF_3$ has been determined to be $1.8.10^7 sec^{-1}$. The lifetime data require that kinetically bimolecular substitution reactions of *good* nucleophiles such as azide or thiolate anion proceed by a concerted mechanism, because the ion triplet intermediate of this putative stepwise (but kinetically bimolecular) reaction cannot exist for the time of a bond vibration. The iminium ion has a lifetime in the presence of aqueous solvent; it forms in water by simple ionisation of the substrate and undergoes diffusion trapping by added nucleophilic reagents ($k_{Nu}/k_s > 1M^{-1}$ when $k_{Nu} = 5\ 10^9 M^{-1} sec^{-1}$).

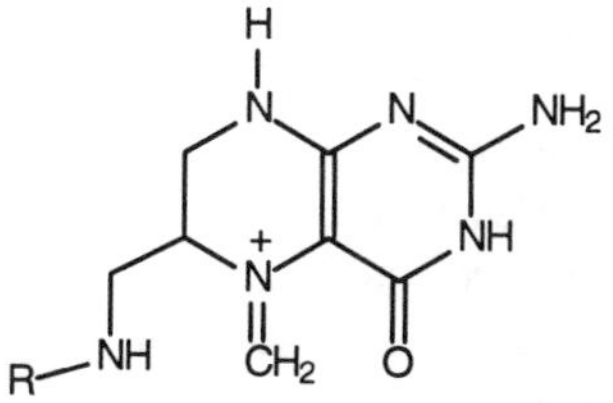

5.9

In the case of aniline thioethers the bimolecular substitution reaction will (by the principle of microscopic reversibility) have a concerted pathway. This reaction is important as an analogue of the thymidylate synthase enzyme reaction[92-94] wherein 2'-deoxyuridylic acid (dUMP) reacts with **5.9** to yield 2'-deoxythymidylic acid (dTMP). The lifetime of N^5,N^{10}-methylene tetrahydrofolate (**5.9**) may be estimated by extrapolation from the Brønsted plot of Figure 5.5 assuming that the pK_a of the N^5 atom in **5.9** is 4.8. The value of $k_w = 4.10^6 sec^{-1}$ is large enough that it is possible that nucleophilic attack from a nucleophile (–NHR) that is in close proximity to the iminium ion would prevent the iminium ion from existing for a significant period. This condition could give rise to a concerted bimolecular substitution mechanism in the conversion of dUMP to dTMP catalysed by thymidylate synthase.

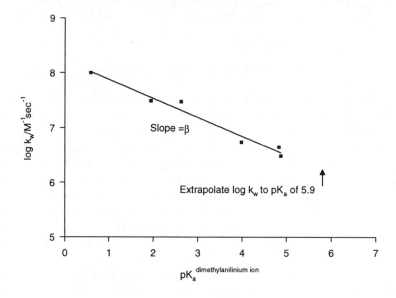

Figure 5.5. Brønsted correlation of the hydration of $H_2C=N^+(CH_3)C_6H_4X$ against the pK_a's of the corresponding N,N-dimethylanilines. Data from *Ref. 90.*

The reaction of azide ion with 4-methoxybenzyl derivatives in aqueous acetone solutions has a mechanism which conforms to that of Scheme 5.2. The reaction has terms both zero-order and first-order in azide ion concentration and this indicates that there is a concurrent D_N+A_N and bimolecular process.[95] The D_N+A_N process must perforce involve an ion pair but the bimolecular reaction is concerted (*via* $N_3^-.ArCH_2Cl$) because (among others) of the following arguments: i) the value of the Grunwald-Winstein m parameter is ~0.5 for the bimolecular reaction implying much less charge change to the transition state compared with that in the D_N+A_N reaction; ii) reactivity in the bimolecular reaction increases for SCN^-, N_3^- and $S_2O_3^{2-}$ is not consistent with a Sneen trapping mechanism which should indicate an identical second-order rate constant for nucleophiles stronger than Cl^-; iii) the rate constant for collapse of the $R^+.N_3^-$ ion pair can be estimated to be ~$10^{15}sec^{-1}$ thus the triple ion pair complex (*cf.* Scheme 5.2) cannot be an intermediate in the reaction of RCl with azide ion.

5.4.2 Isotope effects

Secondary α-deuterium kinetic isotope effects have been widely used to study the mechanism of nucleophilic substitution reactions and to determine how the substituents alter the structure of the S_N2 transition state.[96] Shiner (1970)[97] concluded that solvolyses involving rate-determining dissociation of ion-pairs the α-deuterium isotope effects exhibit maxima independent of carbenium ion structure but dependent on the structure of the leaving group. Leffek[98,99] found very large secondary α-deuterium isotope effects of 1.12 and 1.10 for reaction between bromide ion and phenyldimethylbenzylammonium ion in chloroform and acetone respectively; it was reasoned on the basis of the large α-deuterium isotope effect that this result was due to carbenium ion formation, and a scheme similar to that of Sneen (Scheme 5.3) was proposed involving rate-determining formation of the carbenium ion followed by fast reaction with the nucleophile. On the basis of Leffek's results the mechanism appears to change from S_N2 to an S_N1 process on going from protic to dipolar aprotic solvents. Since there appears to be no obvious rationale for this behaviour Westaway and his co-workers studied the mechanism of the displacement at phenyldimethylbenzylammonium ion by thiophenoxide anion[100,101] in N,N-dimethylformamide; the observation of a substantial kinetic isotope effect for nitrogen ($^{14}k/^{15}k$) of 1.02±0.0007 is not consistent with the formation of a carbenium in a step prior to the rate-determining step. Since second-order kinetics *require* the putative carbenium ion mechanism to have rate-limiting addition of the nucleophile the relatively large ρ value for substituents on the anilino nucleus effectively excludes the stepwise process leaving concerted displacement as the current mechanism (Scheme 5.10).

Westaway and Ali[101] attributed the large $^Hk/^Dk$ effects for α-deuterium isotope substitution to the large steric crowding in the $C_α$-H bonds which is reduced in the S_N2 transition state. The α-deuterium isotope effect cannot therefore be employed as an absolute criterion to distinguish S_N2 from S_N1 mechanisms when applied to substrates with bulky groups.

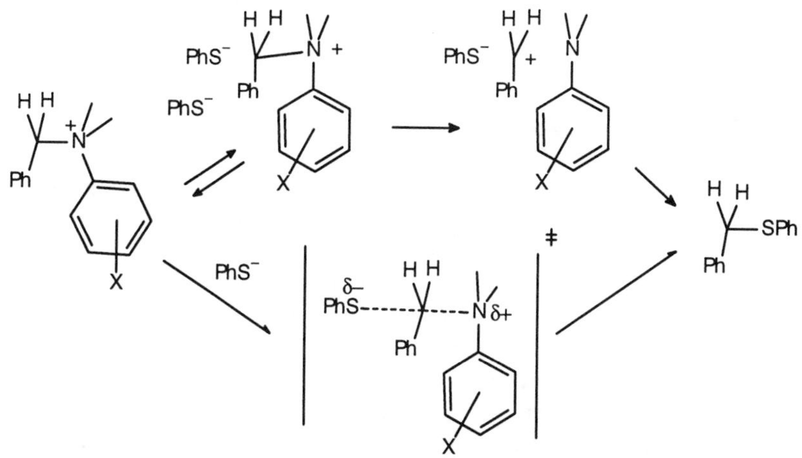

Scheme 5.10. Exclusion of a stepwise process by isotope and substituent effect.

The magnitude of the secondary α-deuterium kinetic isotope effect results from changes that occur in the $C_α$-H(D) out-of-plane bending vibrations when the reactant is converted to the transition state. When the tetrahedral carbon is converted to the trigonal bipyramidal transition state the out-of-plane bending vibrations remain almost constant or become even higher in energy. The S_N2 mechanism should thus be characterised by either a small normal or an inverse isotope effect ($^H k / ^D k < 1.04$). An S_N1 mechanism involves a decrease in energy of the out-of-plane bending vibrations on passage from tetrahedral to trigonal carbon and should therefore be characterised by a normal isotope effect ($^H k / ^D k > 1.07$). This criterion (Shiner, 1970)[97] is complicated by variations in the isotope effect for the S_N2 reaction according to the "looseness" or "tightness" of the transition state. A relatively large normal isotope effect would thus be expected for a loose transition state. It is important to emphasise that secondary α-deuterium isotope effects are often difficult to interpret in the absence of other mechanistic information. For example, there are several examples of reactions which are clearly bimolecular in added nucleophilic reagents and which show secondary α-deuterium isotope effects similar to that observed for stepwise reactions through a carbenium ion intermediate.[70,102]

The earliest application of isotope effects to the question of ion-pair intermediates showed that the reaction of azide ion and solvent with benzyl chlorides possessed high $^{12}k/^{14}k$ ratios and low $^H k / ^D k$ ratios for isotopic substitution of the central carbon and of the α-hydrogens.[43] The results are not consistent with the ion-pair mechanism, thereby excluding the idea that *all* bimolecular substitution reactions involve ion-pairs. The observation that $^{35}k/^{37}k$ chlorine isotope effect for reaction of azide ion with 4-methoxy benzyl chloride increases with azide ion concentration is consistent with a change in rate-limiting step.[62] However, a detailed study of the reaction (see earlier) indicates that it proceeds by competing stepwise SN1 and concerted paths; Sims has shown that variations of heavy atom kinetic isotope effects can be accommodated by the S_N2 mechanism.[103]

5.4.3 Front-side displacement

Two extreme types of concerted mechanisms can be advanced for displacement reactions at saturated carbon ~ the one involving *in-line* stereochemistry[104] and the other *front-side* stereochemistry (Scheme 5.11). The former mechanism requires inversion of configuration and the latter retention of configuration at the central carbon. The trajectories illustrated in Scheme 5.11 are *extrema* and the concerted mechanism would give rise to a racemic product if the trajectory angle, θ, were equal to the tetrahedral angle of 109°. The potential energy of the transition structure of the front-side process depends on van der Waals, orbital symmetry and electrostatic factors. The attack of nucleophiles at saturated carbon will give a transition structure with substantial van der Waals repulsion between entering and leaving groups whatever the nature of nucleophile or substrate. Electrostatic repulsion will be suffered between entering and leaving groups in reaction of neutral with charged species (illustrated in Scheme 5.11 for negative nucleophile and neutral substrate) but will be relieved if both reactants are neutral. Further repulsive forces arise from the orbital dissymmetry of attacking nucleophile (Nu) and leaving nucleofuge (Lg). In order for the HOMO of the attacking nucleophile to overlap with the LUMO of the central carbon it is required to have an out-of-phase, repulsive, overlap with the HOMO of the nucleofuge (Lg). Overlap repulsion, van der Waals repulsion and electrostatic repulsion have reduced significance as the bonding to leaving group and nucleophile both become smaller in the transition structure; under these circumstances, close to the border between A_ND_N and A_N+D_N mechanisms, front-side attack is a possibility.

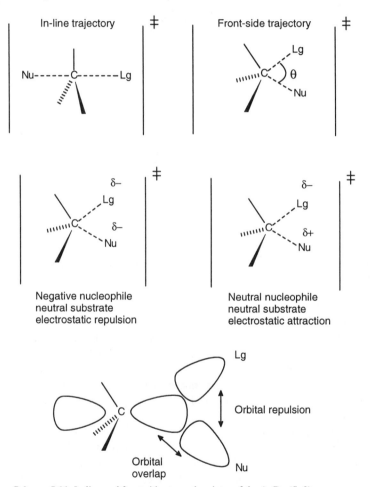

Scheme 5.11. In-line and front-side stereochemistry of the A_ND_N (S_N2) process.

Retention of configuration during nucleophilic displacement reactions has been reported for many systems; the bromination of phenylalkylmethanols by hydrobromic acid involves retention as does nucleophilic aliphatic substitution by phenols.[105] The mechanism was at one time believed to involve a concerted front-side displacement but ion-pair mechanisms can account for the results just as convincingly. The types of mechanism which are proposed to explain retention of configuration include rear-side shielding of the carbenium ion, three-, four- and six-centre cyclic models and solvent separated ion-pair models (Scheme 5.12).

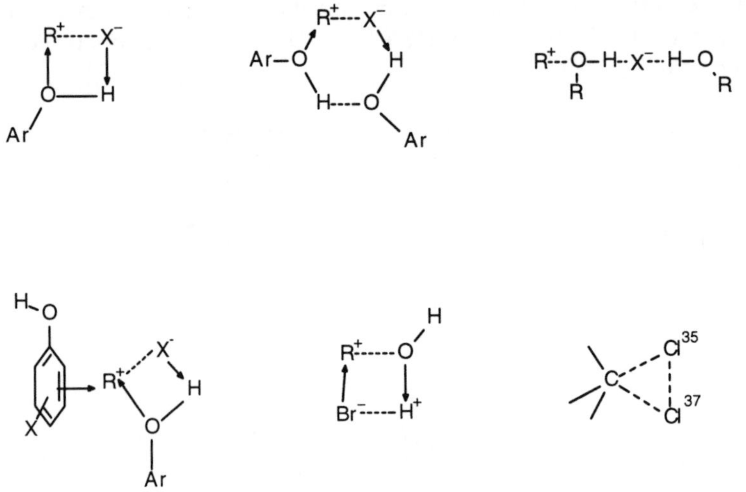

Scheme 5.12. Mechanisms for retention of configuration in bimolecular nucleophilic substitution.

The identity reaction of chloride[39] ion with methyl chloride[37] may be followed in the mass spectrometer[106] and the rate coefficient measured as a function of the centre of mass kinetic energy. An increase in rate coefficient above 0.4 ev corresponds to reaction through a [$CH_3.Cl_2^-$] complex *via* attack of chloride ion on the chlorine of CH_3Cl. This could be interpreted as a front-side process which could involve concerted attack and fission at the carbon as in Scheme 5.12. A study at a high level of computation has investigated this concerted channel.[107]

Halogenation of an alcohol by decomposition of its chlorosulphite or chlorocarbonate esters exhibits complete retention of configuration and this property was originally explained by a *concerted* front-side attack by halogen atom as indicated in Scheme 5.13. This classical concerted mechanism was replaced by one proposed by Lewis and Boozer[108] because it was observed that the reaction is strongly accelerated by polar solvents. Charge development in a concerted mechanism would not be affected by the polarity of the solvent to the same extent as the formation of ion-pair intermediates.

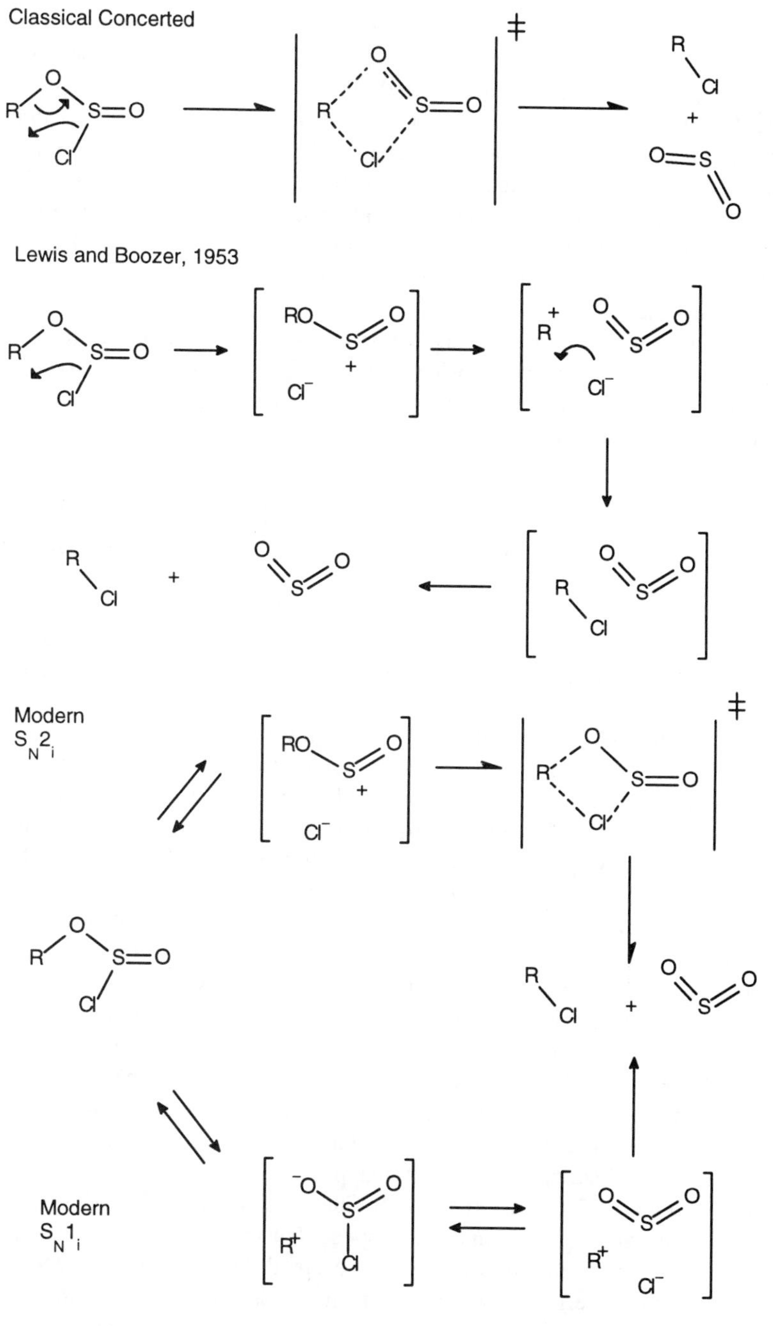

Classical Concerted

Lewis and Boozer, 1953

Modern
S_N2_i

Modern
S_N1_i

Scheme 5.13. Retention of configuration in halogenation by use of halosulphite esters. Similar mechanisms can be written for the analogous chlorocarbonate system (ROCOCl→RCl + CO₂).

The ROSO$^+$ species have been isolated and a crystallographic structure has even been reported for CH$_3$OSO$^+$Sb$_2$F$_{11}$$^-$;[109] it is likely that halogenation *via* the chlorocarbonate esters could involve the ROCO$^+$ ion which undoubtedly has some stability (Chapter 4). It has been commonly assumed that the concerted front-side attack mechanism is symmetry-forbidden but this is not so (Scheme 5.11). Retention mechanisms, where front-side attack is concerted with leaving group departure are possible when the entering and leaving bonds are very extended or there is an interaction between nucleophile and leaving group as in the cyclic transition state described in Schemes 5.12 and 5.13.[110]

The positional isotope exchange experiments could indicate a concerted mechanism if there were no significant barrier for collapse of the carbenium ion - sulphonate ion pair (Scheme 5.14). A similar scheme can be written for carbenium ion-carboxylate ion pairs.

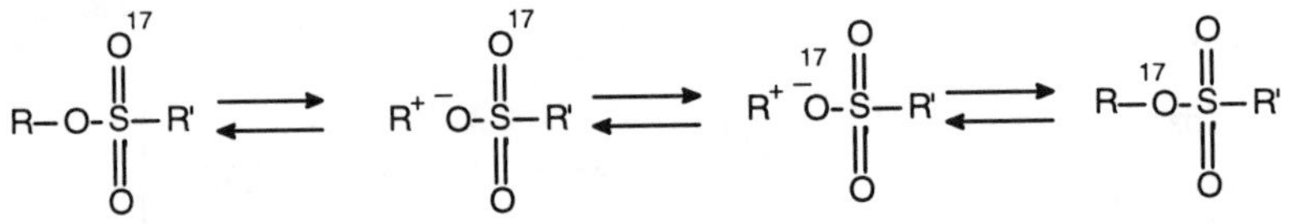

Scheme 5.14. Stepwise positional isotope exchange preceeding nucleophilic displacement.

Dietze and Wojciechowski (1990)[111] showed that positional isotope exchange occurs in the solvolysis of 2-butyl brosylate. Since there is no concurrent racemisation in the reactant the lifetime for a rotation of the butyl carbenium ion in the putative ion pair must exceed that for recombination which thus has little or no barrier. The positional isotope exchange, which must involve front-side displacement, is therefore a concerted process.

Ab-initio studies[109,112,113] indicate that the in-line displacement is preferred (by some 47 kcal/mole) in the model gas phase reaction of fluoride ion with methyl fluoride. Front-side displacement by the fluoride ion of the ion-pair (Li$^+$F$^-$) possesses a 16 kcal/mole advantage over the in-line process due to interactions between the lithium cation and the entering and leaving fluoride ions (Scheme 5.15). The presence of the cationic species (in the above case, lithium ion) appears to enhance the predicted retention mechanism. Calculations carried out on the methyl chlorosulphite decomposition indicate an electrostatic advantage to the front-side process.

The evidence for concerted front-side nucleophilic displacement in solution is sparse at present and we suppose that this is, *in part*, because of the reluctance to challenge the widely accepted *in-line* dogma. However, when the orbital-overlap requirements are met it is not impossible that such processes will occur. The large negative ρ_{XZ} (-0.56) for the reaction of 1-phenylethyl arenesulphonate esters with anilines (Scheme 5.16) suggests that the nucleophile (X) and leaving group (Z) are adjacent in the transition state (Scheme 5.16). The α-deuterium isotope effect (where the hydrogen on the nucleophilic nitrogen is substituted by deuterium) varies between 1.70 and 2.58 in acetonitrile consistent with **5.10**. The hydrogen H_a of Structure **5.10** would give a relatively small primary kinetic isotope effect due to the non-linear and unsymmetrical structure of N-H-O together with an inverse isotope effect for H_b. In contrast, Structure **5.11** is predicted to have a small ρ_{xz} in agreement with the observed value of -0.1(Lee, Shim, and Lee, 1989);[114] the bending vibrations of N-H_a and N-H_b would give an inverse secondary isotope effect as observed.[115]

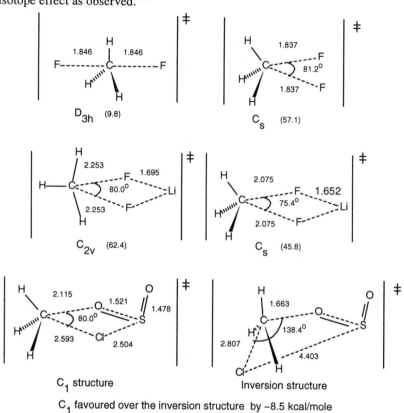

Scheme 5.15. Results of *ab-initio* calculations for model displacements at the methyl centre; bond angles and bond lengths (Ånstroms) are from *Refs. 112* and *113*. Values in parentheses are the calculated enthalpies of formation (kcal/mole) relative to the initial complexes between reactants (F^-:CH_3F and LiF:CH_3F).

Both stereochemistry and cross-correlation effects are not alone diagnostic of concerted front-side displacements. It is conceivable that in-line and front-side concerted displacements could give cross-correlation coefficients sufficiently different for them to be diagnostic provided there were some sort of normalised effect with which the values could be compared. Lee showed that the magnitude of cross-correlation should be inversely proportional to the distance between the reacting centres.[114-120] Utilising the three interaction coefficients ρ_{XY}, ρ_{XZ} and ρ_{YZ} the value of ρ_{XZ} might be expected to be smaller than ρ_{XY} and ρ_{YZ} as seen for reaction of benzyl benzenesulphonate esters with anilines (**5.11**). A large ρ_{XZ} compared with ρ_{XY} and ρ_{YZ} would be consistent with concerted front-side displacement (**5.10**) as seen for the reaction of anilines with substituted 1-phenylethyl benzenesulphonate esters.

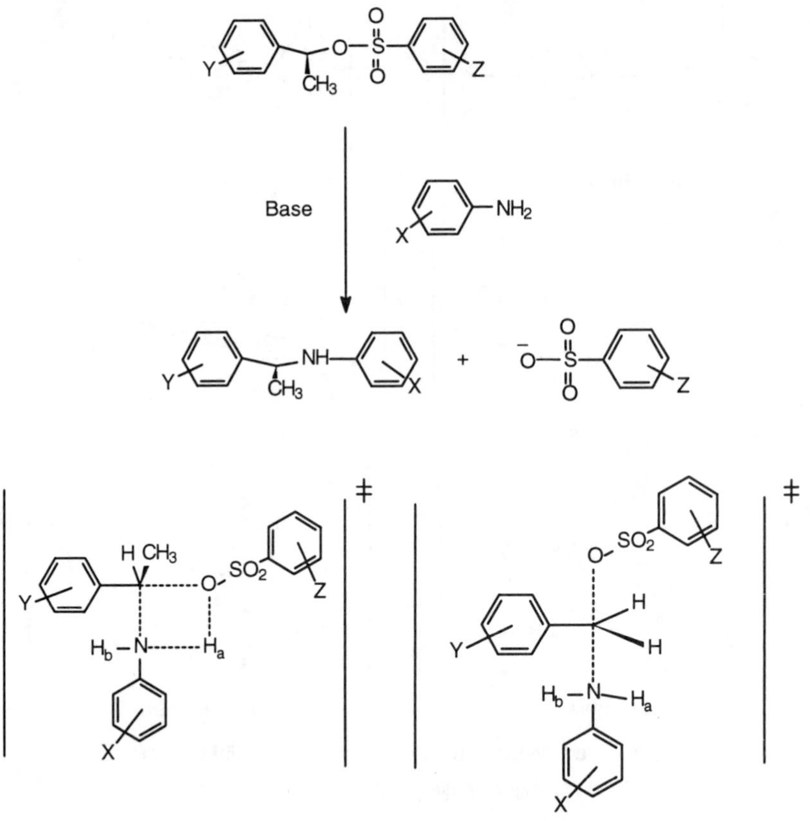

5.10   5.11

Scheme 5.16. Cross interaction for reaction of arenesulphonate esters with anilines.

It is interesting that leaving groups at bridgehead carbon can be displaced even though in-line attack is sterically impossible. This stereochemical retention is not unequivocally in favour of a concerted front-side mechanism as seen in Scheme 5.17 for the displacement of bromine by elemental fluorine from adamantyl bromide.[121] An electrophilic S_E2 (A_ED_E) process (see Scheme 5.17) has not been excluded for either case.

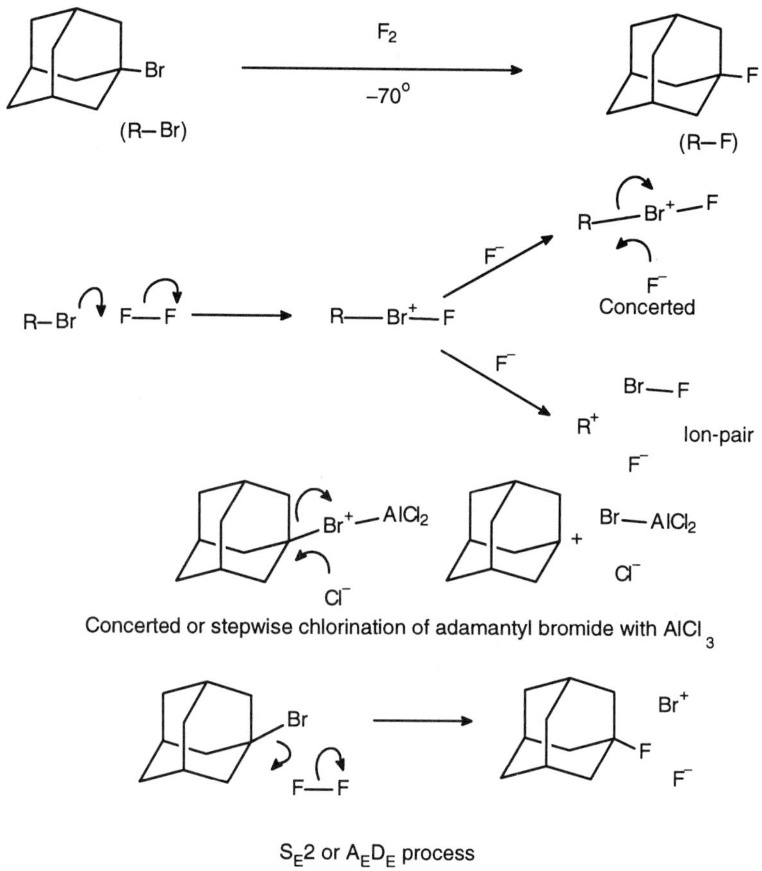

Scheme 5.17. Front-side displacement of bromine by fluorine and chlorine from adamantyl bromide (R-Br).

An experiment favouring front-side concerted displacement was quoted by Schleyer[112] where chlorine, from $AlCl_3$, substitutes for bromine in bridgehead bromoadamantane; the result is not unequivocal, however, for the reasons given for the fluorination results (Scheme 5.17) and, moreover, the results are also consistent with A_ED_E substitution. Contrary to the case of front-side nucleophilic displacement at saturated carbon the corresponding front-side concerted process has achieved general acceptance for electrophilic displacements at saturated carbon (in, for example, metal exchange).[122,123] It is noted that the evidence for this is stereochemical (retention of configuration) and thus carries the same interpretational difficulties as that for the nucleophilic case. The A_ED_E processes known up to now are all systems where the entering and departing groups are tied to each other either covalently or by some weaker bonding types. On electrostatic grounds the configurations for front-side attack in A_ED_E and A_ND_N mechanisms would have identical energies (Scheme 5.18).

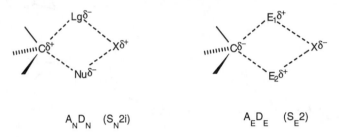

$$A_ND_N \quad (S_N2i) \qquad\qquad A_ED_E \quad (S_E2)$$

Scheme 5.18. Identity of the electrostatic interactions for electrophilic and nucleophilic front-side displacement mechanisms.

FURTHER READING

Amyes, T. L., and Jencks, W. P., Concerted bimolecular substitution reactions of acetal derivatives of propionaldehyde and benzaldehyde, *J. Amer. Chem. Soc., 111,* 7900, 1989.

Anh, N. T., and Minot, C., Conditions favouring retention of configuration in S_N2 reactions. A perturbational study, *J. Amer. Chem. Soc., 102,* 103, 1980.

Arnett, E. M., and Reich, R., Electronic effects on the Menschutkin reaction. A complete kinetic and thermodynamic dissection of alkyl transfer to 3- and 4-substituted pyridines, *J. Amer. Chem. Soc., 102,* 5892, 1980.

Bordwell, F. G., Are nucleophilic bimolecular concerted reactions involving four or more bonds a myth? *Accs. Chem. Res., 3,* 281, 1970.

Bron, J., Model calculations of kinetic isotope effects in nucleophilic substitution reactions, *Can. J. Chem., 52,* 903, 1974.

Buddenbaum, W. E., and Shiner, V. J., ^{13}C kinetic isotope effects and reaction coordinate motions in transition states for S_N2 displacement reactions, *Can. J. Chem., 54,* 1146, 1976.

Chabinyc, M. L., Craig, S. L., Regan, C. K., and Brauman, J. I., Gas-phase ionic reactions: dynamics and mechanism of nucleophilic displacements, *Science, 279,* 1882, 1998.

Chandrasekhar, J., and Jorgensen, W. L., Energy profile for a nonconcerted S_N2 reaction in solution, *J. Amer. Chem. Soc., 107,* 2974, 1985.

Craig, S. L., and Brauman, J. I., Intramolecular microsolvation of thermoneutral gas-phase S_N2 reactions, *J. Amer. Chem. Soc., 118,* 6786, 1996.

Dannenberg, J. J., Theoretical study of nucleophilic substitution on simple alkyl systems, *J. Amer. Chem. Soc., 98,* 6261, 1976.

Dewar, M. J. S., Mechanism of aliphatic substitution, *Annual Reports of the Chem. Soc., 48,* 121, 1951.

Dietze, P. E., and Jencks, W. P., Swain-Scott correlations for reactions of nucleophilic reagents and solvents with secondary substrates, *J. Amer. Chem. Soc., 108,* 4549, 1986.

Dodd, J. A., and Brauman, J. I., On the application of the Marcus equation to methyl transfer, *J. Amer. Chem. Soc., 106,* 5356, 1984.

Evanseck, J. D., Blake, J. F., and Jorgensen, W. L., *Ab initio* study of the S_N2 reactions of OH^- and OOH^- with CH_3Cl, *J. Amer. Chem. Soc., 109,* 2349, 1987.

Fry, A., Application of the successive labelling technique to C, N, and Cl isotope effects studies of organic reaction mechanisms, *Pure and Applied Chem., 8,* 409, 1964.

Hirao, K., and Kebarle, P., S_N2 reactions in the gas phase. Transition states for the reaction: $Cl^- + RBr = ClR + Br^-$, where $R = CH_3$, C_2H_5, and iso-C_3H_7, from *ab initio* calculations and comparison with experiment. Solvent effects, *Can. J. Chem., 67,* 1261, 1989.

Hu, W. P., and Truhlar, D. G., Modelling of transition state solvation, *J. Amer. Chem. Soc.,* *116,* 7797, 1994.

Hu, W. P., and Truhlar, D. G., Factors affecting the competitive ion-molecule reactions ClO⁻ + EtCl and C_2D_5Cl via E2 and S_N2 channels predicts isotope effects, *J. Amer. Chem. Soc.,* *118,* 860, 1996.

Jencks, W. P., When is an intermediate not an intermediate? Enforced mechanisms of general acid-base catalysed, carbocation, carbanion, and ligand exchange reactions, *Accs. Chem. Res., 13,* 161, 1980.

Jencks, W. P., How does a reaction choose its mechanism? *Chem. Soc. Reviews, 10,* 345, 1981.

Katritzky, A. R., and Brycki, B. E., The mechanisms of nucleophilic substitution in aliphatic compounds, *Chem. Soc. Reviews, 19,* 83, 1990.

Kim, H. J., and Hynes, J. T., A theoretical model for S_N1 ionic dissociation in solution. 1. Activation free energetics and transition state structure, *J. Amer. Chem. Soc., 114,* 10508, 1992.

Kim, H. J., and Hynes, J. T., A theoretical model for S_N1 ionic dissociation in solution. 2. Nonequilibrium solvation reaction path and reaction rate, *J. Amer. Chem. Soc., 114,* 10528, 1992.

Lee, I., Characterisation of transition states for reactions in solution by cross-interaction constants, *Chem. Soc. Reviews, 19,* 317, 1990.

Lee, I., Cross-interaction constants and transition state structure in solution, *Adv. Phys. Org. Chem., 27,* 57,1992.

Lipson, M., Deniz, A. A., and Peters, K. S., Nature of the potential energy surfaces for the S_N1 reaction: a picosecond kinetic study of homolysis and heterolysis for diphenylmethyl chlorides, *J. Amer. Chem. Soc., 118,* 2992, 1996.

Lipson, M., Deniz, A. A., and Peters, K. S., Picosecond kinetic study of the dynamics for photoinduced homolysis and heterolysis in diphenylmethylchloride, *J. Phys. Chem., 100,* 3580, 1966.

Martin, J. C., "Frozen" transition states: pentavalent carbon *et al., Science, 221,* 509, 1983.

McClelland, R. A., Flash photolysis generation and reactivities of carbenium ions and nitrenium ions, *Tetrahedron, 52,* 6823, 1996.

McLennan, D. J., A case for the concerted S_N2 mechanism of nucleophilic aliphatic substitution, *Accs. Chem. Res., 9,* 281, 1976.

Melander, L., and Saunders, W. L. Jr., *Reaction Rates of Isotopic Molecules,* John Wiley & Sons, New York, 1980.

Pellerite, M. J. and Brauman, J. I., Intrinsic barriers in nucleophilic displacements. a general model for intrinsic nucleophilicity toward methyl centres model suggests delocalisation effects do not greatly influence nucleophilic reactivity, *J. Amer. Chem. Soc., 105,* 2672, 1983.

Poirier, R. A., Wang, Y., and Westaway, K. C., A theoretical study of the relationship between secondary α-deuterium kinetic isotope effects and the structure of S_N2 transition states, *J. Amer. Chem. Soc., 116,* 2526, 1994.

Richard, J. P., and Jencks, W. P., Concerted S_N2 displacement reactions of 1-phenylethyl chlorides, *J. Amer. Chem. Soc., 104,* 4691, 1982.

Richard, J. P., A consideration of the barrier for carbocation-nucleophile combination reactions, *Tetrahedron, 51,* 1535, 1995.

Rodgers, J., Femec, D. A., and Schowen, R. L., Isotopic mapping of transition state structural features associated with enzymic catalysis of methyl transfer, *J. Amer. Chem. Soc., 104,* 3263, 1982.

Sastry, G. N., Danovich, D., and Shaik, S., Towards the definition of the maximum allowable tightness of an electron transfer transition state in the reactions of radical anions and alkyl halides, *Angew. Chem., Int. Ed. Engl., 35,* 1098, 1996.

Saveant, J-M., Single electron transfer and nucleophilic, *Adv. Phys. Org. Chem., 26,* 1, 1990.

Shi, Z., and Boyd, R. J., Transition state electronic structures in S_N2 reactions, *J. Amer. Chem. Soc., 111,* 1575, 1989.

Shiner, V. J. Jr., Deuterium isotope effects in solvolytic substitution at saturated carbon, in *Isotope Effects in Chemical Reactions*, ACS Monograph 167, Collins, C. J. and Bowman, N. S. eds., Van Nostrand Reinhold Company, New York, 1970, Chapter 2, p. 90.

Sims, L. B., Fry, A., Netherton, L. T., Wilson, J. C., Reppond, K. D., and Crook, S. W., Variations of heavy-atom kinetic isotope effects in S_N2 displacement reactions, *J. Amer. Chem. Soc., 94,* 1364, 1972.

Sneen, R. A., Organic ion pairs as intermediates in nucleophilic substitution and elimination reactions, *Accs. Chem. Res., 6,* 46, 1973.

Speiser, B., Electron transfer and chemical reactions - stepwise or concerted? On the competition between nucleophilic substitution and electron transfer, *Angew. Chem., Int. Ed. Engl., 35,* 2471, 1996.

Streitwieser, A., *Solvolytic Displacement Reactions*, McGraw-Hill, New York, 1962.

Wilbur, J. L., and Brauman, J. I., Intermediates and potential surfaces of gas-phase S_N2 reactions, *J. Amer. Chem. Soc., 113,* 9699, 1991.

Willi, V., Kinetic carbon and other isotope effects in cleavage and formation of bonds to carbon, in *Isotope Effects in Organic Chemistry. Vol. 3. Carbon-13 in Organic Chemistry*, Buncel, E., and Lee, C. C., Elsevier, Amsterdam, 1977, Chapter 5.

Wolfe, S., Mitchell, D. J., and Schlegel, H. B., Theoretical studies of S_N2 transition states. Substituent effects, *Can. J. Chem., 60,* 1291, 1982.

Xhao, X. G., Tucker, S. C., and Truhlar, D. G., Solvent and secondary isotope effect for S_N2 of Cl on MeCl, *J. Amer. Chem. Soc., 113,* 826, 1991.

REFERENCES

1. **Jencks, W. P.,** How does a reaction choose its mechanism? *Chem. Soc. Reviews, 10,* 345, 1981.

2. **Westaway, K. C., and Ali, S. F.,** Isotope effects in nucleophilic substitution reactions. III. The effect of changing the leaving group on transition state structure in S_N2 reactions, *Can. J. Chem., 57,* 1354, 1979.

3. **Harris, J. M., Shafer, S. G., Moffatt, J. R., and Becker, A. R.,** Prediction of S_N2 transition state variation by the use of More O'Ferrall plots, *J. Amer. Chem. Soc., 101,* 3295, 1979.

4. **Westaway, K. C., and Waszcylo, Z.,** Isotope effects in nucleophilic substitution reactions. IV. The effect of changing a substituent at the α-carbon on the structure of S_N2 transition states, *Can. J. Chem., 60,* 2500, 1982.

5. **Jaffé, H. H.,** Does carbon utilise *3d*-orbitals in bonding? *J. Chem. Phys., 21,* 1893, 1953.

6. **Gillespie, R. J.,** The use of *3d*-orbitals in bonding by carbon, *J. Chem. Phys., 21,* 1893, 1953.

7. **Dewar, M. J. S.,** The role of *3d*-electrons in valency states of first-row elements, *J. Chem. Soc.,* 2885, 1953.

8. **Gillespie, R. J.,** The use of *3d*-orbitals in certain valency states of the carbon atom and other first-row elements, *J. Chem. Soc.,* 1002, 1952.

9. **Winstein, S., Darwish, D., and Holness, N. H.,** Merged bimolecular substitution and elimination, *J. Amer. Chem. Soc., 78,* 2915, 1956.

10. **Oae, S., and Uchida, Y.,** Ligand-coupling reactions of hypervalent species, *Accs. Chem. Res., 24,* 202, 1991.

11. **Uhl, W., Layh, M., and Massa, W.,** Synthesis and crystal structure of $Li_2[(Me_3SiCH_2)_3Al-CH_2-Al(CH_2SiMe_3)_3]$ - a dilithiomethane derivative? *Chem. Ber., 124,* 1511, 1991.

12. **Schmidbauer, H., Brachthauser, B., and Steigelmann, O.,** Direct observation of the central atom in $[C\{Au[P(C_6H_5)_2(p\text{-}C_6H_5NMe_2)]\}_6(BF_4)_2$ by ^{13}C NMR spectroscopy, *Angew. Chem., 103,* 1488, 1991.

13. **Poumbga, C. N., Benard, M., and Hyla-Kryspin, I.,** Planar tetracoordinate carbons in dimetallic complexes: quantum chemical investigations, *J. Amer. Chem. Soc., 116,* 8259, 1994.

14. Elsenbroich, C., and Salzer, A., *Organometallics. A Concise Introduction,* VCH., Weinheim, 1992, pp. 407, 408.
15. Albrecht, M., Erker, G., and Kruger, C., The synthesis of stable, isolable planar-tetracoordinate carbon compounds, *Synlett.,* 441, 1993.
16. Martin, J. C., "Frozen" transition states: pentavalent carbon *et al., Science, 221,* 509, 1983.
17. Forbus, T. R., Kahl, J. L., and Martin, J. C., Electrochemical evidence for hypervalent (10-C-5) pentacoordinate carbon, *Heteroatom Chem., 4,* 137, 1993.
18. Forbus, T. R., and Martin, J. C., An NMR-study of a C-13-labeled hypervalent (10-C-5) pentacoordinate carbon species, *Heteroatom Chem., 4,* 129, 1993.
19. Forbus, T. R., and Martin, J. C., An observable model for the S_N2 transition state-hypervalent trigonal bipyramidal carbon (10-C-5), *Heteroatom Chem., 4,* 113, 1993.
20. Forbus, T. R., and Martin, J. C. (1979) Quest for an observable model for the S_N2 transition state. Pentacoordinate carbon, *J. Amer. Chem. Soc., 101,* 5057, 1979.
21. Arduengo, A. J., Carbon's octet rule, *Chem. and Engineering News, November 28, 3,* 1983.
22. Richard, J. P., and Yeary, P. E., On the importance of carbocation intermediates in bimolecular nucleophilic substitution reactions in aqueous solution, *J. Amer. Chem. Soc., 115,* 1739, 1993.
23. Amyes, T. L., and Jencks, W. P., Concerted bimolecular substitution reactions of acetal derivatives of propionaldehyde and benzaldehyde, *J. Amer. Chem. Soc., 111,* 7900, 1989.
24. Andrieux, C. P., Savéant, J-M., Tallec, A., Tardival, R., and Tardy, C., Concerted and stepwise dissociative electron transfers. Oxidability of the leaving group and strength of the breaking bond as mechanism and reactivity governing factors illustrated by electrochemical reduction of α-substituted acetophenones, *J. Amer. Chem. Soc., 119,* 2420, 1997.
25. Andrieux, C. P., Differding, E., Robert, M., and Saveant, J-M., Controlling factors of stepwise versus concerted reductive cleavages. Illustrative examples in the electrochemical reductive breaking of nitrogen-halogen bonds in aromatic N-halosultams, *J. Amer. Chem. Soc., 115,* 6592, 1993.
26. Andrieux, C. P., Robert, M., Saeva, F. D., and Saveant, J-M., Passage from concerted to stepwise dissociative electron transfer as a function of the molecular structure and of the energy of the incoming electron. Electrochemical reduction of aryldialkyl sulphonium cations radicals, *J. Amer. Chem. Soc., 116,* 7864, 1994.
27. Haberfield, P., Trapping the single electron transfer intermediate in an S_N2 reaction, *J. Amer. Chem. Soc., 117,* 3314, 1995.
28. Sastry, G. N., Danovich, D., and Shaik, S., Towards the definition of the maximum allowable tightness of an electron transfer transition state in the reactions of radical anions and alkyl halides, *Angew. Chem., Int. Ed. Engl., 35,* 1098, 1996.
29. Saveant, J-M., Single electron transfer and nucleophilic substitution, *Adv. Phys. Org. Chem., 26,* 1, 1990.
30. Richard, J. P., The extraordinarily long lifetimes and other properties of highly destabilised ring-substituted 1-phenyl-2,2,2-trifluoroethyl carbocations, *J. Amer. Chem. Soc., 111,* 1455, 1989.
31. Richard, J. P., A consideration of the barrier for carbocation-nucleophile combination reactions, *Tetrahedron, 51,* 1535, 1995.
32. Richard, J. P., and Jencks, W. P., A simple relationship between carbocation lifetime and reactivity-selectivity relationships for the solvolysis of ring-substituted 1-phenylethyl derivatives, *J. Amer. Chem. Soc., 104,* 4689, 1982.
33. Young, P. R., and Jencks, W. P., Trapping of the oxocarbenium ion intermediate in the hydrolysis of acetophenone dimethyl ketals, *J. Amer. Chem. Soc., 99,* 8238, 1977.
34. Hammett, L. P., *Physical Organic Chemistry,* McGraw-Hill, New York, 1940, p. 171.
35. Winstein, S., Ion-pairs in solvolysis and exchange, *Chem. Soc. Spec. Publ., 19,* 109, 1965.
36. Weiner, H., and Sneen, R. A., Substitution at a saturated carbon V. A clarification of the mechanism of solvolyses of 2-octyl sulfonates. Kinetic considerations, *J. Amer. Chem. Soc., 87,* 292, 1965.

37. **Sneen, R. A., and Larsen, J. W.,** Substitution at a saturated carbon Part 10. The unification of mechanisms SN1 and SN2, *J. Amer. Chem. Soc., 91*, 362, 1969.

38. **Sneen, R. A., and Larsen, J. W.,** Substitution at a saturated carbon atom. 11. The generality of the ion-pair mechanism of nucleophilic substitution, *J. Amer. Chem. Soc., 91*, 6031, 1969.

39. **Sneen, R. A., and Larsen, J. W.,** Identification of an ion-pair intermediate in an S_N2 reaction, *J. Amer. Chem. Soc., 88*, 2593, 1966.

40. **Sneen, R. A.,** Organic ion pairs as intermediates in nucleophilic substitution and elimination reactions, *Accs. Chem. Res., 6*, 46, 1973.

41. **Gregory, B. J., Kohnstam, G., Paddon-Row, M., and Queen, A.,** Nucleophilic substitution in the mechanistic border-line region, *Chem. Soc. Chem. Commun.*, 1032, 1970.

42. **Gregory, B. J., Kohnstam, G., Queen, A., and Reid, D. J.** Mechanisms of nucleophilic substitution. Kinetics of the reactions of benzyl and diphenylmethyl chlorides in aqueous acetone, *J. Chem. Soc. Chem. Commun.*, 797, 1971.

43. **Hall, R. E., Harris, J. M., Raber, D. J., and Schleyer, P. v. R.,** Comment on the Sneen mechanism, *J. Amer. Chem. Soc., 93*, 4821, 1971.

44. **Kurz, J. L., and Harris, J. C.,** Evidence against the generality of the ion-pair mechanism for nucleophilic substitution, *J. Amer. Chem. Soc., 92*, 4117, 1970.

45. **Raaen, V. F., Juhlke, T., Brown, F. J., and Collins, C. J.,** Do S_N2 reactions go through ion pairs? The isotope effect criterion, *J. Amer. Chem. Soc., 96*, 5928, 1974.

46. **Stein, A. R.,** β-Deuterium kinetic isotope effects for identity processes: bromide ion substitution at 1-bromo-1-arylethanes and 2-bromooctane, *Can. J. Chem.*, 72, 1789, 1994.

47. **Stein, A. R.,** The ion-pair mechanism and bimolecular displacement at saturated carbon. VI. Racemization and *radio*-bromide exchange for substituted 1-phenylbromoethanes; solvent effects, *Can. J. Chem.*, 65, 363, 1987.

48. **Stein, A. R.,** The ion-pair mechanism and bimolecular displacement at saturated carbon. VII. Racemization of substituted 1-phenylbromoethanes; ionic strength effects, *Can. J. Chem.*, 67, 297, 1989.

49. **Harris, J. M., Becker, A., Fagan, J. F., and Walden, F. A.,** Ion-pair identification by means of a stability-selectivity relationship, *J. Amer. Chem. Soc., 96*, 4484, 1974.

50. **Richard, J. P.,** On the importance of reactions of carbocation ion pairs in water: common ion inhibition of solvolysis of 1-(4-methoxyphenyl)-2,2,2-trifluoroethyl bromide and trapping of an ion-pair intermediate by solvent, *J. Org. Chem., 57*, 625, 1992.

51. **Richard, J. P., and Jencks, W. P.,** Concerted bimolecular substitution reactions of 1-phenylethyl derivatives, *J. Amer. Chem. Soc., 106*, 1383, 1984.

52. **Müller, P.,** Glossary of terms used in physical organic chemistry, *Pure and Applied Chemistry, 66*, 1077, 1994.

53. **Doering, W. v. E., and Zeiss, H. H.,** Methanolysis of optically active hydrogen 2,4-dimethyl hexyl-4-phthalate, *J. Amer. Chem. Soc., 75*, 4733, 1953.

54. **Goering, H. L., and Levy, J. F.,** The solvolysis of benzhydryl 4-nitrobenzoate carbonyl-^{18}O. A new method for detecting ion pair return, *Tetrahedron Letters*, 644, 1961.

55. **Goering, H. L., Briody, R. G., and Levy, J. F.,** The stereochemistry of ion pair return associated with solvolysis of 4-chlorobenzhydryl 4-nitrobenzoate, *J. Amer. Chem. Soc., 85*, 3059, 1963.

56. **Tsuji, Y., Kim, S. H., Saeki, Y., Yatsugi, K-I., Fujio, M., and Tsuno, Y.,** Oxygen-18 scrambling studies in the solvolysis of benzyl tosylates, *Tetrahedron Letters, 36*, 1465, 1995.

57. **Tsuji, Y., Yatsugi, K-I., Fujio, M., and Tsuno, Y.,** Oxygen-18 scrambling studies in the solvolysis of α-(t-butyl)benzyl tosylates, *Tetrahedron Letters, 36*, 1461, 1995.

58. **Chang, S., and le Noble, W. J.,** Study of ion-pair return in 2-norbornyl brosylate by means of ^{17}O nmr, *J. Amer. Chem. Soc., 105*, 3708, 1983.

59. **Allen, A. D., Fujio, M., Tee, O. S., Tidwell, T. T., Tsuji, Y., Tsuno, Y., and Yatsugi, K-I.,** Ion pairs in the solvolysis of secondary systems. Salt effect, ^{18}O-labelling, and polarimetric studies of 1-(4'-tolyl)-2,2,2-trifluoroethyl tosylate, *J. Amer. Chem. Soc., 117*, 8974, 1995.

60. Streitwieser, A. Jr., Walsh, T. D., and Wolfe, J. R. Jr., Stereochemistry of acetolysis of alkyl sulfonates, *J. Amer. Chem. Soc., 87,* 3682, 1965.

61. Streitwieser, A. Jr., and Walsh, T. D., Salt effects on the stereochemistry of acetolysis of 2-octyl p-toluenesulfonate, *J. Amer. Chem. Soc., 87,* 3686, 1965.

62. Graczyk, D. G., and Taylor, J. W., Chlorine kinetic isotope effects in nucleophilic substitution reactions. Support for the ion-pairs mechanism in the reaction of *p*-methoxybenzyl chloride in 70% aqueous acetone, *J. Amer. Chem. Soc., 96,* 3255, 1974.

63. Raber, D. J., Harris, J. M., Hull, R. E., and Schleyer, P. v. R., The use of added sodium azide as a mechanistic probe for solvolysis reactions, *J. Amer. Chem. Soc., 93,* 4821, 1971.

64. Raber, D. J., Harris, J. M., Hull, R. E., and Schleyer, P. v. R., The effects of sodium azide on the solvolysis of secondary and primary β-arylpropyl derivatives. Azide ion as a probe for k_s processes, *J. Amer. Chem. Soc., 93,* 4829, 1971.

65. McLennan, D. J., A case for the concerted S_N2 mechanism of nucleophilic aliphatic substitution, *Accs. Chem. Res., 9,* 281, 1976.

66. Abraham, M. H., and McLennan, D. J., Comments on the putative ion-pair mechanism for hydrolysis of methyl halides and methyl perchlorate, *J. Chem. Soc. Perkin Trans. 2,* 873, 1977.

67. Abraham, M. H., Thermodynamic parameters for ionisation and dissociation of alkyl halides in water and non-aqueous solvents. Comments on the ion-pair mechanism of nucleophilic substitution, *J. Chem. Soc. Perkin Trans. 2,* 1893, 1973.

68. Amyes, T. L., and Jencks, W. P., Absence of a common ion effect on the hydrolysis of an α-azido ether of an aliphatic aldehyde, *J. Amer. Chem. Soc., 110,* 3677, 1988.

69. Amyes, T. L., and Jencks, W. P., Lifetimes of oxocarbenium ions in aqueous solution from common ion inhibition of the solvolysis of α-azido ethers by added azide ion, *J. Amer. Chem. Soc., 111,* 7888, 1989.

70. Knier, B. L., and Jencks, W. P., Mechanism of reactions of N-(methoxymethyl)-N,N-dimethylanilinium ions with nucleophilic reagents, *J. Amer. Chem. Soc., 102,* 6789, 1980.

71. Dietze, P. E., and Jencks, W. P., Swain-Scott correlations for reactions of nucleophilic reagents and solvents with secondary substrates, *J. Amer. Chem. Soc., 108,* 4549, 1986.

72. McClelland, R. A., Banait, N. S., and Steenken, S., Electrophilic reactivity of the triphenylmethyl cation in aqueous solutions, *J. Amer. Chem. Soc., 108,* 7023, 1986.

73. Richard, J. P., Rothenberg, M. E., and Jencks, W. P., Formation and stability of ring-substituted 1-phenylethyl carbocations, *J. Amer. Chem. Soc., 106,* 1361, 1984.

74. McClelland, R. A., Flash photolysis generation and reactivities of carbenium ions and nitrenium ions, *Tetrahedron, 52,* 6823, 1996.

75. McClelland, R. A., Kanagasabapathy, V. M., and Steenken, S., Nanosecond laser flash photolytic generation and lifetimes in solvolytic media of diarylmethyl and *p*-methoxyphenthyl cations, *J. Amer. Chem. Soc., 110,* 6913, 1988.

76. Craze, G. A., Kirby, A. J., and Osborne, R., Bimolecular nucleophilic substitution on an acetal, *J. Chem. Soc. (Perkin Trans. 2),* 357, 1978.

77. Capon, B., and Thacker, D., The mechanism of anomerisation of methyl D-glucopyranosides and methyl D-glucofuranosides, *J. Chem. Soc. (B),* 1010, 1976.

78. Banait, N. S., and Jencks, W. P., Reactions of anionic nucleophiles with α-D-glucopyranosyl fluoride in aqueous solution through a concerted, A_ND_N (S_N2) mechanism, *J. Amer. Chem. Soc., 113,* 7951, 1991.

79. Banait, N. S., and Jencks, W. P., General-acid and general-base catalysis of the cleavage of α-D-glucopyranosyl fluoride. Concerted mechanism of nucleophilic attack and proton abstraction enforced by the absence of significant lifetime of the glycosyl cation in presence of fluoride ion, *J. Amer. Chem. Soc., 113,* 7958, 1991.

80. Horenstein, B. A., and Bruner, M., The N-acetyl neuraminyl oxecarbenium ion is an intermediate in the prsence of anionic nucleophiles, *J. Amer. Chem. Soc., 120,* 1357, 1998.

81. Huang, X., Surry, C., Hiebert, T., and Bennet, A. J., Hydrolysis of (2-deoxy-β-δ-glucopyranosyl)pyridinium salts, *J. Amer. Chem. Soc., 117,* 10614, 1995.

82. Huang, X., Surry, C., Hiebert, T., and Bennet, A. J., Hydrolysis of (2-ethoxy-β-D-glucopyranosyl)pyridinium salts, *J. Amer. Chem. Soc., 117,* 10614, 1995.

83. **Sinnott, M. L., and Jencks, W. P.,** Solvolysis of D-glucopyranosyl derivatives in mixtures of ethanol and 2,2,2-trifluroethanol, *J. Amer. Chem. Soc., 102,* 2026, 1980.

84. **Zhu, J., and Bennet, A. J.,** Hydrolysis of (2-deoxy-β-D-glucopyranosyl)pyridinium salts: the 2-deoxyglucosyl oxocarbenium is not solvent-equilibrated in water, *J. Amer. Chem. Soc., 120,* 3887, 1998.

85. **Jencks, W. P.,** When is an intermediate not an intermediate? Enforced mechanisms of general acid-base catalysed, carbocation, carbanion, and ligand exchange reactions, *Accs. Chem. Res., 13,* 161, 1980.

86. **Ritchie, C. D.,** Cation-anion combination reactions. XIII. Correlation of the reactions of nucleophiles with esters, *J. Amer. Chem. Soc., 97,* 1170, 1975.

87. **Ritchie, C. D., and Gandler, J.,** Cation-anion combination reactions. 17. Reactivity of alkylthiolate anions in aqueous solution, *J. Amer. Chem. Soc., 101,* 7318, 1979.

88. **Eldin, S., and Jencks, W. P.,** Lifetimes of iminium ions in aqueous solution, *J. Amer. Chem. Soc., 117, 4851,* 1995.

89. **Eldin, S., Digits, A. J., Huang, S-T., and Jencks, W. P.,** Lifetime of an aliphatic iminium ion in aqueous solution, *J. Amer. Chem. Soc., 117,* 6631, 1995.

90. **Ritchie, C. D.,** Cation-anion combination reactions. 26. A review, *Can. J. Chem., 64,* 2239, 1986.

91. **Eldin, S., and Jencks, W. P.,** Concerted bimolecular substitution reactions of anilino thioethers, *J. Amer. Chem. Soc., 117,* 9415, 1995.

92. **Bruice, T. W., and Santi, D. V.,** Isotope effects in reactions catalysed by thymidylate synthase, in *Enzyme Mechanisms from Isotope Effects,* Cook, P. F., ed., CRC Press, Boca Raton, 1991, p. 457.

93. **Benkovic, S. J., and Bullard, W. P.,** Mechanism of action of folic acid cofactors, *Prog. Bioorg. Chem., 2,* 135, 1973.

94. **Kallen, R. G., and Jencks, W. P.,** The mechanism of the condensation of formaldehyde with tetrahydrofolic acid, *J. Biol. Chem., 241,* 5851, 1966.

95. **Amyes, T. L., and Jencks, W. P.,** Concurrent stepwise and concerted substitution reactions of 4-methoxybenzyl derivatives and the lifetime of the 4-methoxybenzyl carbocation, *J. Amer. Chem. Soc., 112,* 9507, 1990.

96. **Westaway, K. C., Pham, V. T., and Fang, Y-r.,** Using secondary α-deuterium kinetic isotope effects to determine the symmetry of S_N2 transition states, *J. Amer. Chem. Soc., 119,* 3670, 1997.

97. **Shiner, V. J. Jr., Rapp, M. W., and Pinnick, H. R.,** α-Deuterium isotope effects in S_N2 reactions with solvents, *J. Amer. Chem. Soc., 92,* 232, 1970.

98. **Ko, E. C. F., and Leffek, K. T.,** Studies on the decomposition of tetra-alkylammonium salts in solution. Part IV. The α-deuterium secondary isotope effect, *Can. J. Chem., 49,* 129, 1971.

99. **Leffek, K. T., and Tsao, F. H. C.,** Studies on the decomposition of tetra-alkylammonium salts in solution. Part I, *Can. J. Chem., 46,* 1215, 1968.

100. **Westaway, K. C., and Poirier, R. A.,** Isotope effects in nucleophilic substitution reactions. I. The mechanism of the reaction of phenylbenzyldimethylammonium ion with thiophenoxide ion, *Can. J. Chem., 53,* 3216, 1975.

101. **Westaway, K. C., and Ali, S. F.,** Isotope effects in nucleophilic substitution reactions. II. Secondary α-deuterium kinetic isotope effects: a criterion of mechanism? *Can. J. Chem., 57,* 1089, 1979.

102. **Vitullo, V. P., Grabowski, J., and Sridhara, S.,** α-Deuterium isotope effects in benzyl halides. 2. Reaction of nucleophiles with substituted benzyl bromides. Evidence for a change in transition state structure with electron donating substituents, *J. Amer. Chem. Soc., 102,* 6463, 1980.

103. **Sims, L. B., Fry, A., Netherton, L. T., Wilson, J. C., Reppond, K. D., and Crook, S. W.,** Variations of heavy-atom kinetic isotope effects in S_N2 displacement reactions, *J. Amer. Chem. Soc., 94,* 1364, 1972.

104. **Salem, L.,** Orbital interactions and reaction paths, *Chem. Brit.,* 449, 1969.

105. **Okamoto, T., Takeuchi, K., and Inoue, J.** Retentive solvolysis. Part 12. Mechanism of the reaction of optically active 1-p-methoxyphenylethyl trifluoroacetate with phenol and

methanol in benzene and cyclohexane Front-side attack, phenolyses of optically active secondary and tertiary systems, *J. Chem. Soc. Perkin Trans. 2*, 842, 1980.

106. **Barlow, S. E., Van Doren, J. M., and Bierbaum, V. M.,** The gas-phase displacement reaction of chloride ion with methyl chloride as a function of kinetic energy, *J. Amer. Chem. Soc., 110*, 7240, 1988.

107. **Deng, L., Branchadell, V., and Ziegler, T.,** Potential energy surfaces of the gas-phase S_N2 reactions $X^- + CH_3Cl = XCH_3 + X^-$ ($X = F$, Cl, Br, I): A comparative study by density functional theory and ab initio methods, *J. Amer. Chem. Soc., 116*, 10645, 1994.

108. **Lewis, E. S., and Boozer, C. E.,** The decomposition of secondary alkyl chlorosulphites. II. Solvent effects and mechanism, *J. Amer. Chem. Soc., 75*, 3182, 1953.

109. **Schreiner, P., Schleyer, P. v. R., and Hill, R. K.,** Reinvestigation of the S_Ni reaction. The ionisation of chlorosulphites, *J. Org. Chem., 58*, 2822, 1993.

110. **Anh, N. T., and Minot, C.,** Conditions favouring retention of configuration in S_N2 reactions. A perturbational study, *J. Amer. Chem. Soc., 102*, 103, 1980.

111. **Dietze, P. E., and Wojciechowski, M.,** Oxygen scrambling and stereochemistry during the trifluoroethanolysis of optically-active 2-butyl-4-bromobenzenesulfonate, *J. Amer. Chem. Soc., 112*, 5240, 1990.

112. **Harder, S., Streitwieser, A., Petty, J. T., and Schleyer, P. v. R.,** Ion-pair S_N2 reactions: theoretical study of inversion and retention mechanisms, *J. Amer. Chem. Soc., 117*, 3253, 1995.

113. **Schreiner, P., Schleyer, P. v. R., and Hill, R. K.,** Mechanisms of front-side substitutions. The transition states for the S_Ni decomposition of methyl and ethyl chlorosulphite in the gas phase and in solution, *J. Org. Chem., 59*, 1849, 1994.

114. **Lee, I., Shim, C. S., and Lee, H. W.,** Cross-interaction constants as a measure of the transition state structure. Part 4. Brønsted-type cross-interaction constants, *J. Chem. Soc. Perkin Trans. 2*, 1205, 1989.

115. **Lee, I.,** Secondary kinetic isotope effects involving deuterated nucleophiles, *Chem. Soc. Reviews, 24*, 223, 1995.

116. **Lee, I.,** Characterisation of transition states for reactions in solution by cross-interaction constants, *Chem. Soc. Reviews, 19*, 317, 1990.

117. **Lee, I.,** Mechanistic significance of the magnitude of cross-interaction constants, *J. Phys. Org. Chem., 5*, 736, 1992.

118. **Lee, I., and Sohn, S. C.,** The mechanistic significance of cross-interaction constants, r_{ij}, *J. Chem. Soc., Chem. Commun.*, 1055, 1986.

119. **Lee, I.,** Cross interaction between identical groups, *Bull. Korean Chem. Soc., 8*, 200, 1987.

120. **Lee, I.,** Cross interaction constants in elimination reactions, *Bull. Korean Chem. Soc., 8*, 426, 1987.

121. **Rozen, S., and Brand, M.,** Electrophilic attack of elemental fluorine on organic halogens. Synthesis of fluoroadamantanes, *J. Org. Chem., 46*, 733, 1981.

122. **Ingold, C. K.,** *Structure and Mechanism in Organic Chemistry,* 2nd edn., Cornell University Press, Ithaca, New York, 1969.

123. **Abraham, M. H.,** Electrophilic substitution at a saturated carbon atom, in *Comprehensive Chemical Kinetics, Volume 12*, Bamford, C. H., and Tipper, C. F. H. eds., Elsevier, Amsterdam, 1973.

Chapter 6

Displacement Reactions at Heteroatoms

Displacement reactions at heteroatoms (Scheme 6.1) feature widely in both organic and bio-organic chemistry and they are among the most important processes in metabolism. This chapter is about the transfer of electrophiles between nucleophiles but also includes hydrogen transferring formally with its valency electrons between electrophilic donors (E_2) and acceptors (E_1); hydrogen transferring as an electrophile *without* its valency electrons is discussed in Chapter 3. Fashion in chemical thought has often influenced the acceptance of concerted or stepwise mechanisms and in the cases shown in Scheme 6.1 reactions 1, 2, 5 and 6 were generally accepted to have concerted mechanisms whereas reactions 3, 4 and 7 were regarded as stepwise. In order to overturn these dogmas it was necessary to find *positive* evidence to demonstrate concertedness in reactions 3, 4 and 7 or a stepwise path for reactions 1, 2, 5 and 6. There is still no evidence against concerted mechanisms for any reactions of type 1, 2, 5 and 6 and it has now been demonstrated that reactions 3 and 4 can be concerted under suitable conditions.

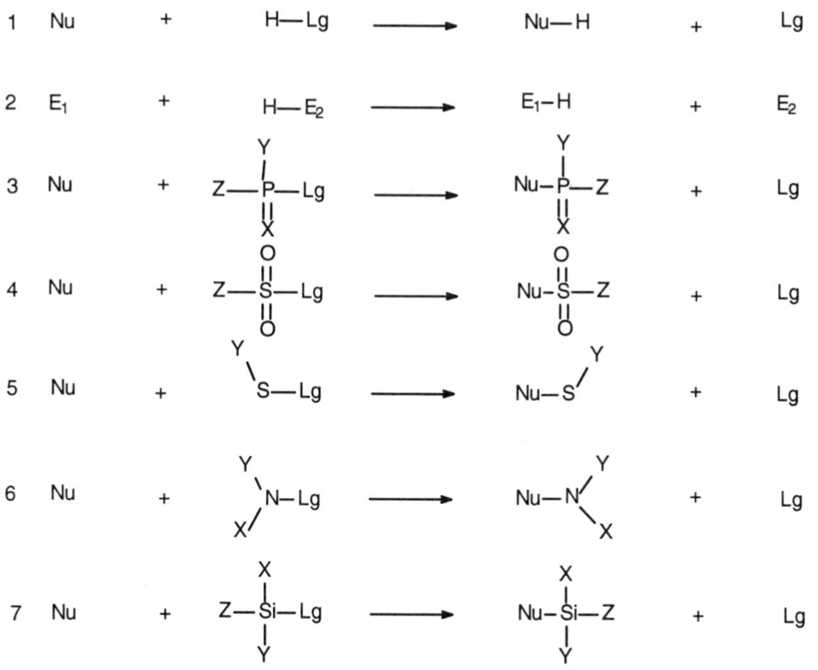

Scheme 6.1. Displacement reactions at heteroatoms. E_1 and Nu are electrophilic and nucleophilic acceptors and E_2 and Lg are the corresponding donors.

6.1   Electrophilic displacement at hydrogen

Transfer of hydrogen with its valency electrons is usually referred to as *hydride ion transfer* and is central in many biological redox reactions which involve the NAD^+ - NADH couple. Hydride ion transfer is a common reaction involved in organic reductions including the action of lithium aluminium hydride, the Meerwein-Pondorff-Verley reaction, the Cannizzaro reaction and hydride shifts in carbocations involved in terpene biosynthesis and similar bio-organic systems.

Kreevoy and Lee (1984)[1-6] made extensive studies of hydride transfer between NAD^+ analogues (Scheme 6.2). A Leffler plot (log k_{ij} versus log K_{ij}) for systems with different identities of R (Figure 6.1a) is linear with a slope $\alpha_{ij} = 0.37$. Moreover, the plot is linear over a range of log K_{ij} (5 log units) where a break would be expected for a stepwise mechanism (at zero log K_{ij}) due to a change in rate-limiting step (the expected correlation is shown in Figure 6.1b). The linearity of the plot is not consistent with the intervention of an intermediate of the type **6.1** and **6.2** and expression of the hydride ion as a *discrete* intermediate is excluded by the second-order rate-law.

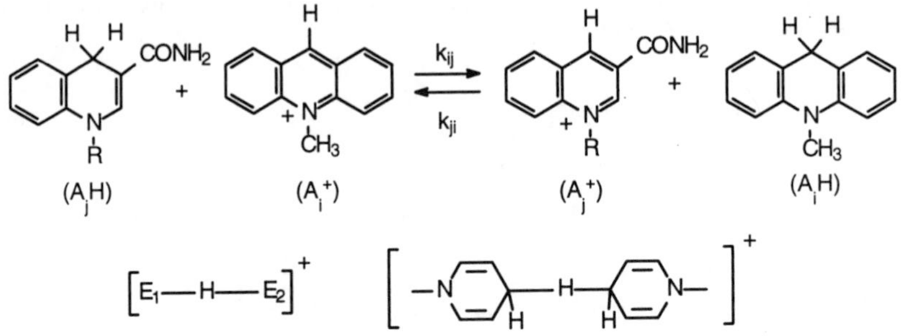

Scheme 6.2. Hydride ion transfer between NAD^+ analogues.

Scheme 6.3. A mechanism involving an hydride ion intermediate in an encounter complex; k_a and k_d are diffusion processes.

A pre-association mechanism, consistent with the kinetics, can be formulated (Scheme 6.3) which would also yield a non-linear Leffler plot with a break-point when E_1^+ and E_2^+ are identical ($K_{ij} = 1$). The reaction is essentially symmetrical; at $K_{ij} > 1$, $k_{-1} < k_2$, and k_1 is rate-limiting. Since E_2^+ (the variant electrophile) is only a spectator in this step the overall rate constant should be independent of E_2 when $K_{ij} > 1$. At $K_{ij} < 1$, $k_{-1} > k_2$, and the overall rate constant should vary with the structure of E_2^+ as shown in Figure 6.1c.

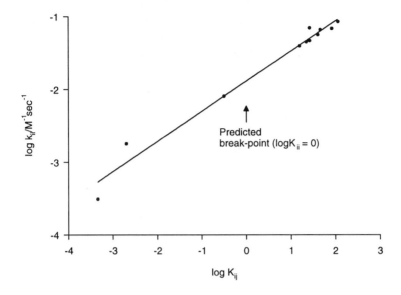

Figure 6.1a

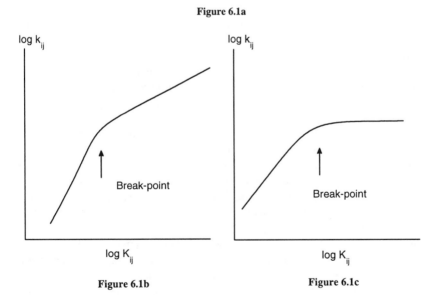

Figure 6.1b   **Figure 6.1c**

Figure 6.1. a). Hydride ion transfer between NAD$^+$ analogues. Data is from *Ref. 1* and the line is fitted to a linear equation (slope $\alpha = 0.37$). The Leffler plot is linear for *the quasi-symmetrical reaction* consistent with a concerted mechanism; the arrow indicates the expected break-point for a stepwise process. b). Correlation expected for an A_E+D_E mechanism. c). Correlation expected for a pre-associative D_E+A_E mechanism.

The tightness (τ) of the transition state structure in the concerted reaction may be deduced from Leffler's α value. Kreevoy and Lee[1] proposed a relationship (Equation 6.1) where δ is the effective charge on the transferring nucleus, τ is the "tightness" parameter (this is the sum of the bond orders of breaking and forming bonds, η_{lg} and η_{nuc} respectively)[7] and β_{ii} and β_{eq} have the meanings given in Chapter 4 (Equations 4.2 to 4.6). The value of τ may be deduced from Equation (6.2).[7]

$$\delta = \tau - 1 = \partial\log k_{ij}/\partial\log K_{ij} = \beta_{ii}/\beta_{eq} \qquad 6.1)$$

$$\tau = (\eta_{lg} + \eta_{nuc}) = \delta + 1 = (\beta_{ii}/\beta_{eq}) + 1 \qquad (6.2)$$

Since $\beta_{eq} = \beta_{nuc} - \beta_{lg}$ and $\beta_{ii} = \beta_{nuc} + \beta_{lg}$

$$\tau = (\beta_{nuc} + \beta_{lg} + \beta_{nuc} - \beta_{lg})/(\beta_{nuc} + \beta_{lg})$$

and Equation (6.3) follows:

$$\tau = 2\,\beta_{nuc}/\beta_{eq} = 2\,\alpha_{nuc} \qquad 6.3)$$

Substituting the value of 0.37 (the slope of the line in Figure 6.1a) yields a τ value of 0.74 in very good agreement with that (0.77) calculated from the Marcus equation. The calculated effective charge on the in-flight hydrogen (δ) is -0.23. Considerations of molecular orbital symmetry predict a linear transition state structure as for proton transfer. *Ab initio* calculations lead to a slightly non-linear C...H...C configuration in the transition state.[8-13] Only two electrons are involved in the bonding of the three centre system (E_1:H E_2) whereas proton transfer involves four electrons (Nu_1:H:Nu_2). Hydride ion would therefore be bound covalently and more tightly than the proton where the extra pair of electrons would be in an anti-bonding orbital in the transition state structure. The transition state for hydride ion transfer is expected to be tighter than that for proton transfer. The argument has an intuitive appeal but it calls for a general conclusion that the relative destabilisation of the C-H bond caused by partial occupancy of the anti-bonding orbital in a transition state for proton transfer will always be greater than the loss of stabilisation arising from loss of electron occupancy of a bonding orbital in the transition state for hydride ion transfer. This may not always be the case. The geometry is in agreement with what might be expected intuitively and also from isotope effects. The linear C-H-C configuration in the transition-structure is supported by large $^Hk/^Dk$ isotope effects of between 7 and 9.[14,15]

The hydride shift is very rapid in the example illustrated (Scheme 6.4,)[16] for an intramoleuclar reaction but there is no evidence for a *synchronous* process, although the reaction must be concerted.

Scheme 6.4. An intramolecular hydride shift.

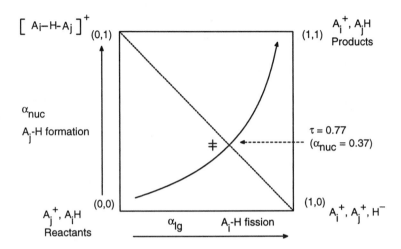

Figure 6.2. The More O'Ferrall-Jencks diagram for the quasi-symmetrical transfer of hydride ion between pyridinium ions A_i^+ and A_j^+ (Scheme 6.2).

6.2 Nucleophilic displacement at saturated nitrogen

Despite its possessing a lone pair of electrons, nitrogen, in its trivalent level of oxidation can act as an electrophile. Displacement reactions at nitrogen, analogous to substitution at sp^3 carbon, have been studied (Scheme 6.5). Although there appears to be no well-known biological analogue of transfer of saturated nitrogen the reaction type is important in synthesis[17-19] including the synthesis of heterocycles, amination processes in reaction umpolung[20] and probably in carcinogenesis by aromatic amines.[21,22] The reaction type is also involved in the Raschig synthesis of hydrazine where reaction (2) in Scheme 6.5 is an important step.[23] Although nucleophilic displacements at nitrogen have not been investigated kinetically to any great extent those studies which are available indicate second-order kinetics for the intermolecular reaction.[24-28]

Ab initio molecular orbital calculations for the reaction of fluoride ion with fluoramine in the gas phase indicate that the fluoride ion initially forms a hydrogen-bond with the substrate (F$^-$...H-N) and the displacement reaction is concerted, involving an essentially linear F...H...N arrangement ($\angle$FNF ~ 162°)[29,30] in the transition state.

The endocyclic restriction test has been applied to the displacement at neutral nitrogen[31] and it was shown that reaction (1) of Scheme 6.5 exhibits *no* isotopic scrambling in the products in a double-labelling experiment involving the two reactants (**6.3, 6.4**). The intermediacy of an ion-pair is also unlikely since positional isotopic exchange (PIX) is not observed in the reactant labelled with ^{18}O at the phosphoryl group even after 80% completion in the intermolecular reaction (Scheme 6.6) between the phosphoryl oxime and phenylacetonitrile. The ion-pair mechanism would require the nucleophile within the encounter complex as a spectator because the lifetime of the ion-pair would be too short for diffusion of the nucleophile to compete with the return reaction or reaction with solvent. A lower limit can be placed on the ion-pair mechanism as the nucleophile would need to react with a rate constant greater than $10^{12}sec^{-1}$ with the nitrenium ion for scrambling to be excluded. The retention of the nitrogen within the endocyclic transition-structure of the intramolecular reaction is consistent with an $\angle ONC$ angle of ~60° (**6.5**) indicating that there is a fairly large "cone of opportunity" for nucleophilic displacement at neutral nitrogen. Displacement at sp^3 carbon does not appear to possess such a wide "cone" because endocyclic nucleophilic displacement does *not* occur in similar substrates with carbon replacing the electrophilic nitrogen.[32] The experimental error in the scrambling and double labelling experiments does not completely exclude the expression of a nitrenium ion (R_2N^+)[33] in the encounter complex.

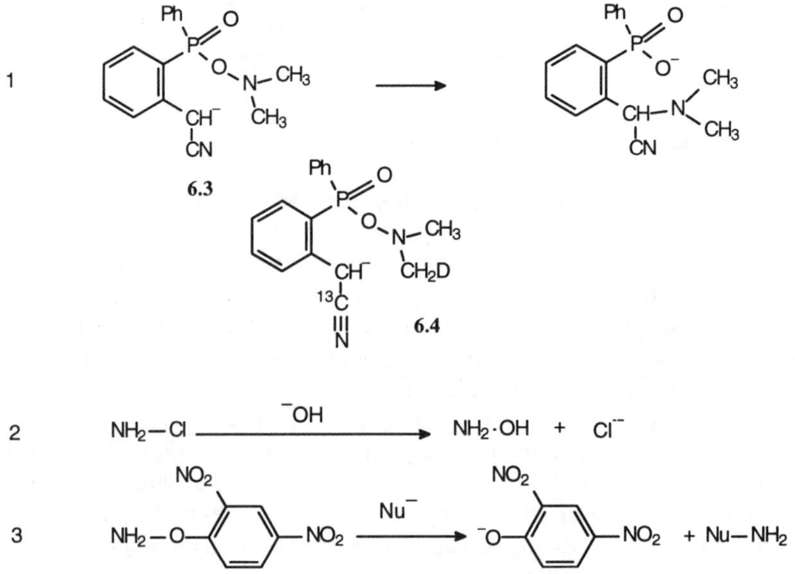

Scheme 6.5. Some nucleophilic substitution reactions at saturated nitrogen.

A nitrenium ion intermediate is possible in a reaction which has kinetics first-order in substrate and first-order in nucleophile (Scheme 6.7). However, little work has been done to elucidate pathways of this nature in displacements at nitrogen. The observation of substantial nucleophilic substituent effects would not be consistent with Scheme 6.7.

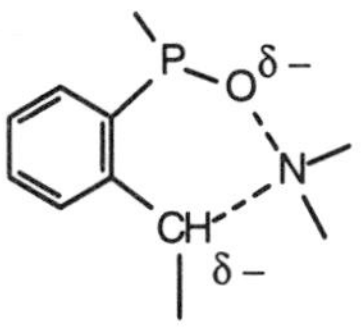

6.5

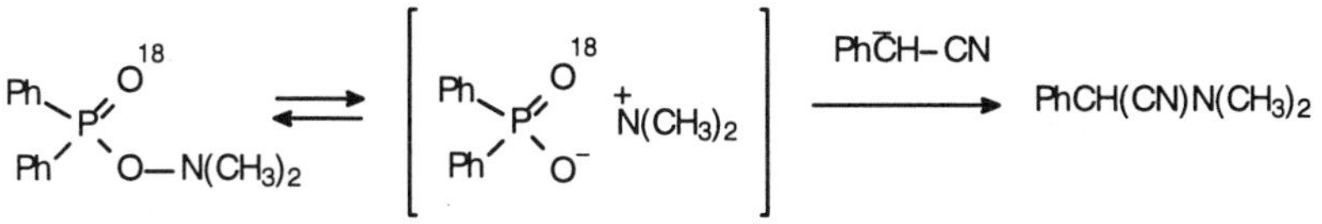

PhC̃H–CN

PhCH(CN)N(CH₃)₂

Scheme 6.6. An ion-pair mechanism would yield isotopic "scrambling" unless its lifetime were too short for diffusion of the nucleophile.

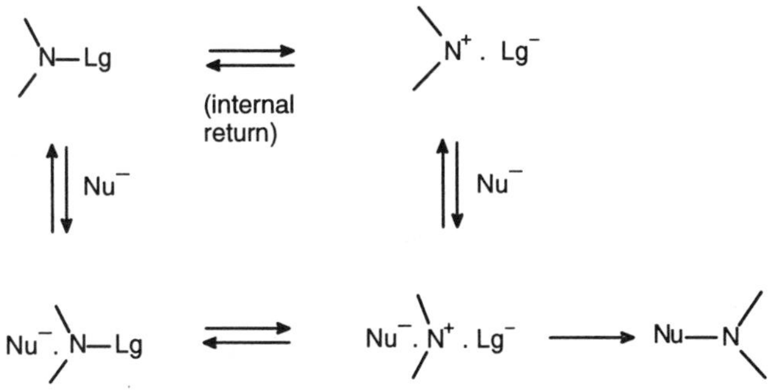

Scheme 6.7. A nitrenium ion involvement in displacements at nitrogen.

Fishbein and McClelland (1987)[34] estimated the lifetime of the 2,6-dimethylphenylnitrenium ion to be *ca.* 10^{-9} sec in the Bamberger rearrangement of N-arylhydroxylamines in aqueous solution by use of an azide ion trapping technique (Scheme 6.8). The reaction of Scheme 6.8 gives classical evidence for the nitrenium ion intermediate comprising a rate constant (k_{obs}, the *net* rate constant for disappearance of reactant) independent of azide ion concentration over the range where the yields of phenol product (**6.6**) and azide product (**6.7**) undergo substantial changes (Figure 6.3).

6.3   Nucleophilic displacement at silicon

Silicon possesses d-orbitals of energy sufficiently close to that of the valence electrons thus enabling it to readily increase its coordination number.[35] Displacements at silicon are important in synthetic organic chemistry where silyl compounds are employed largely as protective groups (Fleming).[36]

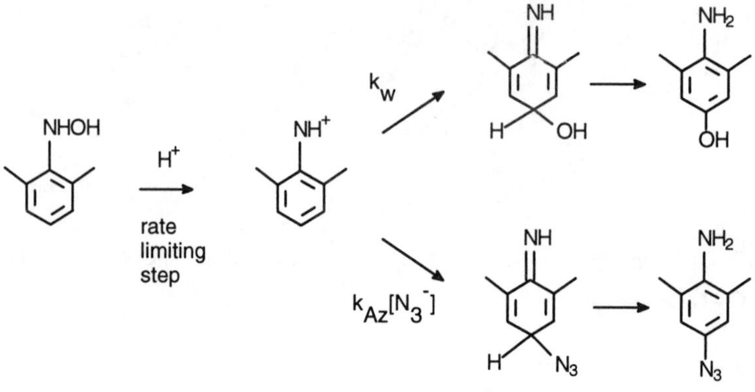

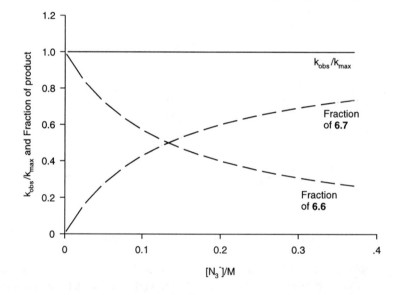

Scheme 6.8. Azide ion trapping of an arylnitrenium ion in the Bamberger rearrangement.

Figure 6.3. Kinetics and product analysis as a function of azide ion concentration for the Bamberger reaction; lines calculated from data in *Ref. 34*.

The hydrolysis of phenyl dimethylsilyl ethers[37-39] catalysed by hydroxide ion and the displacement by trifluoroethoxide ion are thought to involve an A_ND_N rather than an A_N+D_N process. Both nucleophiles give Brønsted lines of similar slope for the rate constant versus the pK_a of the leaving group.[39] In this system it was not possible to span the pK_a of the 2,2,2-trifluoroethoxide ion with a range of leaving groups of similar structure and indeed the overall plot exhibits two parallel lines (one for phenolate ions and one for alcoholate ions) although there is a certain degree of scatter due to different steric requirements. The stepwise process would require a different selectivity to the leaving group at pK_a's above and below that corresponding to the change in rate-limiting step (Chapter 2). The observation of identical sensitivities is not consistent with a stepwise process. A good test of the concerted mechanism is the linear plot of log k_{TFE} versus log k_{OH} where k_{OH} would not be expected to involve a change in rate-limiting step whereas k_{TFE} *would* be expected to have a change in rate-limiting step at the position shown by the arrow in Figure 6.4 corresponding to the identity reaction of trifluoroethoxide ion with the trifluoroethyl ether. The upper limit, β_{eq}, for complete bond fission of the leaving group is unknown so that the value of $\beta_{lg} = 0.7$ cannot be used to determine the Leffler parameter for the reaction or a comparison of effective charges.

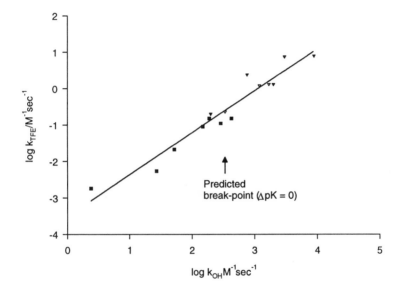

Figure 6.4. Displacement of phenoxy (∇) and alkoxy ($\blacksquare$) leaving groups from phenyl dimethyl silyl ethers by trifluoroethoxide ion (k_{TFE}) and hydroxide ion (k_{OH}); a break-point is expected at the position shown for a stepwise mechanism. Data are from *Ref. 40* and the line is a linear fit.

Pentacoordinate species have been isolated (Holmes, 1990)[41] in silicon chemistry and the geometry of these ranges from trigonal bipyramidal (**6.8, 6.9**) to square pyramidal (**6.10**) giving good precedent for the intervention of a pentacoordinate species in displacement reactions at silicon. The equatorial bond lengths are less than those of the axial in the trigonal bipyramidal structures **6.8** and **6.9** whereas they have similar lengths in the square pyramidal structure **6.10**. The acyclic SiX_5^- complexes of trigonal bipyramidal shape undergo pseudorotation with energy barriers of between 9 and 12 kcal/mole.

Both inversion and retention of configuration have been observed[41-43] in nucleophilic substitution at silicon and while these results are consistent with *in-line* and *front-side* concerted mechanisms they are equally explained by the intervention of pentacoordinate intermediates (Scheme 6.9).

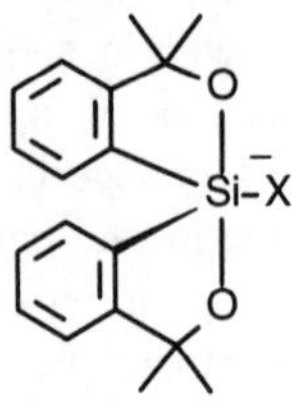

6.8 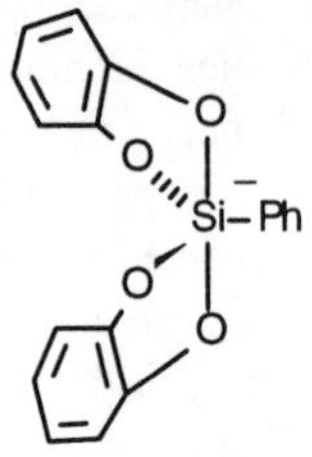   6.9 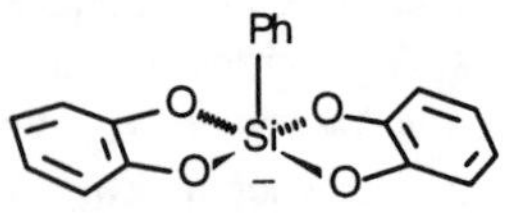   6.10

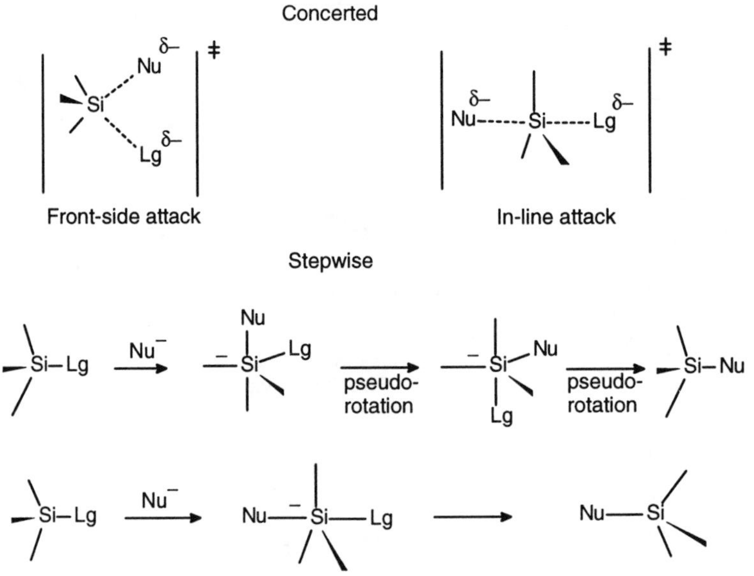

Concerted

Front-side attack   In-line attack

Stepwise

pseudo-   pseudo-
rotation   rotation

Nu——Si——Lg

Nu——Si

Scheme 6.9. Stereochemistry of displacement at silicon *via* stepwise and concerted mechanisms.

6.4 Nucleophilic displacement at phosphorus (V)

Displacements at the phosphorus (V) centre comprise examples from many aspects of chemistry and biochemistry ranging from organic synthesis through to membrane transport processes. The richness of phosphorus chemistry is due to the variety of valence types that the element can assume (Scheme 6.9) and the great interest in phosphorus chemistry over that of other elements which have similar valence properties stems from its relevance to biological chemistry and from its usefulness in synthesis. It is because of the multiplicity of phosphorus types that the general way of referring to these centres as *phosph*orus *acyl* that the term *phosphyl* was introduced. Within the oxygenated species (**6.11, 6.12, 6.14** and **6.15**) the systems can be even more complicated due to ionisation as seen in Scheme 6.11 and these comprise those phosphyl groups which are of biological significance (**6.17** and **6.18**).

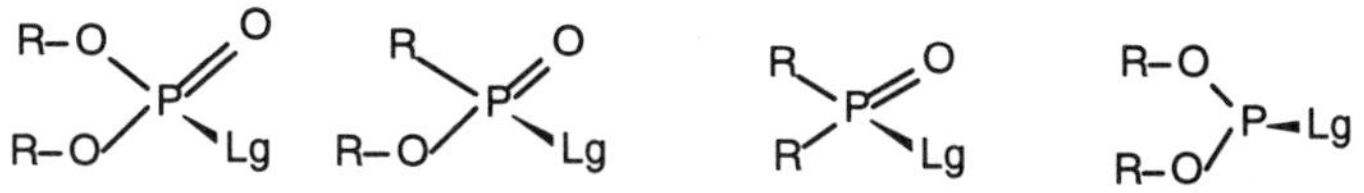

Scheme 6.10. Some phosphyl centres.

Scheme 6.11. Ionisation of phosphorus oxyacids.

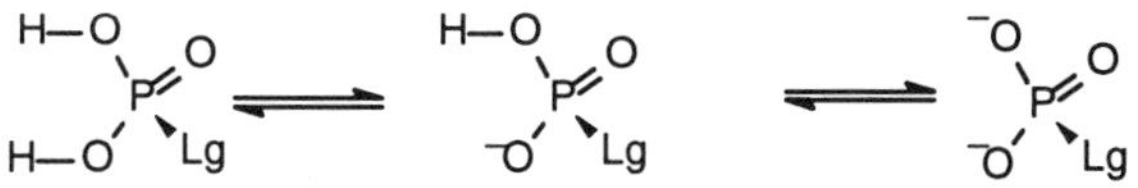

It is reasonable to say that three main types of displacement process are involved generally in phosphyl group transfer. These are the stepwise *addition-elimination*, the stepwise *elimination-addition* and the *concerted displacement* mechanisms.[44] In order to discuss the concerted displacement mechanism at phosphorus it is necessary to delineate the mechanistic *extrema* namely the stepwise (D_N+A_N) and (A_N+D_N) processes.

6.4.1 The D_N+A_N mechanism

The question central to many biological processes involving ATP as a phosphorylating agent is whether the metaphosphate ion[45-49] is a discrete intermediate (Todd, 1959)[50] or if the transfer mechanism is concerted (Herschlag and Jencks, 1989,[51] Scheme 6.12). This question has never been satisfactorily answered as evidence for the free metaphosphate ion *is very elusive indeed.*[52-60] Analogous unsaturated phospho intermediates have been demonstrated (Scheme 6.13) by unequivocal experiments involving trapping, loss of stereochemical integrity of a central chiral phosphorus atom, measurements of effective charge in transition-structures, comparisons of rate constants with reactions not able to participate in D_N+A_N processes and isotopic and entropy studies.

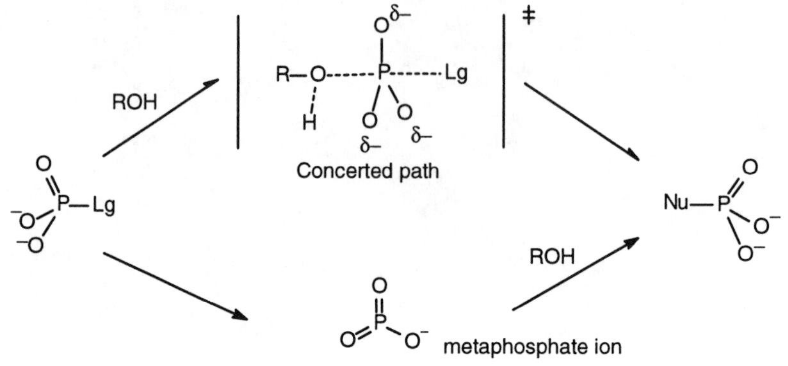

Scheme 6.12 Concerted and metaphosphate pathways for phosphorylation.

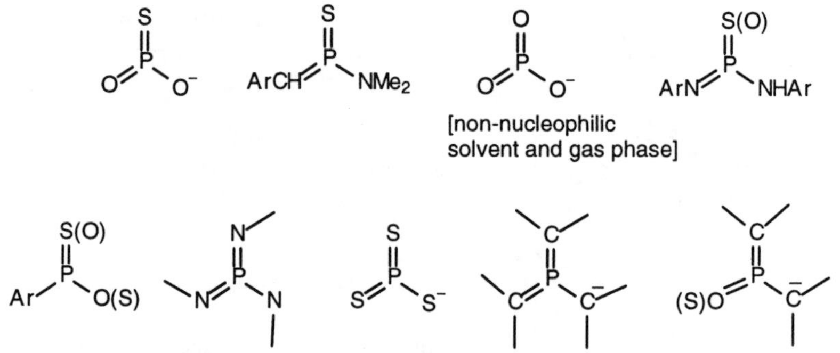

Scheme 6.13. Metaphosphate ion and analogues which have been demonstrated in some phosphyl transfer reactions or directly detected.[45,50-59,61-68]

In the case of the metaphosphate ion its extreme reactivity may be moderated by the use of solvents which are only weakly nucleophilic. Ramirez[69-75] found that methanol and *tert*-butanol had very similar reactivities as nucleophiles towards 2,4-dinitrophenylphosphate in acetonitrile solution whereas in water ethanol, methanol and isopropanol have *widely* differing reactivities. Racemisation at chiral phosphorus in the phosphate ion is also observed. Acetonitrile itself is a weakly nucleophilic solvent and could react with the ester to give a reactive phospho-species which could undergo several interchanges with acetonitrile solvent to lead to racemisation. There is good precedent for a phospho-solvent species because sulphur trioxide, isoelectronic with the metaphosphate ion, interacts with such weakly nucleophilic solvents as dioxane. Diagnosis is complicated by the fact that any solvent for the phospho esters must, of necessity, possess some form of nucleophilic character through its solubilising heteroatoms. There is a further complication of interpreting such stereochemical results because concerted alcoholysis has an initially formed species which is protonated (**6.19** in Scheme 6.14)[76] and can either lose a proton or undergo multiple transfers with the solvent *tert*-butanol. The racemisation at the phosphorus can then only be diagnostic if the proton transfer to solvent is faster than the displacement by the alcohol. The finding that the entropy of activation for the reaction of 4-nitrophenyl phosphate in *tert*-butanol (+24.5 e.u/mole at 25°) is much more positive than that for the reaction in aqueous solution (+3.5 e.u/mole at 25°) is consistent with a metaphosphate ion intermediate in the alcoholic solution.[78]

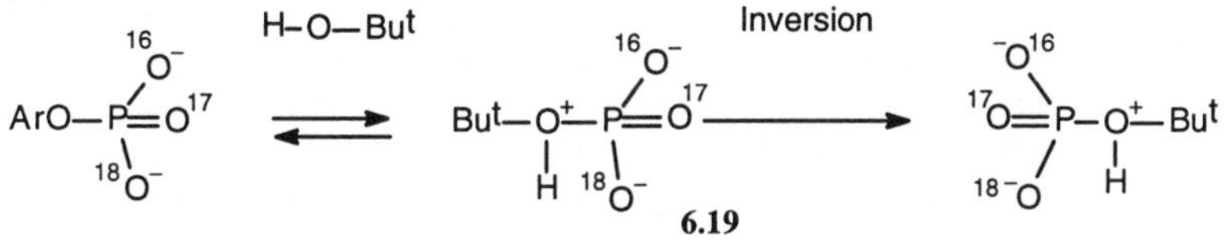

Scheme 6.14. Racemisation at phosphorus is possible during concerted displacement by solvent *tert*-butanol.

Positional isotopic exchange (PIX)[77] is a good diagnostic tool for the demonstration of a metaphosphate ion intermediate and has been employed extensively. The reaction of ADP in acetonitrile or acetonitrile-butanol solvents (Scheme 6.15) involves interchange of the bridging oxygens consistent with an encounter complex possessing a metaphosphate ion component.

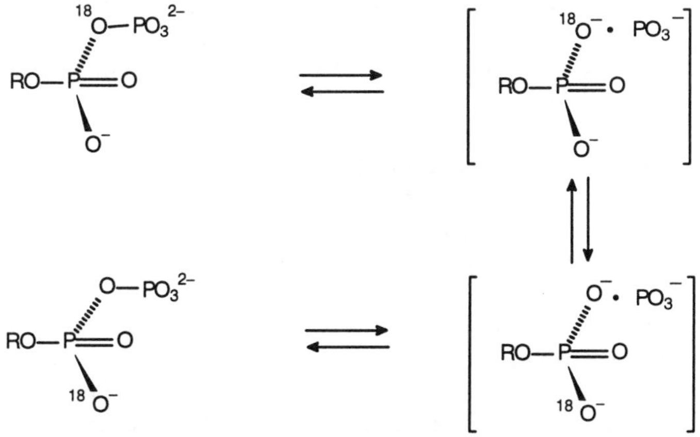

Scheme 6.15. Positional isotopic exchange within an encounter complex (RO- = adenosine).[77]

Replacing the oxygens by other atoms such as sulphur or nitrogen leads to much more stable analogues of the metaphosphate ion (Scheme 6.13) and these can be readily demonstrated.

Positional isotopic exchange experiments have also been carried out on a number of enzymatic phosphoryl group transfer systems. In these and in the solution reactions the assumption that the lifetimes of the postulated intermediate are sufficiently long for tumbling to occur deserves mention. In the enzyme systems there is also the question of the degree of rotational freedom of the hypothetical intermediate. Positional isotopic exchange has been used to probe a number of kinase reactions and, for example, the absence of PIX in the presence of substrate analogues which cannot undergo phosphorylation argues against a free metaphosphate ion in these mechanisms.[79-83] The absence of PIX is not definitive as it could be due to interactions of catalytic groups or coordination to metal ions inhibiting rotation of the metaphosphate ion.

6.4.2 The A_N+D_N mechanism

The evidence for the A_N+D_N mechanism is largely from analogy with the mechanism of nucleophilic displacement at the carbonyl centre. Pentacoordinate oxyphosphoranes have been isolated and some of these are illustrated in Structures **6.20** and **6.21**;[84-86] positional isotopic exchange experiments demonstrate the existence of this type of intermediate in some cases of phosphyl group transfer.

The stereochemistries of most of the displacements at phosphyl(V)[87] can be accommodated by the intervention of pentacoordinated intermediates wherein permutation isomerisation of the ligands (pseudorotation) can occur (Scheme 6.16).[88,89] Although stereochemistry is not diagnostic of either a stepwise or a concerted process the stereochemistries are still *consistent* with either in-line or adjacent concerted displacements (Scheme 6.16).

6.4.3   Concerted mechanisms
6.4.3.1   Dissociative process

Effective charge studies.   Skoog   and   Jencks[90,91]   and   Bourne   and Williams[92,93] independently utilised the *quasi symmetrical* technique to distinguish between a concerted displacement of pyridines from N-phosphopyridines and the pre-association mechanism. The latter mechanism involves a metaposphate intermediate which is formed only in an encounter complex and does not survive long enough to enable it to diffuse into the bulk solution (Scheme 6.17). The reaction exhibits bimolecular kinetics thus excluding any mechanism involving metaphosphate ion as a *free* intermediate.

A Brønsted plot of the overall second-order rate constant (log k_{xpy}) for attack by substituted pyridines (xpy) against the pK_a of the attacking pyridine is linear with a low slope (β_{nuc} ~0.15). The linearity extends over a range of pK_a's well above and well below that of the leaving pyridine (Figure 6.5). The stepwise mechanism shown in Scheme 6.17 would result in a non-linear Brønsted plot with a break at the pK_a of the leaving pyridine group. The two steps k_a and k_d in the stepwise mechanism (Scheme 6.17) involve formation and breakdown of the two encounter complexes (**6.22**) and (**6.24**) and would not be susceptible to changes in the substituent X in the attacking pyridine nucleophile; the k_1 step, where xpy is a spectator would also be independent of structural change in the attacking pyridine (xpy). When pK_a^{xpy} is greater than that of the leaving pyridine the rate-limiting step would be the formation of (**6.23**) from (**6.22**) because $k_2 >$ k_{-1}. Under these conditions the overall rate constant $k_{xpy} = k_1 k_a/k_{-a}$ and is thus independent of the structure of the nucleophilic pyridine (xpy). The rate-limiting step for the condition $pK_a^{xpy} < pK_a^{ypy}$ is k_2 and since $k_{xpy} = (k_1/k_{-1}) (k_a/k_{-a}) k_2$ the value of k_{xpy} is dependent on the structure of the attacking pyridine (xpy).

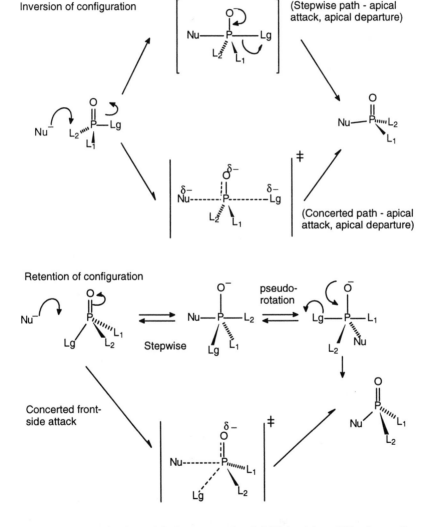

Scheme 6.16. Stereochemistry of displacement at phosphyl (V) involving addition intermediates and concerted mechanisms.

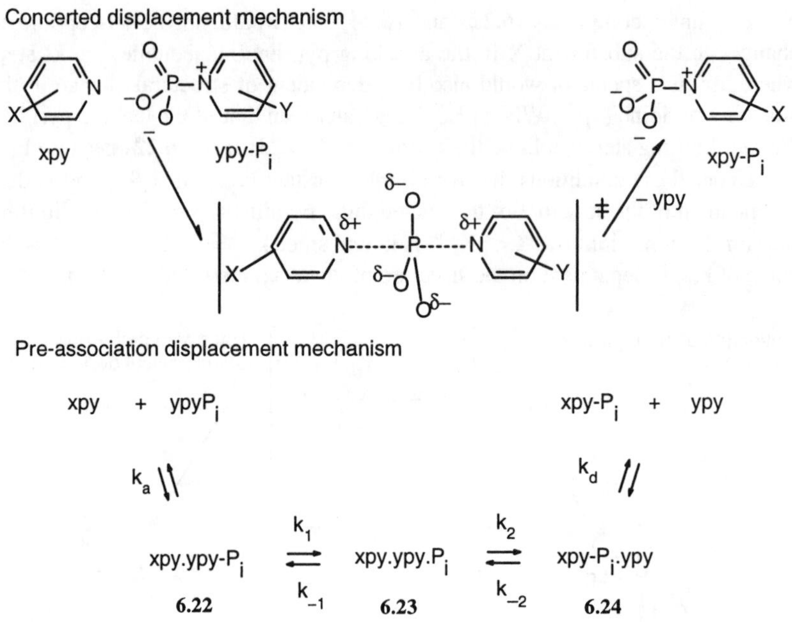

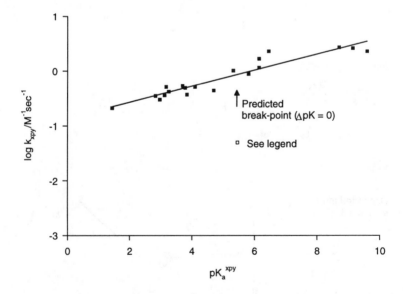

Concerted displacement mechanism

Pre-association displacement mechanism

xpy   +   ypyP$_i$ xpy-P$_i$ +   ypy

k_a k_d

xpy.ypy-P$_i$ $\rightleftharpoons$ xpy.ypy.P$_i$ $\rightleftharpoons$ xpy-P$_i$.ypy

$\quad\quad\quad k_1$ $\quad k_2$

6.22 k_{-1} 6.23 k_{-2} 6.24

Scheme 6.17. Concerted displacement at N-phosphopyridines; xpy-P$_i$ is an N-phospho-pyridinium ion.

Figure 6.5. Brønsted dependence of the reaction of substituted pyridines (xpy) with N-phospho-isoquinolinium ion; expected break-point is shown; open square is that for general base-catalysed hydrolysis of the phospho-isoquinolinium ion by isoquinoline (Data are from *Refs. 92* and *93*).

The stepwise mechanism of Scheme 6.17 can be analysed by use of the Bodenstein steady-state approximation to yield the rate-law dependence on the pK_a of the nucleophile (Equation (6.4)) and is valid provided $k_1 < k_{-a}$ and $k_d > k_{-2}$.

$$k_{xpy} = (k_a/k_{-a})\ k_1\ k_2/(k_{-1} + k_2) \tag{6.4}$$

$$k_{xpy} = k_o\ 10^{\ \beta\ \Delta pK}\ /(1 + 10^{-\Delta pK\ \Delta\beta}) \tag{6.5}$$

The individual rate constants (excluding the diffusion rate constants) can be written as linear Brønsted relationships and this leads to Equation (6.5) as described in Chapter 2. When $\Delta\beta = 0$, Equation (6.5) becomes linear with slope β. The data of the plot (Figure 6.5) can be analysed according to Equation (6.5) to yield a value of $\Delta\beta$ by use of a curve fitting programme. In the case of the displacement at the N-phospho-isoquinolinium ion the confidence limit on $\Delta\beta$ is greater than the parameter itself ($\Delta\beta$ is ~0.1 units). The difference in effective charge on the nitrogen atom of the attacking nucleophile between the two putative transition structures corresponding to k_1 and k_2 would be less than the confidence limit in $\Delta\beta$ (~0.1). The change in effective charge on the attacking nitrogen between reactant and product is 1.07 and the difference between the charges in the two putative transition structures is therefore about 10% of the total change. If the stepwise path prevails then the change in effective charge between the nitrogen in the intermediate and that in one of the transition structures must be less than half of the difference in effective charges between the two putative transition states; in this case it is 5% of the total change in the formation of the xpy-Pi bond and it is unlikely that a significant barrier could be accommodated within this small range. The effective charge in the reaction is mapped in Scheme 6.18 which reveals that there is an imbalance of change in effective charge which must be compensated for by charge uptake in the PO_3 group of atoms equivalent to +0.84 effective charge units (2.07 − 1.23).

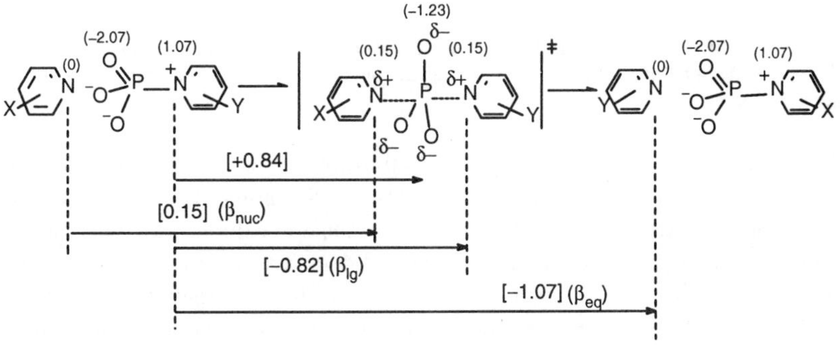

Scheme 6.18. Map of effective charge (in parentheses) for the transfer of the phosphoryl group between pyridine nucleophiles; values in square brackets refer to changes in effective charge.

The imbalance of effective charge in the transfer of the phosphoryl group between pyridine nucleophiles (Scheme 6.18) is due to the bond fission process being advanced over that for bond formation and the resultant transition structure with relatively weak entering and departing bonds is referred to as an "exploded" transition structure. The position of the transition structure can be located in the More O'Ferrall-Jencks map (Figure 6.6) by simple application of β_{nuc} and β_{eq} since the reaction is quasi-symmetrical. The shape of the surface at the transition structure may be determined by the technique elaborated by Jencks and Jencks (1977)[94] (see Chapter 2) from the values of the p_{xy}, p_x and p_y coefficients of 0.014, 0.006 and 0.006 respectively. The angle between the two "level lines" is 64.6° and the reaction coordinate, bisecting these lines, is 45° relative to the β_{nuc} and β_{lg} coordinate axes. Thus the transition structure is at a col with steeper upward curvature of the walls perpendicular to the reaction coordinate compared with that possessed by the downward slopes to reactant or product along the reaction path.

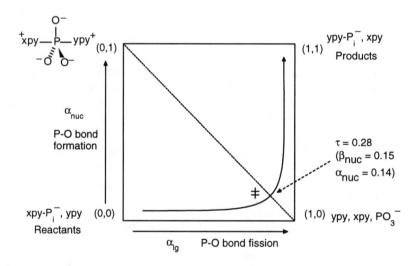

Figure 6.6. More O'Ferrall-Jencks map for the displacement reactions of pyridines at N-phosphopyridinium ions; transition state structure indicated is for the identity reaction with isoquinoline.

Kinetic isotope effects. Primary heavy atom isotope effects for ^{18}O-replacements indicate that the P-O bond fission is rate-limiting in reactions with phosphomonoesters. 2,4-Dinitrophenyl phosphate exhibits kinetic isotope effects of 1.5 to 2.5% over the pH-range 4.4 to 8 when the bridging ^{16}O is replaced by ^{18}O.[95] This data is consistent with that obtained for the 4-nitrophenyl ester by Hengge and Cleland using their very accurate indirect technique namely the secondary ^{15}N-effect in the nitro function of the leaving group.[95-98] It is necessary to estimate the extent of bond formation and this can be done by observing the secondary isotope effects on the non-bridging oxygens which give an estimate of the bonding order to these atoms.[96] The hydrolysis of the dianion of 4-nitrophenyl phosphate exhibits a very small $^{18}O/^{16}O$-isotope effect in the non-bridging oxygens corresponding to that expected for no change in bond-order from ground to transition state for the PO_3 group of atoms. In formal terms there is a change in bond order from 5/3 in the oxygen atoms of the metaphosphate ion to 4/3 for the same atoms in the phosphate ground-state. This change might be expected to give a larger isotope effect than observed; however, computational studies[96] indicate that there is very little bond order difference between phosphate mono ester dianion and metaphosphate ion and this predicts small inverse isotope effects for dissociative mechanisms. Coupled with the isotope effect for P-O bond fission the results for the non-bridging oxygens are consistent with the "exploded" transition state structure.

6.4.3.2 Associative process

Effective charge studies. Theories of displacement at neutral phospho esters were dominated by ideas of pentacoordinate intermediates until the late 1980's when the *quasi symmetrical* technique was applied to the problem. The reaction of phenolate ions with aryl diphenylphosphinate esters (Scheme 6.19) was studied in aqueous solution.[99] The kinetics obey second-order rate-laws and the value of log k_{OAr} is linear against the pK_a of the attacking phenolate ion (Figure 6.7) over a wide range where ΔpK is both positive and negative indicating a single transition state and a concerted displacement mechanism for this reaction.

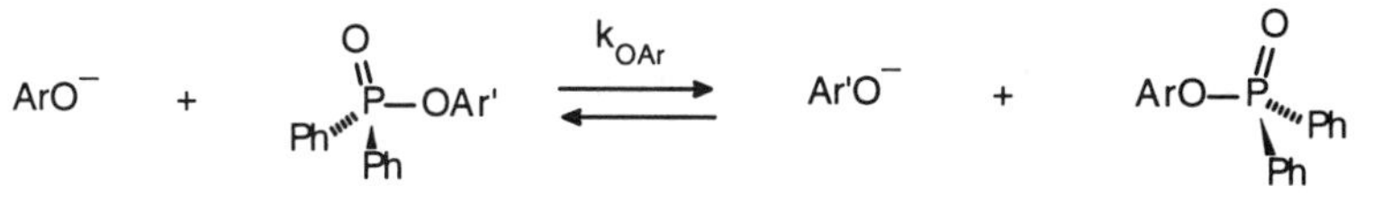

Scheme 6.19. Reaction of aryl oxide ions with aryldiphenylphosphinate esters.

The upper limit on $\Delta\beta$ (0.2) for the putative stepwise process (Equation 6.5) indicates that the two transition structures of the putative stepwise process would lie less than 0.2 effective charge units apart on the reaction map (Figure 6.8). There would be too little space on the energy surface for an intermediate if it were to have a significant barrier for its collapse to reactant or product[90,91,100] because the structure would be less than 0.1 units of effective charge from one of its transition structures in a total change of 1.25 units for full bond formation.

If the reaction were stepwise (Scheme 6.20) the k_2 step would be rate-limiting at $\Delta pK < 0$ and in this region the value of β_{nuc} is $\beta_1 - \beta_{-1} + \beta_2$. The value of β_{nuc} in this region is no greater than 0.6 thus $\beta_{eq(1)}$ for formation of the pentacoordinate intermediate would be <0.6. The value of β_{-2} (lg) would be −0.79 thus $\beta_{eq(2)}$ for formation of products from the pentacoordinate intermediate is >0.79. Since $\beta_{eq(1)}$ corresponds to a full bond formation and $\beta_{eq(2)}$ to only a hybridisation change from P-O the observed inequalities are *not* compatible with the stepwise process but may be explained by a concerted mechanism.

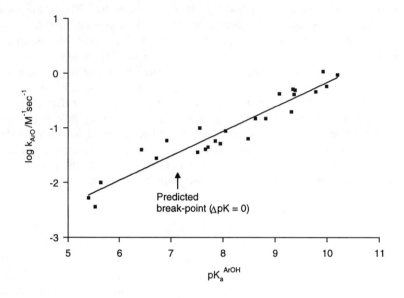

Figure 6.7. Brønsted dependence for the quasi-symmetrical displacement reaction of aryl oxide ions with 4-nitrophenyl diphenylphosphinate; data is taken from *Ref. 99* and line is a linear fit.

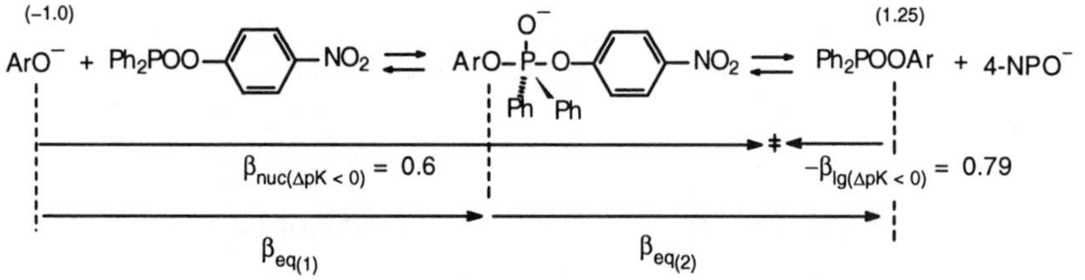

Scheme 6.20. Effective charge map for the displacement at the diphenylphosphinate centre in a putative stepwise process; 4-NPO = 4-nitrophenolate ion.

The variation of β_{lg} as a function of the pK_a of the nucleophile leads to an estimate of β_{eq} for transfer of the phosphinoyl group between phenolate ions (1.25). The value of β_{eq} enables the calculation of the position of the transition structure of the concerted mechanism in the More O'Ferrall-Jencks map (Figure 6.8).

Analogues of the putative pentacoordinate intermediate for example (**6.27**) have been isolated and characterised[101,102] and have weak nucleophilic oxygen ligands in the apical positions. The stability of this species is enhanced by the equatorial oxygen existing in its neutral form and this provides little driving force for decomposition. A chelate effect will also contribute to stability and constraints imposed by the 5-membered rings could minimise additional steric

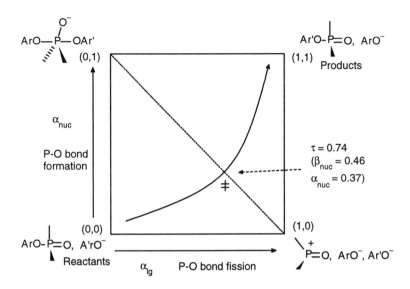

Figure 6.8. More O'Ferrall-Jencks map for the reactions of aryl oxide ions with aryl diphenylphosphinate esters.

repulsions. Ramirez and his coworkers have also studied such analogues as well as the equilibrium between the 4-coordinated reactant (**6.28**) and its pentacoordinated isomer.[84-86]

A good linear Brønsted plot has also been observed for *the quasi symmetrical* reaction at a neutral phosphate centre (Scheme 6.21) and provides good evidence that the neutral triester undergoes concerted displacement reactions.[103-105]

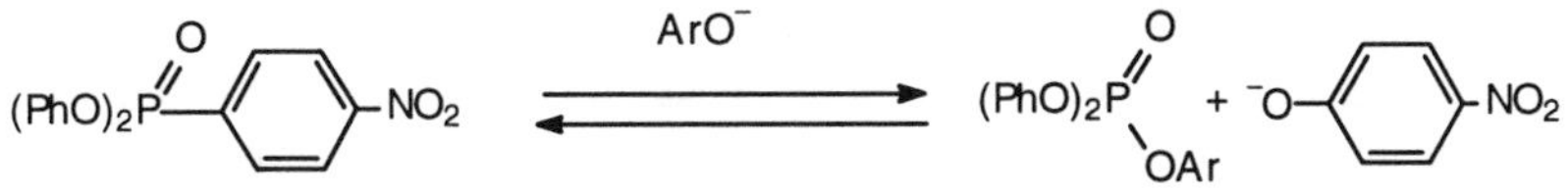

Scheme 6.21. Displacement reaction of aryl oxide ions with 4-nitrophenyl diphenyl phosphate.

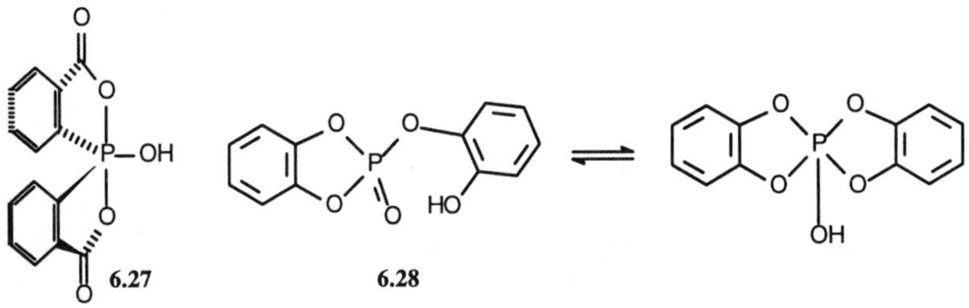

6.27   6.28

Reactions at the phosphodiester level of substitution have direct relevance to the fission and formation of RNA and DNA type biomolecules. The phosphodiester has a structure intermediate between that of the monophosphate dianion and the triphosphate. Since the mono- and tri-phosphate esters undergo concerted displacement reactions the phosphodiester is also be expected to suffer displacement reactions with a concerted pathway if the leaving and displacing groups have sufficiently low nucleophilicity. Provided this is the case the base-catalysed alcoholysis of aryl phosphodiesters can be analysed according to the Scheme 6.22.[106-108]

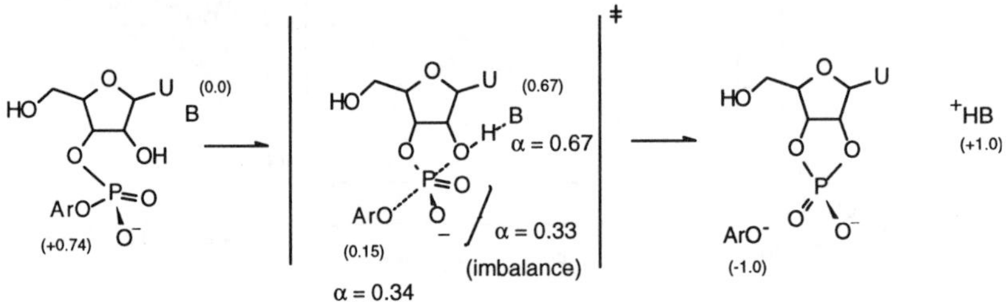

Scheme 6.22. Effective charges and Leffler α-values for cyclisation of aryl uridine-3'-phosphates.

It is not possible to compare directly the β value for the effect of the general base structure with the β_{lg} for the variation of leaving group in the alcoholysis of aryl phosphodiesters. Recourse must be made to Leffler α-values if measures of imbalance are required for this non-symmetrical system. The observed *imbalance* of 0.33 in the Leffler α-parameters indicates that there is a build-up of negative charge on the central O_2PO group of atoms (Scheme 6.22) but there is no indication of where the extra charge resides. Data of Kirby and Younas (1970)[109] indicate a Leffler α-parameter of 0.27 for formation of the ArO–P bond (**6.29**); since there is considerable similarity between the above intermolecular reaction and the intramolecular case only weak bonding is assigned to the forming O-P bond in the cyclisation reaction and the excess charge will thus reside mainly on the attacking oxygen as in (**6.30**) and the reaction is mapped in Figure 6.9.

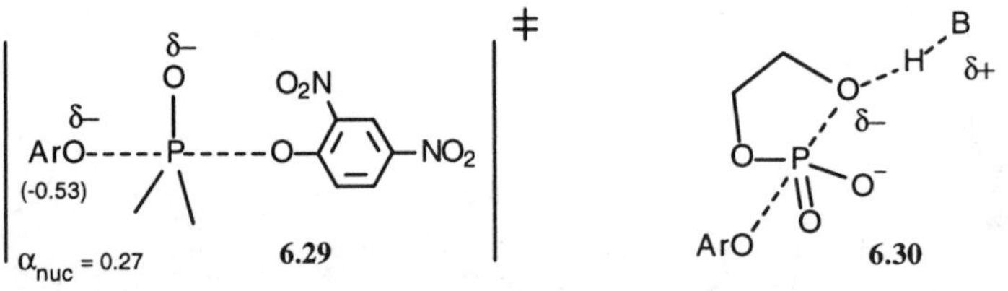

Kirby and Younas (1970)[109] find that the β_{nuc} for reaction of substituted pyridines with methyl aryl phosphates decreases with increasing basicity of the leaving aryl oxide ion indicating coupling between bond fission and bond formation. This is consistent with a concerted mechanism for this reaction.

Kinetic isotope effects. The transfer reactions of phosphodiesters of 4-nitrophenol may be studied by $^{18}O/^{16}O$ secondary isotope effects on the non-bridging oxygens and the $^{15}N/^{14}N$ secondary isotope effect measures the extent of P-O fission to the leaving 4-nitrophenol.[110,111] The value of 0.16% for the ^{15}N-isotope effect on alkaline hydrolysis indicates that the P-O bond is some 57% broken in the transition state. The observation of normal $^{18}O/^{16}O$ isotope effects in the non-bridging oxygens indicates that there is slightly higher bond order to these oxygens consistent with a slightly dissociative process. These data and the ^{15}N isotope effect both indicate less transition state bond fission in the diester consistent[110] with the picture from structure/reactivity analysis.[109] Reaction of hydroxide ion with the neutral phosphotriester exhibits a ^{15}N-isotope effect consistent with only 25% bond fission but the secondary ^{18}O-isotope effect on the non-bridging oxygens indicates a slightly associative mechanism.[110-112]

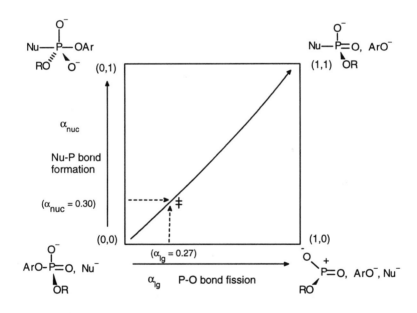

Figure 6.9. Nucleophilic displacement of aryl oxide ions from phosphodiesters.

6.5 Displacement at sulphur

Sulphur and phosphorus, while occupying different groups in the periodic table, appear to have similar chemistry and stereochemistry for many of their oxyacids. Transfer reactions comprise nucleophilic displacements at a number of sulphur centres (Scheme 6.23) and the multiplicity of such centres has lead to the use of the general term "sulphyl" (*sulph*ur ac*yl*) analogous to the "phosphyl" term.

6.5.1 Sulphonyl group transfer

The transfer of the sulphonyl group between nucleophiles could involve a stepwise associative (A_N+D_N), a stepwise dissociative (D_N+A_N), or concerted mechanisms (A_ND_N) as illustrated in Scheme 6.24.[113] The mechanistic types are closely similar to those of the phosphorus series and all types shown in Scheme 6.24 have been observed.

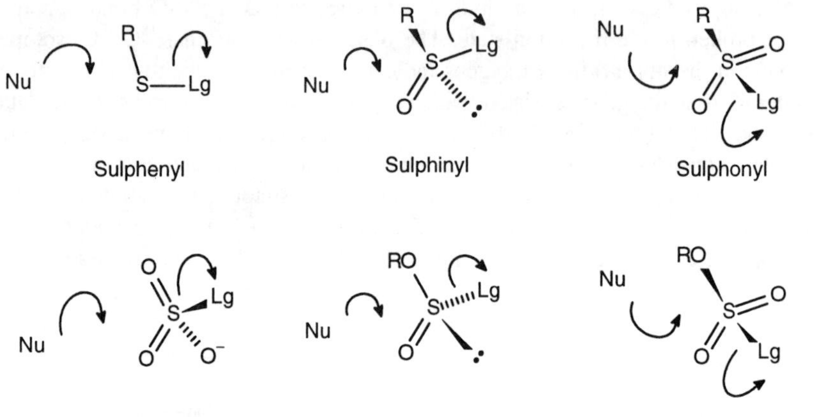

Scheme 6.23. Concerted nucleophilic displacements at sulphyl centres.

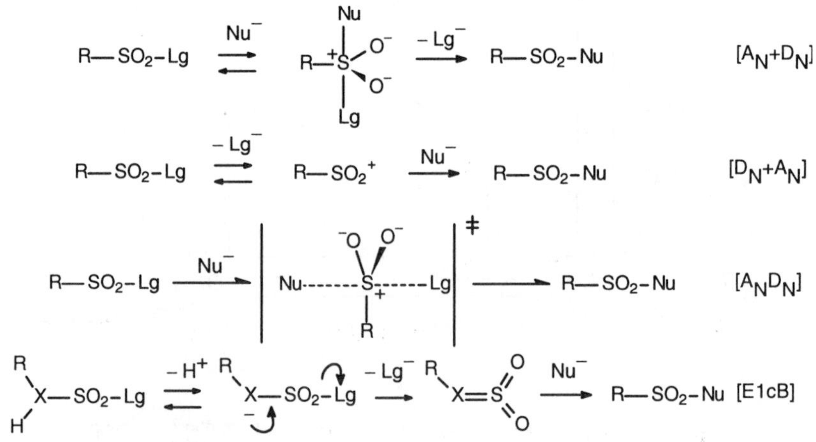

Scheme 6.24. Stepwise and concerted mechanisms for sulphonyl group transfer (E1cB process is named $A_{xh}D_H+D_N$).

6.5.1.1   Bimolecular displacement

The most direct evidence against a pentacoordinate intermediate in sulphonyl group transfer comes from the application of the *quasi-symmetrical* technique to the reaction of oxyanions with 4-nitrophenyl 2,4-dinitrobenzenesulphonate[114]

(Equation 6.6). The Brønsted type plot of log k_{RO} versus the pK_a of the alcohol is linear over a pK_a-range at least 2 units above and below the pK_a of the leaving phenol.

$$O_2N-\underset{NO_2}{\overset{O}{\underset{\|}{\overset{\|}{S}}}}-O-\underset{}{}-NO_2 \quad \xrightarrow[RO^-]{-O-\underset{}{}-NO_2} \quad O_2N-\underset{NO_2}{\overset{O}{\underset{\|}{\overset{\|}{S}}}}-OR \qquad (6.6)$$

Reaction of phenolate ions with sultones (Equation (6.7))[115] has been studied in both directions and β_{eq}, β_{nuc} and β_{lg} have been determined for variation of X and Ar. The value of Leffler's α_{nuc} for the ArO-S bond-forming step in Equation (6.7) is 0.45, measured from the structural effect of the attacking phenolate ion. The value of α_{lg} (0.5) for the bond fission step in Equation (6.7) is determined from the effect of the substituent (X) in the aromatic ring of the substrate. If the displacement were concerted then it would be almost synchronous and the observation of substantial values of α for both bonding changes places the transition structure centrally in the More O'Ferrall-Jencks diagram (Figure 6.10). The absence of imbalance in the system implies that there is no charge taken by the sulphonyl group ($-CH_2SO_2-$) which must therefore retain its integrity through the reaction coordinate. The strong coupling between bond fission and bond formation, in accord with the *subsidiary* definition, is consistent with a concerted process.

$$\underset{O}{\overset{X}{\underset{}{}}}SO_2 \quad \xrightleftharpoons{ArO^-} \quad \underset{O^-}{\overset{X}{\underset{}{}}}SO_2OAr \qquad (6.7)$$

It is of very great interest that Martin's group[116-118] have isolated and characterised compounds with structures analogous to the putative addition complex (coordinates 0,1 in the More O'Ferrall-Jencks diagrams) in displacement at sulphur (analogous to structures **6.31** to **6.35**). The apical bonds in these structures are longer than the equatorial. It is possible that the structures derive from a pair of rapidly interconverting isomers (Equation 6.8) but this is not consistent with the observation of only one carbonyl frequency in the infrared spectrum.[119]

The absence of oxygen exchange into sulphonate esters during hydrolysis in ^{18}O-enriched water (Scheme 6.25) may be due to a slower rate of interconversion of the pentacoordinate or to a concerted displacement mechanism.

6.5.1.2 The D_N+A_N mechanism

A mechanism involving an unsaturated intermediate (Scheme 6.24) has been unambiguously proved for both sulphene[120-124] and iminosulphonate[125-128] intermediates (Scheme 6.24 X = CR and N respectively). Scheme 6.26

illustrates some of the unsaturated sulphur intermediates that have been demonstrated for this mechanism.

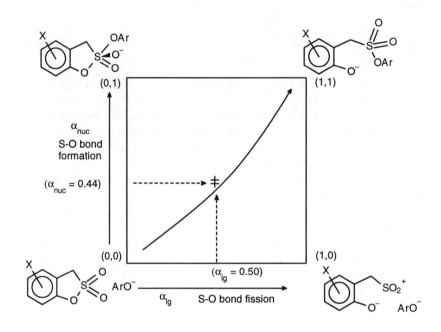

Figure 6.10. More O'Ferrall-Jencks diagram for the fission of sultones (Equation 6.4) by substituted phenolate ions.

6.31   6.32   6.33   6.34   6.35

(6.8)

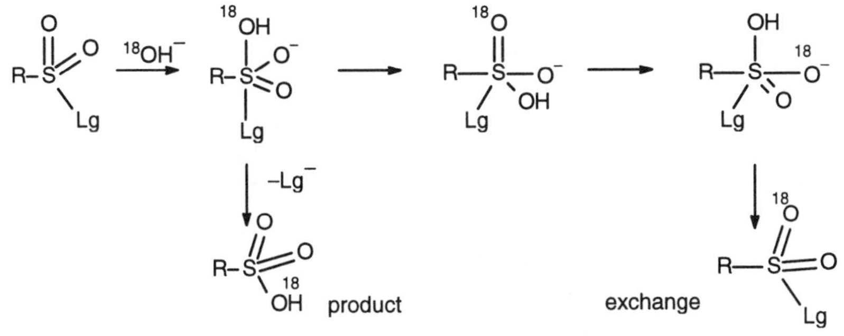

Scheme 6.25. Pseudorotation of the pentacoordinate intermediates in sulphonate ester hydrolysis.

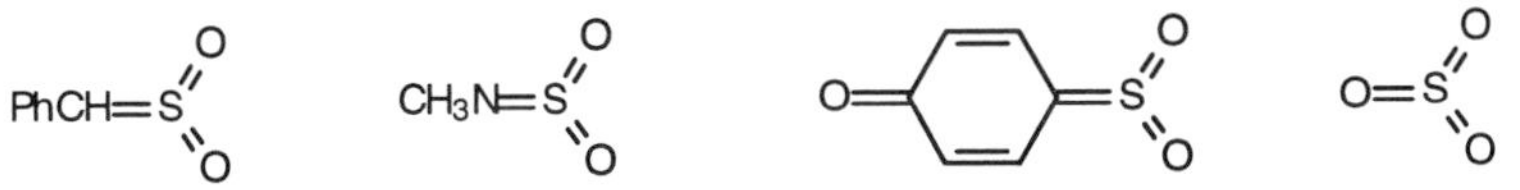

Scheme 6.26. Some putative unsaturated sulphur intermediates for sulphyl group transfer.

The D_N+A_N mechanism is important because sulphur trioxide is a putative intermediate in sulphuryl group transfer and is an analogue of the metaphosphate ion species discussed as a putative intermediate in phosphoryl group transfer. Sulphenes[129-131] and iminosulphonates have been demonstrated as intermediates using classical techniques, and since they are involved in reactions possessing an essential proton transfer (Scheme 6.27) they are discussed in Chapter 3. The function of the proton transfer reaction is to release an anion which effectively stabilises the cationic sulphur of the intermediate. Sulphur trioxide,[132-134] like the metaphosphate anion, has never been unequivocally demonstrated in sulphonyl group transfer reactions in aqueous solution. Displacement reactions with nucleophiles other than water possess rate-laws dependent on the concentration of nucleophile and sulphonate and thus the SO_3 is excluded as a *free* entity; it may intervene as a discrete component of an encounter complex (Scheme 6.28).

Effective charge studies. The pre-association mechanism (Scheme 6.28) can be excluded by use of the technique of *quasi-symmetrical reactions* if nucleophile and leaving group are of similar structure and the nucleophile is varied so that it can be more and less reactive than the leaving group. The reaction of substituted pyridines with N-sulphonato-isoquinolinium ion possesses second-order rate constants which obey a linear Brønsted dependence over a range of pyridine pK_a's some 3 units larger and less than that of isoquinoline (Figure 6.11).[135,136]

The pre-association mechanism is analogous to that for transfer of the PO_3^- group between pyridines (Scheme 6.17) and the kinetic rate-law and associated free energy relationship are those of Equations (6.4) and (6.5). The data may be fit to Equation (6.5) and the value of $\Delta\beta$ is less than its confidence limit of 0.1

indicating that the putative stepwise process could have two transition structures with near identical effective charge on nitrogen for steps k_1 and k_2. The upper limit of $\Delta\beta = 0.1$ indicates that the maximum effective charge difference between the two transition structures in the putative stepwise process is 0.1 and that between the putative intermediate and one of the transition structures (<0.05) is $\leq 5\%$ of the total change in effective charge (1.25) on the nucleophilic nitrogen from reactant to product. Such a small difference in effective charge indicates that there is only a very small effective charge difference between the putative transition structures and that there is very little space on the energy surface to support a significant barrier.

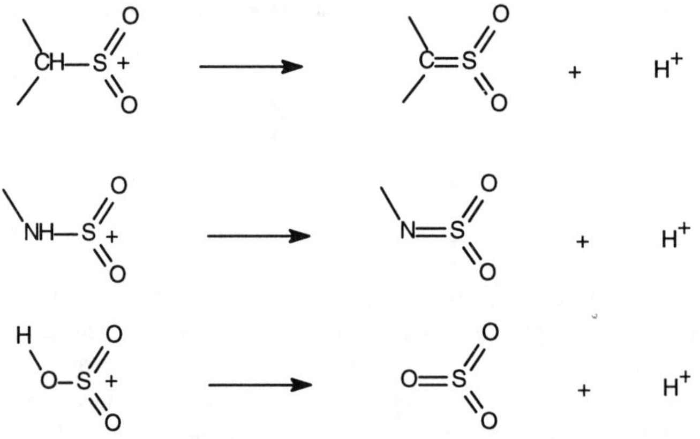

Scheme 6.27. Stabilisation of the cationic sulphur intermediate in a D_N+A_N mechanism.

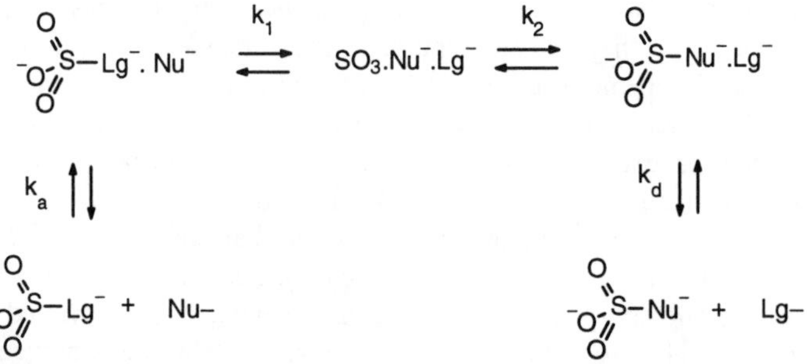

Scheme 6.28. Pre-association mechanism involving expression of sulphur trioxide within an encounter complex.

It is interesting to compare the observed changes in effective charge on the central SO_3 and the entering and leaving pyridines with the model where the transition structure resembles a pentacoordinate intermediate with full bonding of the apical ligands (Scheme 6.29). Scheme 6.29 also illustrates the analogous case for phosphoryl group transfer. The comparisons between charges on transition state and those expected for the pentacoordinate species are markedly different for both sulphuryl and phosphoryl group transfer reactions. The k_1 step in Scheme 6.28 has no dependence on the structure of the nucleophile as this, although it is present in the encounter complex, is merely a spectator of the bond fission reaction. The resultant linearity in the plot in Figure 6.11 is thus excellent evidence that a concerted rather than a stepwise process is involved. The effective charge map for this reaction is closely similar to that for the transfer of

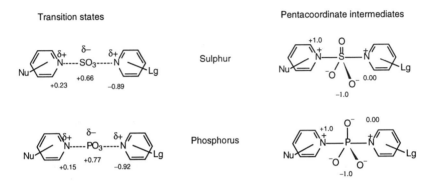

Scheme 6.29. Comparison of observed *changes* in effective charge (between reactants and transition state) with those predicted for the model pentacoordinate intermediate for sulphuryl and phosphoryl group transfer.

of the phosphoryl group and it is reasonable to suppose that, since there is only a small sensitivity to the variation of the pyridine nucleophile, the bond formation is only slightly advanced whereas bond *fission* is significantly advanced in the transition structure. The transition structure is thus an "exploded" structure, like that for the transfer of the phosphoryl group. The More O'Ferrall-Jencks map is similar to that for the phosphoryl group transfer (Figure 6.6), with the transition state located at $\alpha_{nuc} = 0.23/1.25$ for the identity reaction with isoquinoline and N-sulphuryl-isoquinolinium ion (at $\tau = 0.37$ and coordinates 0.816,0.184).[7] The values of β_{nuc} and β_{lg} for the pyridinolysis of N-sulphonatopyridinium ions vary according to the value of pK_{lg} and pK_{nuc} respectively. The coupling between bond fission and bond formation is consistent with the concerted mechanism.

Early work on the effects of amine structure on reactivity of amines with 4-nitrophenyl sulphate indicated that β_{nuc} is very small and at that time the data were interpreted in terms of a completely D_N+A_N process involving the intermediacy of sulphur trioxide.[137] The reactivity of pyridine with aryl sulphates exhibits a very large negative β_{lg} consistent with extensive bond fission in the transition state in this reaction.[134]

Stereochemistry. The stereochemistry of sulphuryl group transfer has been studied using isotopically labelled sulphate esters. Lowe[138,139] demonstrated that the alcoholysis of phenyl sulphate (Scheme 6.30) involves inversion of configuration at the sulphur. The stereochemical results are consistent with concerted displacement but are not sufficient on their own to exclude a process involving the intervention of sulphur trioxide as a component of an encounter complex. If the sulphur trioxide is sufficiently reactive it will be attacked by the nucleophile shortly after it is formed thus precluding tumbling to yield a racemic product.

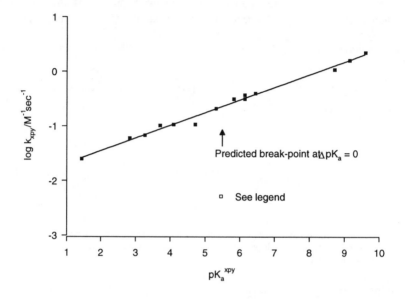

Figure 6.11. Quasi-symmetrical displacement reactions between pyridines and N-sulphonato-isoquinolinium ion; data is from *Ref. 136* and the line is a linear fit. The open square represents the hydrolysis of the sulphonate catalysed by isoquinoline *via* a general base mechanism.

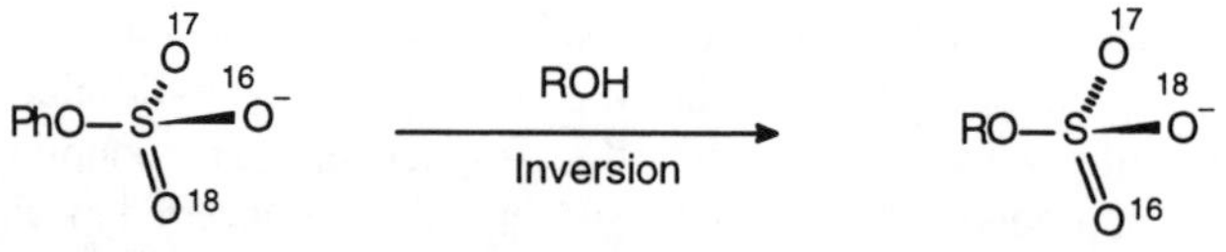

Scheme 6.30. Inversion of configuration during sulphuryl group transfer.

The stereochemical course of the sulphuryl group transfer can be utilised to demonstrate intermediates in sulphotransferase-catalysed reactions. Configuration at sulphur is retained in the sulphuryl transfer catalysed by the sulphotransferase from *aspergillus oryzae* and *Eubacterium* A-44.[140,141] The retention of configuration is consistent with a mechanism involving a sulphuryl-enzyme intermediate formed by a concerted displacement mechanism from the substrate; concerted displacement by the acceptor nucleophile at the sulphuryl-enzyme effects inversion at the sulphur leading to overall retention (Scheme 6.31).

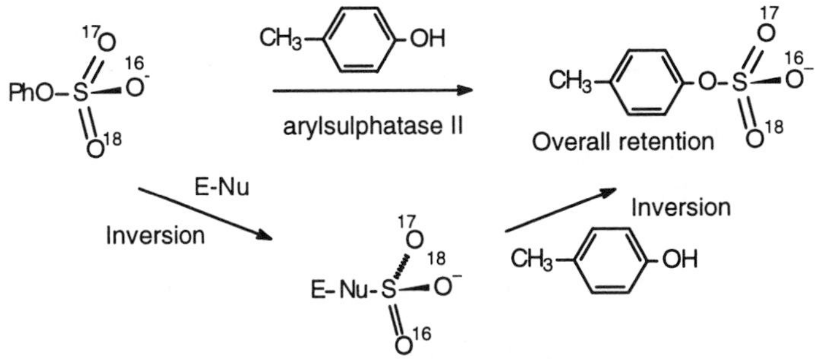

Scheme 6.31. Double inversion in arylsulphatase catalysed displacements at sulphur.

6.5.2 Transfer of sulphur (IV) and sulphur (II) acyl functions

Transfer of the S(IV) group at the RS(O)- level of oxidation is exemplified by transfer of the sulphite and sulphinite groups and stereochemical probes have indicated that displacements occur with inversion of configuration.[142-145] The stereochemical method at the S(IV) level is often difficult to interpret because in some cases racemisation occurs in the product (Scheme 6.32). The rate constant for racemisation = 2 x rate constant for exchange in the reaction (Scheme 6.32) carried out with ^{14}C-labelled methanol.[146] The result indicates that every exchange must occur with an inversion of configuration at the sulphinyl sulphur. Structure **6.34** (see back) is a chemical model of a putative addition adduct in S(IV) transfer between nucleophiles (coordinates 0,1 in a More O'Ferrall-Jencks diagram). Front-side displacement occurs in special cases where the in-line displacement geometry is constrained not to occur by chelate-like interactions (Scheme 6.33).[147]

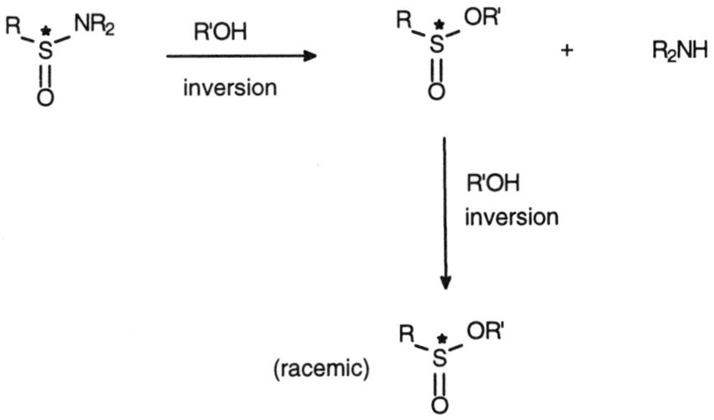

Scheme 6.32. Interpretation of stereochemistry in reactions of S(IV) compounds due to racemisation of product;* indicates a chiral centre.

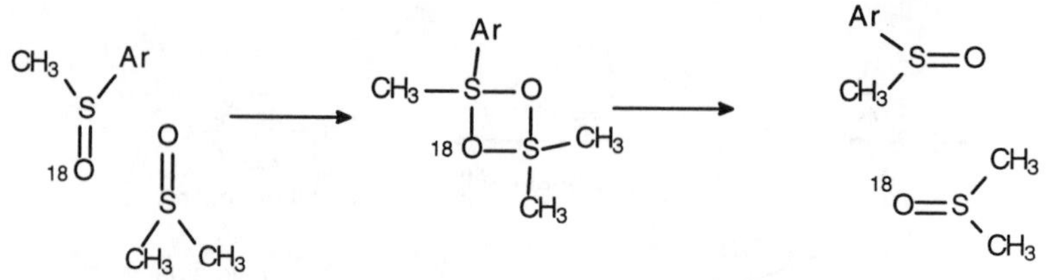

Scheme 6.33. Front-side attack at sulphinyl sulphur indicating scrambling of the oxygen label.

Sulphyl group transfer at the divalent oxidation level is most important in biological systems and involves methyl group transfers (S-adenosyl-methionine chemistry) and acyl transfer reactions (CoA chemistry). Sulphur transfer at the divalent level of oxidation is involved in disulphide cross-linking interchange processes which control the expression of newly formed protein. Disulphide links are formed by nucleophilic substitution by cysteine at disulphide groups followed by further nucleophilic substitution at the sulphur by the thiol of another cysteine residue (Scheme 6.34). The sulphur electrophile in the disulphide link is the sulphyl component of a sulphenic acid (RSOH). Nucleophilic displacement reactions at the sulphenyl group resemble those at tetrahedral carbon (Scheme 6.35) and Hupe and his co-workers showed[148-150] that the *quasi-symmetrical* attack of thiolate anions on aryl methyl disulphides exhibits a linear Brønsted dependence. The Brønsted dependence does not possess a break over the range of nucleophiles spanning the pK_a where $\Delta pK_a = 0$. The effective charge map of the reaction (Structure **6.36**) indicates that there is a relatively small imbalance in the Leffler α-parameters (α_{nuc} and α_{lg}) consistent with a small build-up of negative effective charge on the central sulphur atom ($\Delta \varepsilon = -0.1$) and an almost synchronous concerted mechanism. A chemical model of the putative addition adduct for sulphenyl group transfer is illustrated in Structure **6.35** (see p. 184).

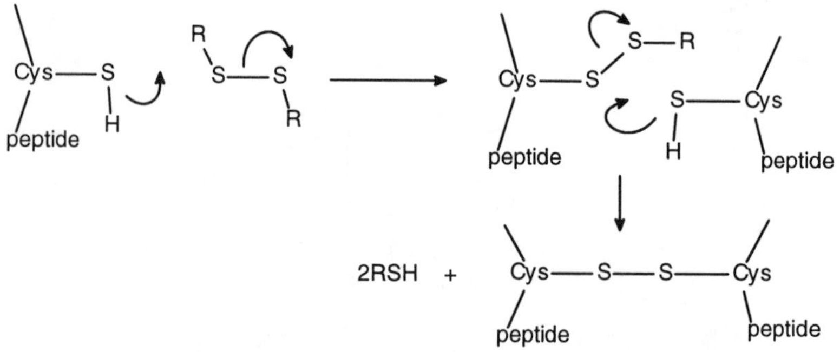

Scheme 6.34. The cross-linking process in the processing of proteins.

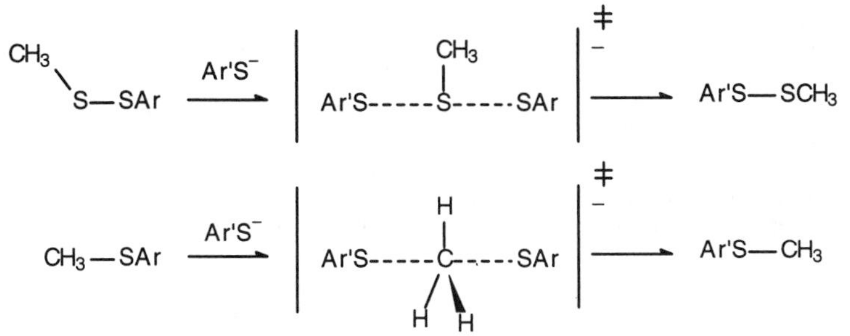

Scheme 6.35. Comparison of nucleophilic displacement at sulphenyl and aliphatic carbon centres.

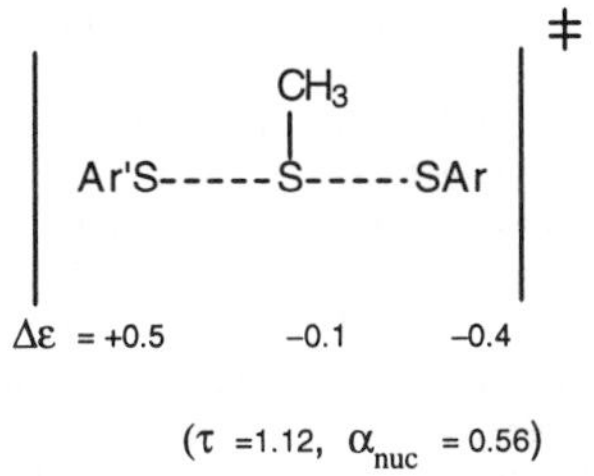

$$Ar'S\text{-----}\underset{\underset{SAr}{|}}{\overset{\overset{CH_3}{|}}{S}}\text{-----}SAr \qquad \ddagger \qquad 6.36$$

$$\Delta\varepsilon = +0.5 \qquad -0.1 \qquad -0.4$$

$$(\tau = 1.12, \ \alpha_{nuc} = 0.56)$$

FURTHER READING

Beak, P., Determination of transition state geometries by the endocyclic restriction test: mechanisms of substitution at non-stereogenic atoms, *Accs. Chem. Res., 25*, 215, 1992.

Brook, A. G., Some molecular rearrangements of organosilicon compounds, *Accs. Chem. Res., 7*, 77, 1974.

Caldwell, S. R., Raushel, F. M., Weiss, P. M., and Cleland, W. W., Transition state structures for enzymatic and alkaline phosphotriester hydrolysis, *Biochemistry, 30*, 7444, 1991.

Erdik, E., and Ng, M., Electrophilic amination of carbanions, *Chem. Rev., 89*, 1947, 1989.

Frey, P. A., Stereochemistry of enzymatic reactions of phosphates, *Tetrahedron, 38*, 1541, 1982.

Guthrie, J. P., Free energy of formation of sulfur trioxide in aqueous solution. Methods of determining the energy level of an unobservable intermediate, *J. Amer. Chem. Soc., 102*, 5177, 1980.

Hall, C., and Inch, T. D., Phosphorus stereochemistry, *Tetrahedron, 36*, 2059, 1980.

Herschlag, D., and Jencks, W. P., Evidence that metaphosphate ion is not an intermediate in solvolysis reactions in aqueous solution, *J. Amer. Chem. Soc., 111*, 7579, 1989.

Holmes, R. R., The stereochemistry of nucleophilic substitution at tetracoordinated silicon, *Chem. Rev., 90*, 17, 1990.

Hudson, R. F., and Brown, C., Reactivity of heterocyclic phosphorus compounds, *Accs. Chem. Res., 5*, 204, 1972.

Kice, J. L., Mechanism and reactivity of organic oxyacids of sulphur and their anhydrides, *Adv. Phys. Org.Chem., 17*, 65, 1980.

Kice, J. L., and Kustateladze, A. G., Mechanism of nucleophilic substitution reactions of *o*-nitrobenzenesulfonamides: evidence for a substitution proceeding through a sulfuranide intermediate, *J. Org. Chem., 57*, 3293, 1992.

King, J. F., Return of sulfenes, *Accs. Chem. Res., 8*, 10, 1975.

Lowe, G., Mechanisms of sulphate activation and transfer, *Phil. Trans. Roy. Soc. Lond.* B, *332*, 141, 1991.

McClelland, R. A., Flash photolysis generation and reactivities of carbenium ions and nitrenium ions, *Tetrahedron, 52*, 6823, 1996.

Okuyama, T., Nakamura, T., and Fueno, T., Acid- and nucleophile-catalysed cleavage of ethyl benzenesulfenate in aqueous solution. Kinetic evidence for a hypervalent intermediate, *J. Amer. Chem. Soc.*, *112*, 9345, 1990.

Okuyama, T., Nakamura, T., and Fueno, T., Hypervalent intermediates in hydrolysis of a sulfenate ester, *Chem. Lett.*, 1133, 1990.

Opitz, G., Wiehn, W., Ziegler, M. L., and Nuber, B., Kristallstrukturanalyse von Bis(trimethylammoniosulfonyl)methanidtetraphenylborat. - n-σ*-wechselwirkungen (Hyperkonjugation und Homohyperkonjugation) in Sulfen-amin-S,N-addukten, *Chem. Ber.*, *125*, 1621, 1992.

Ramirez, F., Oxyphosphoranes, *Accs. Chem. Res.*, *1*, 168, 1968.

Senatore, L., Ciuffarin, E., and Fava, A., Timing of bond formation and breaking in nucleophilic substitution at dicoordinate sulfur. Effect of the basicity of entering and leaving groups in the reaction of oxygen nucleophiles with *para*-substituted phenyl sulfenate esters, *J. Amer. Chem. Soc.*, *92*, 3035, 1970.

Sommer, L. H., *Stereochemistry, Mechanism and Silicon*, McGraw-Hill, New York, 1965.

Stewart, R., The mechanisms of permanganate oxidation. III. The oxidation of benzhydrol, *J. Amer. Chem. Soc.*, *79*, 3057, 1957.

Thatcher, G. R. J., and Kluger, R., Mechanism and catalysis of nucleophilic substitution in phosphate esters, *Adv. Phys. Org. Chem.*, *25*, 99, 1989.

Tillett, J. G., Nucleophilic substitution at tricoordinate sulfur, *Chem. Revs.*, *76*, 747, 1976.

Todd, A. R., Some aspects of phosphate chemistry, *Proc. Natl. Acad. Sci. USA.*, *45*, 1389, 1959.

Tollefson, M. B., Li, J. J., and Beak, P., The endocyclic restriction test: investigation of the geometries of nucleophilic substitution at phosphorus (III) and phosphorus (IV), *J. Amer. Chem. Soc.*, *118*, 9052, 1996.

Waters, W. A., Mechanisms of oxidation by compounds of chromium and manganese, *Quart. Rev. Chem. Soc.*, *12*, 277, 1958.

Westheimer, F. H., Pseudo-rotation in the Hydrolysis of Phosphate Esters, *Accs. Chem. Res.*, *1*, 70, 1968.

Wilkie, J., and Williams, I. H., Geometrical preferences for general acid-catalysed hydride transfer: comparative theoretical study of transition structures for reduction of formaldehyde, *J. Chem. Soc. Perkin Trans. 2*, 1559, 1995.

Young, P. R., and McMahon, P. E., Iodide reduction of sulfilimines. Evidence for the partitioning of sulfurane intermediates, *J. Amer. Chem. Soc.*, *107*, 7572, 1985.

Young, P. R., Zygas, A. P., and Lee, I-w. E., Evidence for concerted general acid catalysis in an S$_N$2-like transition state for the reduction of sulfilimines by thiols, *J. Amer. Chem. Soc.*, *107*, 7578, 1985.

Young, P. R., and Reid, K. J., Thiolate reduction of sulphilimines. 2. Further evidence for a highly coupled concerted transition state, *J. Org. Chem.*, *52*, 2695, 1987.

REFERENCES

1. Kreevoy, M. M., and Lee, I-s. H., Marcus theory of a perpendicular effect on α for hydride transfer between NAD$^+$ analogues, *J. Amer. Chem. Soc.*, *106*, 2550, 1984.

2. Lee, I-s. H., Jeoung, E. H., and Kreevoy, M. M., Marcus theory of a parallel effect on α for hydride transfer reaction between NAD$^+$ analogues, *J. Amer. Chem. Soc.*, *119*, 2722, 1997.

3. Lee, I-s. H., Ostovic, D., and Kreevoy, M. M., Marcus theory of hydride transfer from an anionic reduced deazaflavin to NAD$^+$ analogues, *J. Amer. Chem. Soc.*, *110*, 3989, 1988.

4. Ostovic, D., Lee, I-s., H., Roberts, R. M. G., and Kreevoy, M. M., Hydride transfer and oxyanion addition equilibria of NAD$^+$ analogues, *J. Org. Chem.*, *50*, 4206, 1985.

5. Ostovic, D., Roberts, R. M. G., and Kreevoy, M. M., Isotope effects on hydride transfer between NAD$^+$ analogues, *J. Amer. Chem. Soc.*, *105*, 7629, 1983.

6. Roberts, R. M. G., Ostovic, D., and Kreevoy, M. M., Hydride transfer between NAD$^+$ analogues, *Faraday Discuss. Chem. Soc.*, *74*, 257, 1982.

7.   **Williams, A.,** Bonding in phosphoryl (-PO$_3$$^{2-}$) and sulphuryl (-SO$_3$$^-$) group transfer between nitrogen nucleophiles as determined from rate constants for identity reactions, *J. Amer. Chem. Soc., 107,* 6335, 1985.

8.   **Huskey, W. P., and Schowen, R. L.,** Reaction-coordinate tunneling in hydride-transfer reactions, *J. Amer. Chem. Soc., 105,* 5704, 1983.

9.   **Kim, Y., Truhlar, D. G., and Kreevoy, M. M.,** An experimentally based family of potential-energy surfaces for hydride transfer between NAD$^+$ analogues, *J. Amer. Chem. Soc., 113,* 7837, 1991.

10. **Verhoeven, J. W., Gerresheim, N. van., Martens, F. M., and Kerk, S. M. van der,** Mechanism and transition state studies of hydride-transfer reactions mediated by NAD(P)H-models, *Tetrahedron, 42,* 975, 1986.

11. **Sherrod, M. J.,** Empirically optimised "transition state models", *Tetrahedron Letters, 31,* 5085, 1990.

12. **Wu, Y-d., Lai, D. K. W., and Houk, K. N.,** Transition structures of hydride transfer reactions of protonated pyridinium ion with 1,4-dihydropyridine and protonated nicotinamide with 1,4-dihydronicotinamide, *J. Amer. Chem. Soc., 117,* 4100, 1995.

13. **Donkersloot, M. C. A., and Buck, H. M.,** The hydride-donation reaction of reduced nicotinamide adenine dinucleotide. 1. MINDO/3 and STO-3G calculations on analogue reactions with cyclopropene, trop[ilidene, and 1,4-dihydropyridine as hydride donors and the cyclopropenium cation as acceptor, *J. Amer. Chem. Soc., 103,* 6549, 1981.

14. **More O'Ferrall, R. A.,** Model calculations of hydrogen isotope effects for non-linear transition states, *J. Chem. Soc. (B),* 785, 1970.

15. **Karabatsos, G. J., and Tornaritis, M.,** Structures of the transition states of some intermolecular hydride transfer reactions isotope effects support linear and not non-linear transition states, *Tetrahedron Letters, 30,* 5733, 1989.

16. **Henry, R. S., Riddell, F. G., Parker, W., and Watt, C. I. F.,** The energetics of a readily occurring intramolecular hydride shift., *J. Chem. Soc. Perkin Trans. 2,* 1549, 1976.

17. **Castellino, A. J., and Rapoport, H.,** Synthesis of phenoxyamines, *J. Org. Chem., 49,* 1348, 1984.

18. **Blanchet, P. F., and Krueger, J. H.,** Nucleophilic substitution on nitrogen. Reactions of hydroxylamine-O-sulfonate with thiosulfate and thiourea, *Inorganic Chem., 13,* 719, 1974.

19. **Yamamoto, F., and Oae, S.,** Nucleophilic substitution on trivalent nitrogen atom. Menschutkin type reaction of O-2,4-dinitrophenylhydroxylamine with uncharged nucleophiles pyridine and sulphide nucleophiles, *Bull. Chem. Soc. Japan, 48,* 77, 1975.

20. **Erdik, E., and Ng, M.,** Electrophilic amination of carbanions, *Chem. Rev., 89,* 1947, 1989.

21. **Panda, M., Novak, M., and Magonski, J.,** Hydrolysis kinetics of the ultimateheptacarcinogen N-(sulfonatooxy)-2-(acetylamino)fluorene: detection of long-lived hydrolysis products, *J. Amer. Chem. Soc., 111,* 4524, 1989.

22. **Ulbrick, R., Famulok, M., Busold, F., and Boche, G.,** S$_N$2 at nitrogen: the reaction of N-(4-cyanophenyl)-O-diphenylphosphinoylhydroxylamine with N-methylaniline. A model for the reactions of ultimate carcinogens of aromatic amines with (bio)nucleophiles, *Tetrahedron Letters, 31,* 1689, 1990.

23. **le Noble, W. N.,** Chemical reactions under high pressure. IX. The activation volumes and mechanism of hydrolysis of chloramine, *Tetrahedron Letters, 7,* 727, 1966.

24. **Helmick, J. S., Martin, K. A., Heinrich, J. L., and Novak, M.,** Mechanism of the reaction of carbon and nitrogen nucleophiles with the model carcinogens O-pivaloyl-N-arylhydroxylamines: competing S$_N$2 substitution and S$_N$1 hydrolysis, *J. Amer. Chem. Soc., 113,* 3459, 1991.

25. **Helmick, J. S., and Novak, M.,** Nucleophilic substitution at nitrogen and carboxyl carbon of N-aryl-O-pivaloylhydroxylamines in aqueous solution: competition with S$_N$1 solvolysis of model carcinogens, *J. Amer. Chem. Soc., 56,* 2925, 1991.

26. **Krueger, J. H., Blanchet, P. F., Lee, A. P., and Sudbury, B. A.,** Nucleophilic substitution on nitrogen. Kinetics of reactions of hydroxylamine-O-sulfonate ion, *Inorganic Chem., 12,* 2714, 1973.

27. **Krueger, J. H., Sudbury, B. A., and Blanchet, P. F.,** Correlations of nucleophiles toward trivalent nitrogen reactions of hydroxylamine-O-sulphonate with ethanethiolate, hydroxylamine, and hydroxide, *J. Amer. Chem. Soc., 96,* 5733, 1974.

28. **Sudbury, B. A., and Krueger, J. H.,** Nucleophilic substitution on nitrogen. Kinetics of reactions of hydroxylamine-O-sulfonic acid in dimethyl sulfoxide-water solvents, *Inorganic Chem., 13,* 1736, 1974.

29. **Bühl, M., and Schaefer III, H. F.,** Theoretical characterisation of the transition structure for an S_N2 reaction at neutral nitrogen, *J. Amer. Chem. Soc., 115,* 364, 1993.

30. **Bühl, M., and Schaefer III, H. F.,** S_N2 Reaction at neutral nitrogen: transition state geometries and intrinsic barriers, *J. Amer. Chem. Soc., 115,* 9143, 1993.

31. **Beak, P., and Li, J.,** The endocyclic restriction test: experimental evaluation of transition state structure geometry for a nucleophilic displacement at neutral nitrogen, *J. Amer. Chem. Soc., 113,* 2796, 1991.

32. **Tenud, L., Farooq, S., Seibl, J., and Eschenmoser, A.,** Endocyclische S_N-Reaktionen an gesättigten Kohlenstoff? *Helv. Chim. Acta, 53,* 2059, 1970.

33. **Ren, D., and McClelland, R. A.,** Carbocation-like reactivity patterns in X'-substituted-4-biphenylnitrenium ions, *Can. J. Chem., 76,* 78, 1998.

34. **Fishbein, J. C., and McClelland, R. A.,** Azide ion trapping of the intermediate in the Bamberger rearrangement - lifetime of a free nitrenium ion in aqueous solution, *J. Amer. Chem. Soc., 109,* 2824, 1987.

35. **Greenwood, N. N., and Earnshaw, A.,** *Chemistry of the Elements,* 2nd edition, Butterworth-Heinemann, Oxford, 1997.

36. **Fleming, I., Barbero, A., and Walter, D.,** Stereochemical control in organic synthesis using silicon-containing compounds, *Chem. Rev., 97,* 2063, 1997.

37. **Dietze, P. E.,** Buffer catalysis of the trifluoroethanolysis of phenoxydimethylphenylsilane, *J. Org. Chem., 57,* 6843, 1992.

38. **Dietze, P. E., Khattak, J., and Fickus, E.,** The general acid catalysed trifluoroethanolysis of ethoxydimethylphenylsilane, *Tetrahedron Letters, 32,* 307, 1991.

39. **Dietze, P. E., and Xu, Y.,** Mechanism for the general base catalysed solvolysis of silyl ethers, *J. Org. Chem., 59,* 5010, 1994.

40. **Xu, Y., and Dietze, P. E.,** Evidence for a concerted mechanism in the solvolysis of phenyldimethylsilyl ethers, *J. Amer. Chem. Soc., 115,* 10722, 1993.

41. **Holmes, R. R.,** The stereochemistry of nucleophilic substitution at tetracoordinated silicon, *Chem. Rev., 90,* 17, 1990.

42. **Corriu, R., Guerin, C., and Masse, J.,** Nucleophilic substitution at a silicon atom: evidence for an equatorial attack, *J. Chem. Res. (S),* 160, 1977.

43. **Brook, A. G.,** Some molecular rearrangements of organosilicon compounds, *Accs. Chem. Res., 7,* 77, 1974.

44. **Thatcher, G. R. J., and Kluger, R.,** Mechanism and catalysis of nucleophilic substitution in phosphate esters, *Adv. Phys. Org. Chem., 25,* 99, 1989.

45. **Henchman, M., Viggiano, A. A., Paulson, J. F., Freedman, A., and Wormhoudt, J.,** Thermodynamic and kinetic properties of the metaphosphate anion, PO_3^-, in the gas phase, *J. Amer. Chem. Soc., 107,* 1453, 1985.

46. **Friedman, J. M., Freeman, S., and Knowles, J. R.,** The quest for free metaphosphate in solution: racemization at phosphorus in the transfer of the phospho group from aryl phosphate monoesters to *tert*-butyl alcohol in acetonitrile or in *tert*-butyl alcohol, *J. Amer. Chem. Soc., 110,* 1268, 1988.

47. **Loew, L. M.,** The reactivity of monomeric metaphosphate, *J. Amer. Chem. Soc., 98,* 1630, 1976.

48. **Westheimer, F. H.,** Monomeric metaphosphate, *Chem. Revs., 81,* 313, 1981.

49. **Westheimer, F. H.,** Why nature chose phosphates, *Science, 235,* 1173, 1987.

50. **Todd, A. R.,** Some aspects of phosphate chemistry, *Proc. Natl. Acad. Sci. USA, 45,* 1389, 1959.

51. **Herschlag, D., and Jencks, W. P.,** Evidence that metaphosphate ion is not an intermediate in solvolysis reactions in aqueous solution, *J. Amer. Chem. Soc., 111,* 7579, 1989.

52. **Buchwald, S. L., Friedman, J. M., and Knowles, J. R.,** Stereochemistry of nucleophilic displacement on two phosphoric monoesters and a phosphoguanidine: the role of metaphosphate, *J. Amer. Chem. Soc., 106*, 4911, 1984.

53. **Calvo, K. C.,** The Conant-Swan fragmentation reaction - stereochemistry of phosphate ester formation, *J. Amer. Chem. Soc., 107*, 3690, 1985.

54. **Harnett, S. P., and Lowe, G.,** Stereochemical evidence for monomeric thiometaphosphate as an intermediate in the hydrolysis of (R_P) and (S_P)-deoxyadenosine 5'-[β-^{17}O]-β-thiodiphosphate, *J. Chem. Soc. Chem. Commun.*, 1416, 1987.

55. **Keesee, R. G., and Castleman, A. W.,** Hydration of monomeric metaphosphate anion in the gas phase, *J. Amer. Chem. Soc., 111*, 9015, 1989.

56. **Cullis, P. M., and Misra, R.,** A chemical model for phosphomutases: a dissociative thiophosphoryl transfer reaction proceeding with retention of configuration, *J. Amer. Chem. Soc., 113*, 9679, 1991.

57. **Cullis, P. M., and Rous, A. J.,** Stereochemical course of phosphoryl transfer reactions of P^1,P^1-disubstituted pyrophosphate in aprotic solvent. A model for the enzyme-catalysed "dissociative" phosphoryl transfer, *J. Amer. Chem. Soc., 107*, 6721, 1985.

58. **Cullis, P. M., and Rous, A. J.,** Stereochemical course of the phosphoryl transfer from adenosine 5'-diphosphate to alcohols in acetonitrile and the possible role of monomeric metaphosphate, *J. Amer. Chem. Soc., 108*, 1298, 1986.

59. **Cullis, P. M., Iagrossi, A., and Rous, A. J.,** Phosphoryl and thiophosphoryl transfer reactions: stereochemical imperatives for metaphosphate and thiometaphosphate, *Phosphorus and Sulfur, 30*, 559, 1987.

60. **Lowe, G., and Tuck, S. P.,** Positional isotope exchange in adenosine 5'-[β-^{18}O$_4$]diphosphate and the possible role of monomeric metaphosphate, *J. Amer. Chem. Soc., 108*, 1300, 1986.

61. **Harger, M. J. P., and Hurman, B. T.,** An alkylphosphonyl nucleophilic substitution reaction that proceeds by an elimination-addition mechanism with an alkylidineoxophosphorane (phosphene) intermediate, *J. Chem. Soc. Chem. Commun.*, 1701, 1995.

62. **Gerrard, A. F., and Hamer, N. K.,** Evidence for a planar intermediate in alkaline solvolysis of methyl N-cyclohexylphosphoramidothioic chloride, *J. Chem. Soc. (B)*, 539, 1968.

63. **Williams, A., and Douglas, K. T.,** E1cB mechanisms Part II. Base hydrolysis of substituted phenyl phosphorodiamidates, *J. Chem. Soc. Perkin Trans. 2*, 1454, 1972.

64. **Williams, A., and Douglas, K. T.,** E1cB mechanisms Part IV Base hydrolysis of substituted phenyl phosphoro- and phosphoprothio-diamidates, *J. Chem. Soc. Perkin Trans. 2*, 318, 1973.

65. **Williams, A., and Douglas, K. T.,** Elimination-addition mechanisms of acyl transfer reactions, *Chem. Rev., 75*, 627, 1976.

66. **Freeman, S., and Harger, M. J. P.,** Stereochemistry of reaction of a phosphonamidic chloride with t-butylamine, *J. Chem. Soc. Chem. Commun.*, 1394, 1985.

67. **Guthrie, J. P.,** Hydration and dehydration of phosphoric acid derivatives: free energies of formation of the pentacoordinate intermediates for phosphate ester hydrolysis and of monomeric metaphosphate, *J. Amer. Chem. Soc., 99*, 3991, 1977.

68. **Traylor, P. S., and Westheimer, F. H.,** Mechanisms in the hydrolysis of phosphorodiamidic chlorides, *J. Amer. Chem. Soc., 87*, 553, 1965.

69. **Meyerson, S., Kuhn. E. S., Ramirez, F., Maracek, J. F., and Okazaki, H.,** Electron Impact and field ionisation mass spectrometry of -ketol phosphate salts. Gas phase thermolysis of phosphodiester to monomeric alkyl metaphosphate, *J. Amer. Chem. Soc., 100*, 4062, 1978.

70. **Ramirez, F., Maracek, J. F., and Yemul, S. S.,** Reactions of the monomeric metaphosphate anion generated from different sources, *J. Amer. Chem. Soc., 104*, 1345, 1982.

71. **Meyerson, S., Harvan, D. J., Hass, J. R., Ramirez, F., and Maracek, J. F.,** The monomeric metaphosphate anion in negative-ion chemical-ionization mass spectra of phosphotriesters, *J. Amer. Chem. Soc., 106*, 6877, 1984.

72. **Ramirez, F., and Maracek, J. F.,** Synthesis of complex phosphomonoesters *via* monomeric metaphosphate: the phosphatidic acids, *Synthesis*, 917, 1984.

73. **Ramirez, F., and Maracek, J. F.,** Oxyphosphorane and monomeric metaphosphate ion intermediates in phosphoryl transfer from 2,4-dinitrophenyl phosphate in aprotic and protic solvents, *J. Amer. Chem. Soc., 101*, 1460, 1979.

74. **Ramirez, F., Maracek, J. F., and Yemul, S. S.,** Selectivity of monomeric metaphosphate reactions with alcohols in polar aprotic solvents, *Tetrahedron Letters, 23,* 1515, 1982.

75. **Ramirez, F., Maracek, J. F., Minore, J., Srivastava, S., and le Noble, W.,** On the freeness of the metaphosphate anion in aqueous solution, *J. Amer. Chem. Soc., 108,* 348, 1986.

76. **Cullis, P. M., and Nicholls, D.,** The existence of monomeric metaphosphate in hydroxylic solvent: a positional isotope exchange study, *J. Chem. Soc. Chem. Commun.,* 783, 1987.

77. **Page, M., and Williams, A.,** *Organic and Bio-organic Mechanisms,* Addison Wesley Longman, Harlow, 1997, p. 49.

78. **Hoff, R. H., and Hengge, A. C.,** Entropy and enthalpy contributions to solvent effects on phosphate monoester solvolysis. The importance of entropy effects in the dissociative transition state, *J. Org. Chem., 63,* 6680, 1998.

79. **Rose, I. A.,** Positional isotope exchange studies of enzyme mechanisms, *Adv. Enzymol., 50,* 361, 1979.

80. **Rose, I. A.,** Mechanism of phosphoryl transfer by hexokinase, *Biochem. Biophys. Res. Commun., 94,* 573, 1980.

81. **Lowe, G., and Sproat, B. S.,** Evidence of an $S_N1(P)$ mechanism of phosphoryl transfer by rabbit muscle pyruvate kinase, *J. Chem. Soc. Perkin Trans. 1,* 1622, 1978.

82. **Lowe, G., and Sproat, B. S.,** A synthesis of adenosine 5'-β-^{18}O(2)triphosphate, *J. Chem. Soc. Perkin Trans. 1,* 1874, 1981.

83. **Hassett, A., Blättler, W., and Knowles, J. R.,** Pyruvate kinase - is the mechanism of phospho transfer associative or dissociative? *Biochemistry, 21,* 6335, 1982.

84. **Ramirez, F., Nowakowski, M., and Maracek, J. F.,** Direct observation of a hydroxyphosphorane in equilibrium with a phosphate ester, *J. Amer. Chem. Soc., 99,* 4515, 1977.

85. **Ramirez, F.,** Oxyphosphoranes, *Accs. Chem. Res., 1,* 168, 1968.

86. **Sarma, R., Ramirez, F., McKeever, B., Nowakowski, M., and Maracek, J. F.,** Crystal and molecular structure of *o*-hydroxyphenyl-*o*-phenylene phosphate, (*o*-HOC$_6$H$_4$)(C$_6$H$_4$)PO$_4$. Equilibrium between pentavalent and tetravalent phosphorus in solution, *J. Amer. Chem. Soc., 100,* 5391, 1978.

87. **Hall, C., and Inch, T. D.,** Phosphorus stereochemistry, *Tetrahedron, 36,* 2059, 1980.

88. **Westheimer, F. H.,** Pseudo-rotation in the Hydrolysis of Phosphate Esters, *Accs. Chem. Res., 1,* 70, 1968.

89. **Hudson, R. F., and Brown, C.,** Reactivity of Heterocyclic Phosphorus Compounds, *Accs. Chem. Res., 5,* 204, 1972.

90. **Skoog, M. T., and Jencks, W. P.,** Phosphoryl transfer between pyridines, *J. Amer. Chem. Soc., 105,* 3356, 1983.

91. **Skoog, M. T., and Jencks, W. P.,** Reactions of pyridines and primary amines with N-phosphorylated pyridines, *J. Amer. Chem. Soc., 106,* 7597, 1984.

92. **Bourne, N., and Williams, A.,** The question of concerted or stepwise mechanisms in the nucleophilic displacement of phosphate by pyridines from isoquinoline-N-phosphonate, *J. Amer. Chem. Soc., 105,* 3357, 1983.

93. **Bourne, N., and Williams, A.,** Evidence for a single transition state in the transfer of the phosphoryl group (-PO$_3$$^{2-}$) to nitrogen nucleophiles from pyridine-N-phosphonates, *J. Amer. Chem. Soc., 106,* 7591, 1984.

94. **Jencks, D. A., and Jencks, W. P.,** On the characterization of transition states by structure-reactivity coefficients, *J. Amer. Chem. Soc., 99,* 7948, 1977.

95. **Gorenstein, D. G.,** Oxygen-18 isotope effect in the hydrolysis of 2,4-dinitrophenyl phosphate. A monomeric metaphosphate mechanism, *J. Amer. Chem. Soc., 94,* 2523, 1972.

96. **Weiss, P. M., Knight, W. B., and Cleland, W. W.,** Secondary ^{18}O isotope effects on the hydrolysis of glucose 6-phosphate, *J. Amer. Chem. Soc., 108,* 2761, 1986.

97. **Hengge, A. C., and Cleland, W. W.,** Direct measurement of transition state bond cleavage in hydrolysis of phosphate esters of p-nitrophenol, *J. Amer. Chem. Soc., 112,* 7421, 1990.

98. **Hengge, A. C., Edens, W. A., and Elsing, H.,** Transition state structures for phosphoryl-transfer reactions of *p*-nitrophenyl phosphate, *J. Amer. Chem. Soc., 116,* 5045, 1994.

99. **Bourne, N., Chrystiuk, E., Davis, A. M., and Williams, A.,** A single transition state in the reaction of aryl diphenylphosphinate esters with phenolate ions, *J. Amer. Chem. Soc., 110,* 1840, 1988.

100. **Dietze, P. E., and Jencks, W. P.,** Swain-Scott correlations for reactions of nucleophilic reagents and solvents with secondary substrates, *J. Amer. Chem. Soc., 108,* 4549, 1986.

101. **Segall, Y., and Granoth, I.,** Syntheses of cyclic acyloxyphosphoranes from phosphine oxides: Spectroscopy, stability, and molecular structure. A stable P-hydroxyphosphorane: 1-hydroxy-1,1'-spirobi[3H-2,1-benzoxaphosphole]-3,3'-dione, *J. Amer. Chem. Soc., 100,* 5130, 1978.

102. **Segall, Y., and Granoth, I.,** Surprising synthesis and reactivity of an acidic hydroxyphosphorane, *J. Amer. Chem. Soc., 101,* 3687, 1979.

103. **Ba-Saif, S. A., Waring, M. A., and Williams, A.,** Single transition state in the transfer of a neutral phosphoryl group between phenoxide ion nucleophiles in aqueous solution, *J. Amer. Chem. Soc., 112,* 8115, 1990.

104. **Ba-Saif, S. A., Waring, M. A., and Williams, A.,** Dependence of transition state structure on nucleophile in the reaction of aryl oxide anions with aryl diphenylphosphate esters, *J. Chem. Soc. Perkin Trans. 2,* 1653, 1991.

105. **Waring, M. A., and Williams, A.,** An "open" transition state in the transfer reaction of a neutral phosphoryl group between phenolate anions, *J. Chem. Soc. Chem. Commun.,* 1742, 1989.

106. **Davis. A. M., Hall, A. D., and Williams, A.,** Charge description of base catalysed alcoholysis of aryl phosphodiesters: a ribonuclease model, *J. Amer. Chem. Soc., 110,* 5105, 1988.

107. **Davis, A. M., Regan, A. C., and Williams, A.,** Exzperimental charge measurement at leaving oxygen in the bovine ribonuclease-A catalysed cyclisation of uridine 3'-phosphate aryl esters, *Biochemistry, 27,* 9042, 1988.

108. **Ba-Saif, S. A., Davis, A. M., and Williams, A.,** Effective charge distribution for attack of phenoxide ion on aryl methyl phosphate monoanions: studies related to the action of ribonuclease, *J. Org. Chem., 54,* 5483, 1989.

109. **Kirby, A. J., and Younas, M.,** The reactivity of phosphate esters. Diester hydrolysis, *J. Chem. Soc. (B),* 510, 1970.

110. **Hengge, A. C., and Cleland, W. W.,** Phosphoryl-transfer reactions of phosphodiesters: characterization of transition states by heavy-atom isotope effects, *J. Amer. Chem. Soc., 113,* 5835, 1991.

111. **Hengge, A. C., Tobin, A. E., and Cleland, W. W.,** Studies of transition state structures in phosphoryl transfer reactions of phosphodiesters of *p*-nitrophenol, *J. Amer. Chem. Soc., 117,* 5919, 1995.

112. **Deal, K. A., Hengge, A. C., and Burstyn, J. N.,** Characterization of transition states in dichloro(1,4,7-triazacyclononane)copper(II)-catalyzed activated phosphate diester hydrolysis, *J. Amer. Chem. Soc., 118,* 1713, 1996.

113. **Kice, J. L.,** Mechanism and reactivity of organic oxyacids of sulphur and their anhydrides, *Adv. Phys Org. Chem., 17,* 65, 1980.

114. **D'Rozario, P., Smyth, R. L., and Williams, A.,** Evidence for a single transition state in the intermolecular transfer of a sulphonyl group between oxyanion donor and acceptor, *J. Amer. Chem. Soc., 106,* 5027, 1984.

115. **Deacon, T., Farrar, C. R., Sikkel, B. J., and Williams, A.,** Reactions of nucleophiles with strained cyclic sulfonate esters. Brønsted relationships for rate and equilibrium constants for variation of phenolate anion nucleophile and leaving group, *J. Amer. Chem. Soc., 100,* 2525, 1978.

116. **Perozzi, E. F., Martin, J. C., and Paul, I. C.,** Crystal and molecular structures of a spirodiaryldialkoxysulfurane oxide and its parent sulfurane, *J. Amer. Chem. Soc., 96,* 6735, 1974.

117. **Paul, I. C., Martin, J. C., and Perozzi, E. F.,** Sulfuranes. VII. The crystal and molecular structure of a diaryldialkoxysulfurane, *J. Amer. Chem. Soc., 94,* 5010, 1972.

118. **Perkins, C. W., Wilson, S. R., and Martin, J. C.,** Ground state analogues of transition states for attack of sulphonyl, sulphinyl and sulphenyl sulphur, *J. Amer. Chem. Soc., 107,* 3209, 1985.

119. **Lau, P. H. W., and Martin, J. C.,** Tricoordinate hypervalent sulfur species. Sulfuranide anions, *J. Amer. Chem. Soc., 100,* 7077, 1978.

120. **Davy, M. B., Douglas, K. T., Loran, J. S., Steltner, A., and Williams, A.,** Elimination-addition mechanisms of acyl group transfer: hydrolysis and aminolysis of aryl phenylmethanesufonates, *J. Amer. Chem. Soc., 99,* 1196, 1977.

121. **Douglas, K. T., Steltner, A., and Williams, A.,** An E2 mechanism for the alkaline hydrolysis of toluene-α-sulphonate esters of acidic phenols, *J. Chem. Soc. Chem. Commun.,* 353, 1975.

122. **King, J. F.,** Return of sulfenes, *Accs. Chem. Res., 8,* 10, 1975.

123. **Thea, S., Harun, M. G., and Williams, A.,** Mechanism in sulphonyl group transfer: effective charge in ground-, transition-, and product-states during base catalysed hydrolysis of aryl sulphonate esters possessing associative (AE) and dissociative (EA) mechanisms, *J. Chem. Soc. Chem. Commun.,* 717, 1979.

124. **Williams, A., Douglas, K. T., and Loran, J. S.,** Evidence consistent with a stepwise elimination-addition process for hydrolysis and aminolysis of aryl toluene-α-sulphonates, *J. Chem. Soc. Chem. Commun.,* 689, 1974.

125. **Spillane, W. J., Hogan, G., McGrath, P., King, J., and Brack, C.,** Aminolysis of sulfamate esters in non-aqueous solvents. Evidence consistent with a concerted E2-type mechanism, *J. Chem. Soc. Perkin Trans.* 2, 2099, 1996.

126. **Spillane, W. J., McHugh, F. A., and Burke, P. O.,** Elimination mechanisms in the anilinolysis of sulfamoyl chlorides in chloroform and acetonitrile, *J. Chem. Soc. Perkin Trans.* 2, 13, 1998.

127. **Thea, S., Cevasco, G., Guanti, G., and Williams, A.,** The anionic sulphonylamine mechanism in the hydrolysis of aryl sulphamates, *J. Chem. Soc. Chem. Commun.,* 1582, 1986.

128. **Douglas, K. T., and Williams, A.,** Hydrolysis of Aryl N-methylaminosulphonates: evidence consistent with an E1cB mechanism, *J. Chem. Soc. Perkin Trans.* 2, 1727, 1974.

129. **Cevasco, G., and Thea, S.,** Structure-reactivity correlations in the dissociative hydrolysis of 2',4'-dinitrophenyl 4-hydroxy-X-benzenesulfonates, *J. Org. Chem., 61,* 6814, 1996.

130. **Cevasco, G., and Thea, S.,** The effect of leaving group variation on reactivity in the dissociative hydrolysis of aryl 3,5-dimethyl-4-hydroxybenzenesulfonates, *J. Chem. Soc. Perkin Trans.* 2, 2215, 1997.

131. **Thea, S., Cevasco, G., Guanti, G., Hopkins, A., Kashefi-Naini, N., and Williams, A.,** Sulfoquinones in the hydrolysis of aryl esters of *o*- and *p*-hydroxyarenesulfonic acids in alkaline aqueous solutions of dioxane, *J. Org. Chem., 50,* 2158, 1985.

132. **Hopkins, A. R., Day, R. A., and Williams, A.,** Sulphate group transfer between nitrogen and oxygen: evidence consistent with an open 'exploded' transition state, *J. Amer. Chem. Soc., 105,* 6062, 1983.

133. **Hopkins, A. R., and Williams, A.,** The absence of free sulphur trioxide as an intermediate in sulphonate group transfers, *J. Chem. Soc. Chem. Commun.,* 37, 1983.

134. **Hopkins, A. R., and Williams, A.,** Carboxy group participation in sulphate and sulphamate group transfer reactions, *J. Org. Chem., 47,* 1745, 1982.

135. **Bourne, N., Hopkins, A., and Williams, A.,** A preassociation concerted mechanism in the transfer of the sulphate group between isoquinoline-N-sulphonate and pyridines, *J. Amer. Chem. Soc., 105,* 3358, 1983.

136. **Bourne, N., Hopkins, A., and Williams, A.,** Single transition state for sulfuryl group (-SO$_3^-$) transfer between pyridine nucleophiles, *J. Amer. Chem. Soc., 107,* 4327, 1985.

137. **Benkovic, S. J., and Benkovic, P. A.,** Studies on sulfate esters. I. Nucleophilic reactions of amines with p-nitrophenyl sulfate, *J. Amer. Chem. Soc., 88,* 5504, 1966.

138. **Lowe, G.,** Mechanisms of sulphate activation and transfer, *Phil. Trans. Roy. Soc. Lond.* B, *332,* 141, 1991.

139. **Lowe, G.,** The stereochemical course of sulfuryl transfer reactions, *Phosphorus, Sulfur, Silicon, 59,* 63, 1991.

140. **Chai, C. L. L., and Lowe, G.,** The mechanism and stereochemical course of sulfuryl transfer catalysed by the aryl sulfatase from *Eubacterium* A-44, *Bioorganic Chem., 20,* 181, 1992.

141. **Chai, C. L. L., Loughlin, W. A., and Lowe, G.,** The stereochemical course of sulfuryl transfer catalysed by aryl sulfatase II from *Aspergillus oryzae, Biochem. J., 287,* 805, 1992.

142. **Tillett, J. G.,** Nucleophilic substitution at tricoordinate sulfur, *Chem. Revs., 76,* 747, 1976.

143. **Mikolajczyk, M., Drabowicz, J., and Bujnicki, B.,** Acid-catalysed conversion of sulphinamides into sulphinates: a new synthesis of optically active sulphinates, *J. Chem. Soc. Chem. Commun.*, 568, 1976.

144. **Andersen, K. K., Gaffield, W., Papanikalou, N. E., Foley, J. W., and Perkins, R. J.,** Optically active sulfoxides. The synthesis and rotatory dispersion of some diaryl sulfoxides, *J. Amer. Chem. Soc., 86,* 5637, 1964.

145. **Andersen, K. K.,** Sulfinic acids and their derivatives, in *Comprehensive Organic Chemistry,* Jones, D. N., ed., Pergamon Press, Oxford, 1979, p. 317.

146. **Mikolajczyk, M., Drabowicz, J., Slebocka-Tilk, H.,** Nucleophilil substitution at sulfur. Kinetic evidence for inversion of configuration at sulfinyl sulfur in acid-catalysed transesterification of sulfinates, *J. Amer. Chem. Soc., 101,* 1302, 1979.

147. **Oae, S., Yokohama, M., Kise, M., and Furakawa, N.,** Oxygen exchange reaction of sulfoxides with dimethyl sulfoxide, *Tetrahedron Letters,* 4131, 1968.

148. **Freter, E., Pohl, E. R., Wilson, J. M., and Hupe, D. J.,** Rôle of the central thiol in determining rates of the thiol-disulphide interchange reaction, *J. Org. Chem., 44,* 1771, 1979.

149. **Hupe, D. J., and Pohl, E. R.,** A demonstration of the reactivity-selectivity principle for the thiol-disulphide interchange reaction, *Israel J. Chem.,* 26, 395, 1985.

150. **Wilson, J. M., Bayer, R. J., and Hupe, D. J.,** Structure-reactivity correlations for the thiol-disulphide interchange reaction, *J. Amer. Chem. Soc., 99,* 7922, 1977.

Chapter 7

Cyclic Reactions

Cyclic reactions such as the Diels-Alder cycloaddition, the Cope and the Claisen rearrangements have substantial synthetic[1] and biological[2] significance (Scheme 7.1). The biosynthesis of lanosterol derivatives and associated biomimetic cyclisations, many fragmentation reactions such as decarboxylation, and extrusion of nitrogen and sulphur dioxide constitute part of this large class of reactions. Prior to the mid-1960's evidence had been accumulating which suggested that some cyclic reactions do not possess intermediates and for this reason they were termed "no mechanism" reactions. The available evidence suffered from a lack of precision when judged against current standards and is summarised in review articles by Sara Jane Rhoads.[3,4]

The orbital-symmetry theories of Hoffmann and Woodward[5] do not require that pericyclic reactions should be concerted[6] but their formulation encouraged renewed efforts in the diagnosis of concerted mechanisms in such systems. The eponymous selection rules[5] originated from strong qualitative arguments based on symmetry and explain the stereochemical course of thermal cyclic reactions. Qualitative and quantitative molecular orbital theory has figured extensively in discussions of transition state structure in cyclic reactions and the original idea of concertedness appears to have arisen because of the high degree of stereospecificity invariably observed in these systems. Stereospecificity is required by, but does not require, concertedness; while it is not a precise mechanistic tool, it can be employed in the diagnosis of concerted mechanisms in certain circumstances.

7.1 INTERMEDIATES

7.1.1 Direct observation

The hydrogen-iodine thermal combination reaction obeys a second-order rate-law (rate = $k_2[H_2][I_2]$)[7] which is consistent with a bimolecular mechanism involving a four-centred transition structure (Scheme 7.2). For many years the mechanism was believed to be concerted (four-centre) until consideration of the Hoffmann and Woodward theories indicated the problem that it is not allowed by the symmetry rules.[8] Photochemical initiation allows the reaction to occur at a lower temperature than that of the thermal reaction but the Arrhenius parameters for both photochemical and thermal reactions are identical.[9,10,11] These results are consistent with a *radical chain* mechanism which can be initiated either thermally or photochemically but they do not fit the simple bimolecular mechanism. The observation of radical or di-radical species in a cyclic reaction may be effected by a number of techniques, few of which were available in the early 1960's.[12,13]

The current interest in very fast reaction dynamics has provided an example of a cyclic reaction where the mass spectrum can be measured as a function of time in the femtosecond range. The *retro*-Diels-Alder reaction of norbornene exhibits peaks at 94 amu and at 66 amu. The signal for the mass at 94 amu increases with a $t_{0.5}$ of 30 fs and decays with a $t_{0.5}$ of 220 fs. The signal for the mass at 66 amu decays exponentially with a $t_{1/2}$ of 160 fs and this indicates that this peak is not the final cyclopentadiene product. The results are consistent with two trajectories on the potential energy surface involving diradicals as shown in Scheme 7.3.[14] Symmetrical motions of the

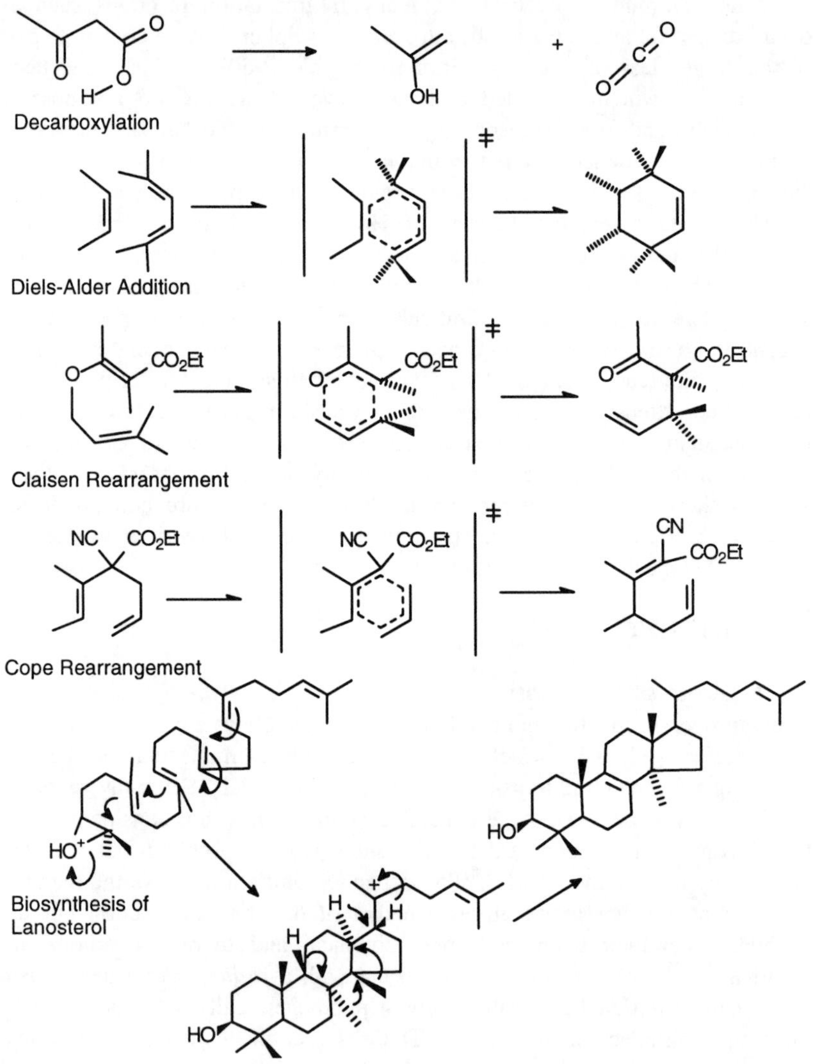

Scheme 7.1. Some cyclic reactions.

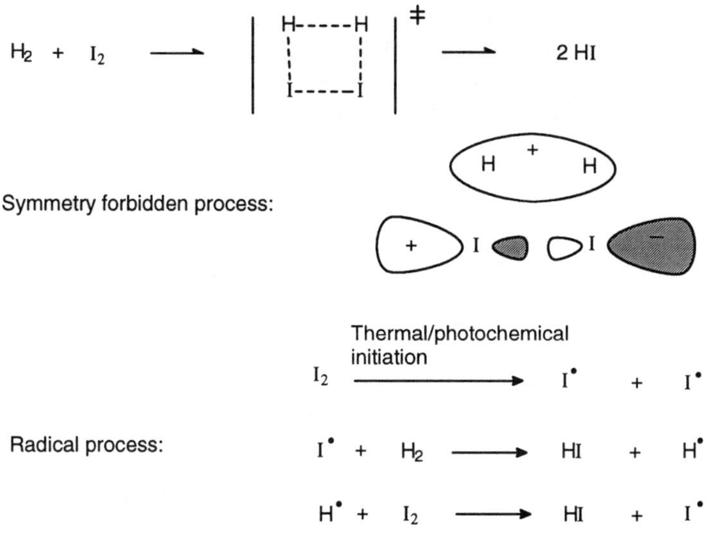

$$H_2 + I_2 \longrightarrow \left| \begin{array}{c} H\text{-----}H \\ \vdots \quad \vdots \\ I\text{-----}I \end{array} \right|^{\ddagger} \longrightarrow 2\ HI$$

Symmetry forbidden process:

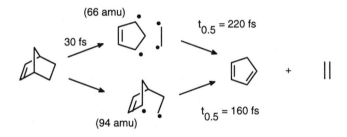

Radical process:

$$I_2 \xrightarrow[\text{initiation}]{\text{Thermal/photochemical}} I^{\bullet} + I^{\bullet}$$

$$I^{\bullet} + H_2 \longrightarrow HI + H^{\bullet}$$

$$H^{\bullet} + I_2 \longrightarrow HI + I^{\bullet}$$

Scheme 7.2. Pericyclic and radical mechanisms for the hydrogen iodine reaction.

Scheme 7.3. *Retro*-Diels-Alder reaction of norbornene studied in the femtosecond range.

two C-C bonds undergoing fission give rise to the concerted trajectory and the observed decay time of 220 fs is due to the vibrationally excited cyclopent-2-en-1,4-diyl decaying to form products. The decay time of 160 fs is the time for interconversion in the second step of the non-concerted route through the diradicaloid structure. The concerted trajectories (by symmetrical bond fission) will produce the vibrationally excited cyclopent-2-en-1,4-diyl which then decays after nuclear rearrangement and electronic hybridisation to form cyclopentadiene and ethylene. The technique is similar to that suggested by Bauer[15] (Chapter 2) whereby the concerted path from symmetrical fission leads to a vibrationally excited product molecule.

An example of the application of conventional kinetic techniques is found in the reaction of *trans*-1-(dimethylamino)-1,3-butadiene with dimethyl dicyanofumarate in acetonitrile at -40°. The kinetics, followed spectroscopically in a stopped flow apparatus,[16] exhibit the rapid disappearance of the reactant absorptions and the appearance of an absorption with a maximum at 400nm. The appearance of the intermediate has a second-order rate constant of $10^7 M^{-1} sec^{-1}$ and it decays with a rate constant of $10.7 sec^{-1}$. The intermediate may be identified as the zwitterionic species given in Scheme 7.4 because the equilibrium constant for its formation is very much larger than that expected for the charge transfer complex.

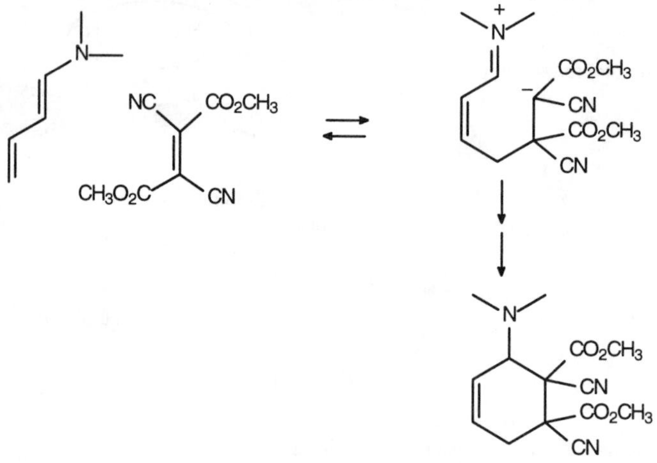

Scheme 7.4. A zwitterionic intermediate (λ_{max} = 400 nm in acetonitrile)) in the Diels-Alder addition of *trans*-1-(dimethylamino)-1,3-butadiene to dimethyl dicyanofumarate.

7.1.2 Indirect exclusion

The rate of the ene reaction of 1,2,3-triphenylcyclopropene (Scheme 7.5) is not affected by addition of di-*tert*-butyl peroxide and both deuteriums from the 1-deuterio species are conserved in the product even when the reaction is carried out in alkali. These classical diagnostic tools exclude the existence of a free radical process and a concerted mechanism can be inferred.[17]

A primary deuterium isotope effect excludes a stepwise process in the solvolytic elimination reaction (Scheme 7.6) where the ester solvolyses[18] to give exclusively the elimination product. The absence of a kinetic isotope effect for the elimination rate constants when the methyl group of the phenylethyl residue is isotopically substituted ($k(CH_3)/k(CD_3)$ = $k(CH_3)/k(CH_2D)$ = 1.00) excludes the mechanism *via* transition state (**7.1**) which requires a proton transfer in the rate-limiting step. The isotope effect, determined from the product yields of the isotopically substituted elimination products indicates that H is lost 3.3-fold faster than D (after taking into account statistical factors). This result indicates that product forming and rate-limiting steps are *not* the same and that the mechanism has a minimum of two steps probably including the decomposition of (**7.2**) in a fast step. It is interesting to note that elimination from cumyl benzoates (see Scheme 7.10) follows a concerted pathway.

The 1,3-rearrangement of bicyclo[2.1.1]hexene-5-d and its phenyl derivatives (Scheme 7.7)[19,20] exhibits largely inversion of configuration at the bridgehead (5-C) carbon. Concerted mechanisms can be expected to exhibit complete stereospecificity but this does not exclude a diradical pathway

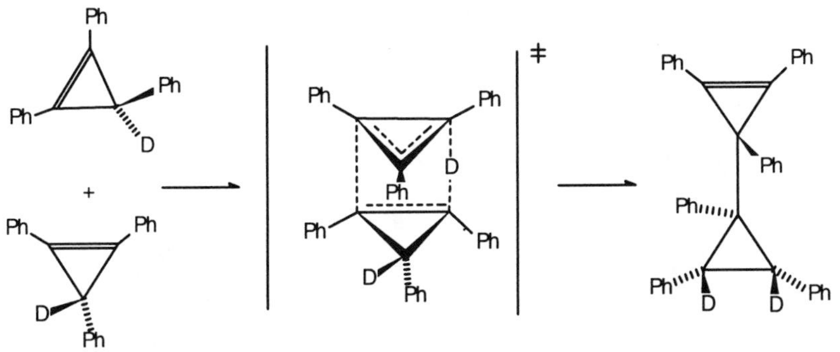

Scheme 7.5. A concerted ene reaction of triphenylcyclopropene.

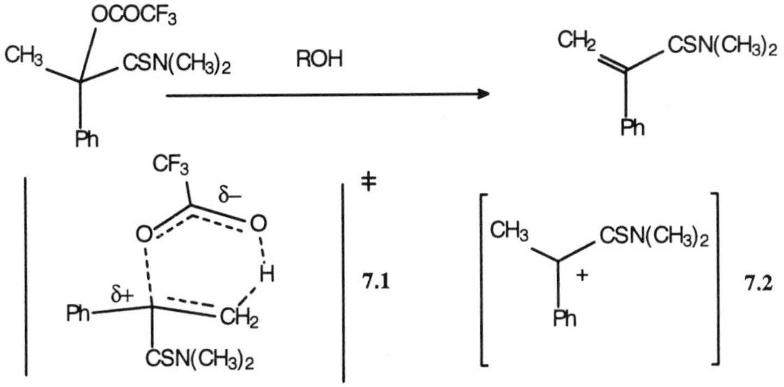

Scheme 7.6. The primary isotope effect demonstrates the existence of two steps in the elimination reaction.

which can retain stereochemical information.[21,22] The recovered reactants after partial reaction show *no* scrambling of the deuterium label but the product for the parent has increasing retention at 5-C as the temperature increases indicating parallel reaction pathways with different activation energies.[23] The phenyl derivatives have temperature *independent* stereospecificities; this is consistent with a single rate-limiting step to give an intermediate (likely to be a singlet diradical) which collapses to products with both possible configurations.

7.2   CLOCK REACTIONS
7.2.1 Conformational change
The stepwise mechanism for the acetylenic Cope rearrangement (Scheme 7.8) involves a diradical intermediate which can undergo a conformational change. The estimated rate constant for the conformational change was utilised by Owens and Berson[24,25] as a "clock" to estimate its decomposition rate constant to the Cope rearrangement products. The acetylenic Cope reaction was chosen as it was considered to be slower than reaction of the regular

Cope olefinic reactant so that the decomposition of the putative diradical

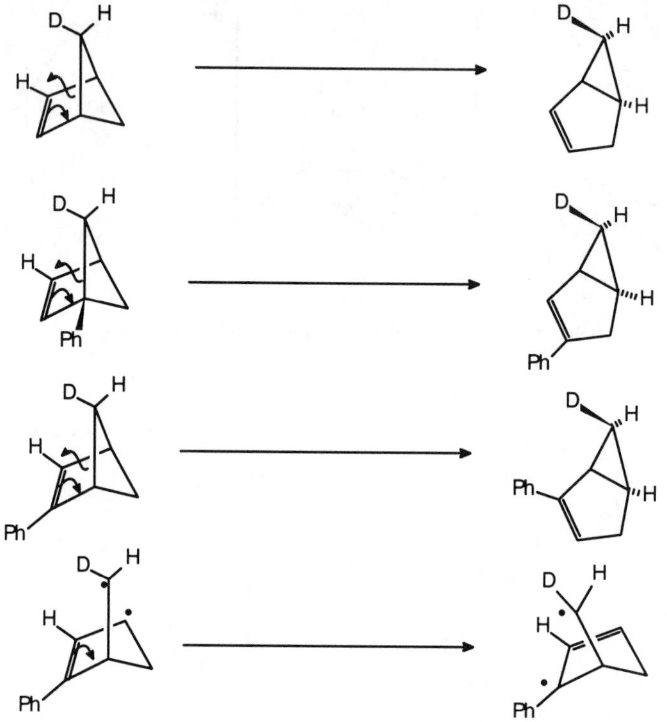

Scheme 7.7. Stereospecificity in rearrangements of bicyclo[2.1.1]hexenes.[12,13]

would give the "clocking" reaction the best opportunity to compete with diradical collapse. Methyl groups were employed as markers and the distribution of product stereochemistry indicated that an antarafacial pathway, orbital symmetry forbidden, cannot exceed 1-2% of the total reactant converted. A model for the conformational inversion of the diradical was taken as for 4-cyclohexenyl radical which, from the hyperfine coupling constants extracted from its ESR spectrum, has a rate constant for the conformational isomerization of $3 \times 10^{11} \mathrm{sec}^{-1}$ at 200°C. The stereochemical results of 94% retention in the reactant (R,E)-4-methyl-5-hepten-1-yne and the 90% (R,E)-4-methyl-1,2,5-heptatriene product coupled with the rate constant of the clock reaction give rate constants for collapse of the diradical of 1.9 and $1.2 \times 10^{13} \mathrm{sec}^{-1}$ respectively with lower limits at 1.1 and $0.83 \times 10^{13} \mathrm{sec}^{-1}$.[24,25] It may be concluded that the concerted mechanism operates for this reaction.

The putative diradical intermediate in the 1,3-dipolar cycloaddition of 4-nitrobenzonitrile oxide to cis- and trans-dideuterioethylene (Scheme 7.9) has a calculated barrier to rotational isomerization (0.1-0.4 kcal/mole); this is very fast relative to cyclization. The absence of stereochemical scrambling in the product (<2%) indicates that the barrier to rotation about the bond derived from the double bond (Scheme 7.9)[26] should be at least 2.3 kcal/mole higher than the barrier to cyclization; since the barrier to rotation is less than 0.4 kcal/mole the only reasonable explanation is that the 1,3-dipolar cycloaddition is concerted.

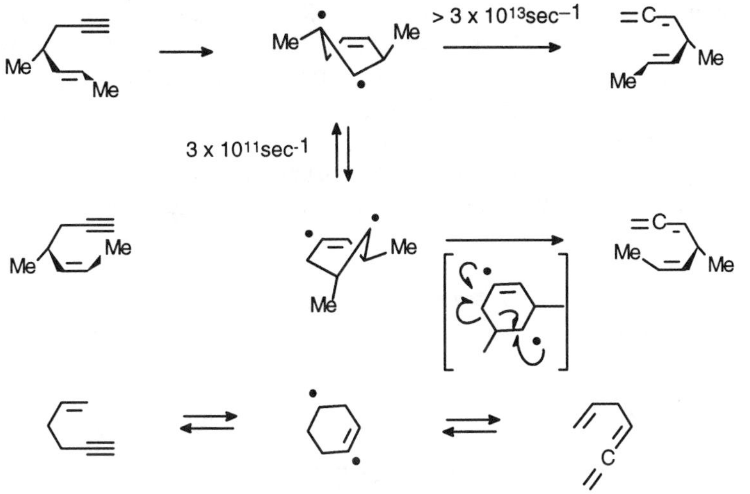

Scheme 7.8. Application of a clock reaction to exclude the stepwise mechanism of the thermal acetylenic Cope rearrangement.

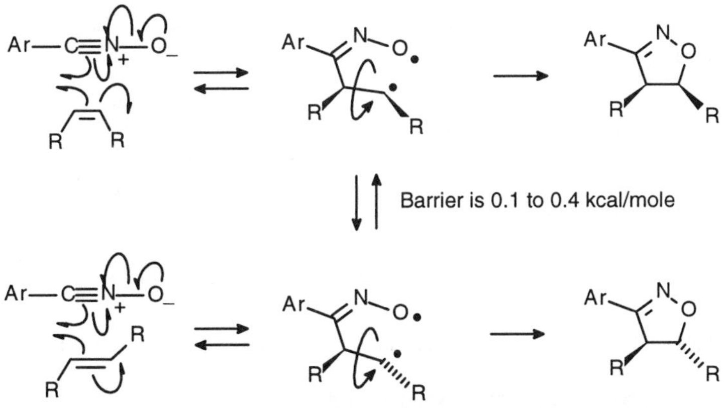

Scheme 7.9. The clock technique applied to exclude the stepwise 1,3-dipolar cycloaddition reaction.

7.2.2 Diffusion processes

Cumyl 4-nitrobenzoates $(ArC(CH_3)_2OCOC_6H_4\text{-}4NO_2)$ (Scheme 7.10) solvolyse in 50/50 v/v trifluoroethanol/water *via* a D_N+A_N mechanism through the carbocation which is captured by solvent at $10^7 - 10^9 sec^{-1}$. Increasing the destabilisation of the carbocation by electron-withdrawing substituents gives rise to increasingly substantial yields of α-methylstyrene *via* an elimination reaction. Addition of NaN_3 increases the yield of the azide adduct, RN_3, but provides no assistance to reaction of the substrate ($k_i = k_c$) and the yield of α-methylstyrene is independent of the $[N_3^-]$ concentration. Thus solvolysis and elimination proceed *via* separate paths. The yield of α-methylstyrene is independent of $[N_3^-]$ concentration and therefore requires that elimination also occurs from the $[N_3^-.RY]$ complex with a rate constant the same as that from RY alone ($k_e = k_e'$).[27] The rate constant for collapse of the ternary complex $[N_3^-.R^+.Y^-]$ to give RN_3 is estimated from that of the diffusion process to be $10^{19} sec^{-1}$; the ternary complex cannot therefore exist as an intermediate. If the ternary complex were to form it would give only RN_3. Since the triplet is too unstable to exist as an intermediate the formation of α-styrene via $[N_3^-.RY]$ *must* be through a concerted mechanism.

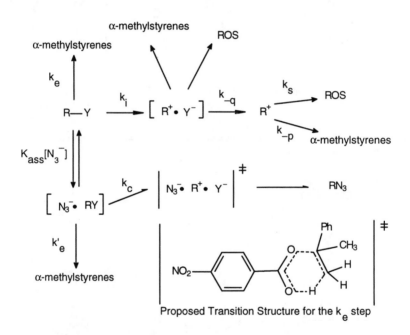

Scheme 7.10. Pericyclic eliminations of cumyl derivatives to give α-methylstyrenes.

7.3   ENERGETIC RELATIONSHIPS

Doering and his coworkers studied the Cope rearrangement of 1,1-dideuteriohexa-1,5-diene (Scheme 7.11)[28,29,30] a reaction analogous to those studied by Berson. The calculated enthalpies of formation given in Scheme 7.11 are determined from Benson's group equivalent method. The stepwise mechanism involving formation of a pair of allylic radicals can be excluded because these are estimated to be of such higher enthalpy than that observed for the transition state. The estimates of the heats of formation of the diradical, cyclohexan-1,4-diyl, have uncertainties which make it difficult to completely exclude as an intermediate. Better estimates of the heats of formation became available but the differentiation between the mechanisms did not improve.

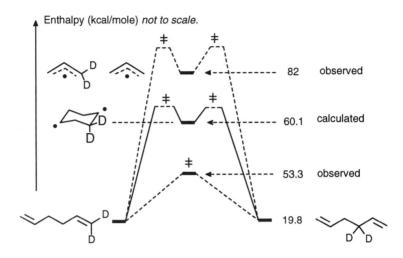

Scheme 7.11. Enthalpies of formation (kcal/mole) for the Cope rearrangement of hexa-1,5-diene through stepwise and concerted routes.

One of the problems referred to in Chapter 2 is that potential energy, even if it is reasonably well approximated by heats of formation, is not an absolute criterion between stepwise and concerted mechanism; the Gibbs' free energy is the appropriate quantity. In the case of the Cope rearrangement of 1,5-hexadienes (Scheme 7.12)[31,32] the Gibbs' free energy of the transition state of the rearrangement is about 12 kcal/mole less than that determined for the transition state for fission of the diradical, cyclohexan-1,4-diyl; this value was obtained experimentally from the kinetics of the rearrangement of bicyclo[2.1.0]hexane (see Scheme 7.12) to hexa-1,5-diene.[33-35] While the free energy of the diyl is not known experimentally the free energy of the transition state (53 kcal/mole at 523°K) is substantially higher than the experimental value for the hexan-1,5-diene rearrangement (41 kcal/mole) thus excluding the diyl as an intermediate.

7.4 NASCENT PRODUCTS

Bauer's proposal that the vibrational energy of nascent products could be employed as a criterion of concertedness[15] (Chapter 2) has been applied to the thermal decomposition of 2,3-diazobicyclo[2.1.1]hex-2-enes (**7.3**) by Chang, Jain and Dougherty (Scheme 7.13).[36] The experimental result is that the bath gas pressure (benzene) affects the yield of bicyclobutane versus cyclobutadiene because benzene acts as a single molecule, hard collision deactivator. Shen and Bergman studied the pyrolysis of pyrazolines[37] (Scheme 7.13) and found that the species **7.4** did not produce an activated hydrocarbon whereas from the species **7.5** the vibrationally activated energy distribution was in the hydrocarbon fragment. These results indicated that the nitrogen (N$_2$) is extruded in a symmetrical fashion from **7.4** (concerted). In the case of **7.5** the cleavage yields a diazenyl diradical leading to a vibrationally activated spiropentane.

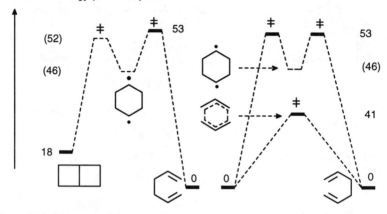

Gibbs' free energy (kcal/mole) at 523 K

Scheme 7.12. Direct use of the Gibbs' free energy of activation (kcal/mole) for diyl fission; numbers refer to experimental free energies; the values in parentheses are estimates.

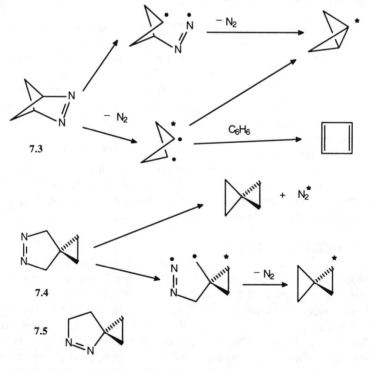

Scheme 7.13. Nascent products in the thermal decomposition of diazabicyclo[2.1.1]hexenes (**7.3**) and pyrazolines (**7.4** and **7.5**); star indicates a vibrationally activated species.

7.5   ISOTOPE EFFECTS

Primary kinetic isotope effects can indicate if a bond is breaking in a rate-limiting step and since more than one bond is involved in a concerted process, diagnosis requires that multiple isotope effects should be examined. Early work on cyclic decarboxylation reactions involved ^{13}C and ^{14}C replacement of the C2 and C1 atoms in malonic acid (**7.6**, Scheme 7.14).[38,39] The heavy atom isotope effects indicate C-C bond fission in the transition structure consistent with, but not required by, a concerted mechanism (**7.7**).

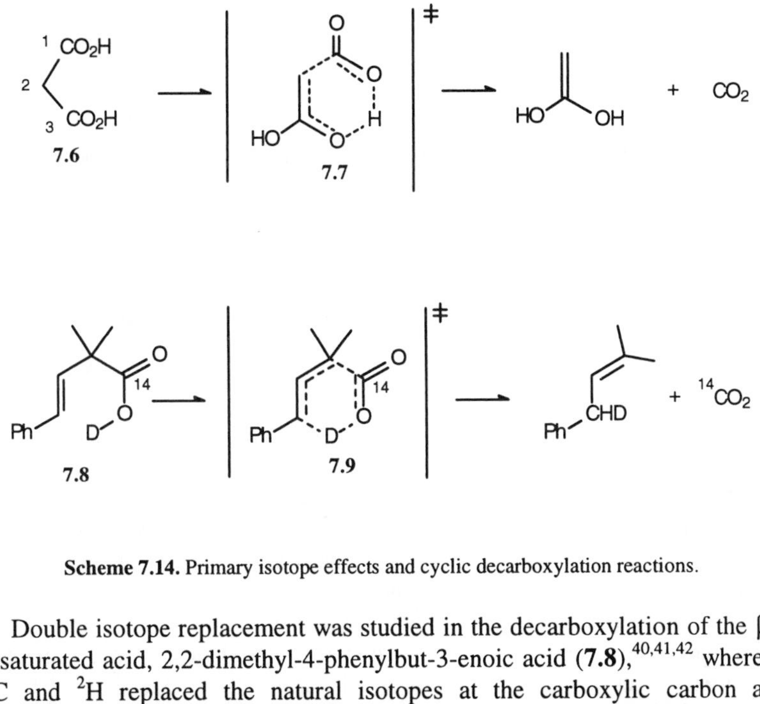

Scheme 7.14. Primary isotope effects and cyclic decarboxylation reactions.

Double isotope replacement was studied in the decarboxylation of the $\beta\gamma$-unsaturated acid, 2,2-dimethyl-4-phenylbut-3-enoic acid (**7.8**),[40,41,42] whereby ^{14}C and ^2H replaced the natural isotopes at the carboxylic carbon and hydrogen. Primary kinetic isotope effects ($^{12}k/^{14}k = 1.035$ and $^1k/^2k = 2.87$) are consistent with a concerted cyclic mechanism as indicated in (**7.9**). The isotope effect data for the decarboxylation reactions can be taken in conjunction with the effect of α, β, and γ substituents on the rate constants indicating that substantial bonding changes take place at these sites in the transition state of the rate-limiting step.

The Tschugaeff reaction of the xanthate ester (Scheme 7.15) can involve transfer of the β-hydrogen to either the sulphur of the thioether or the thion sulphur.[43] The thion sulphur suffers a large bonding change (pathway *via* **7.10**) as evidenced by a large primary isotope effect whereas insignificant isotope effects caused by isotopic substitution at the thioether sulphur and the carbonyl carbon indicate very little bonding change to these atoms. These results are fully consistent with pathway (**7.10**) and exclude the path involving a transition structure with endocyclic thioether sulphur (**7.11**).

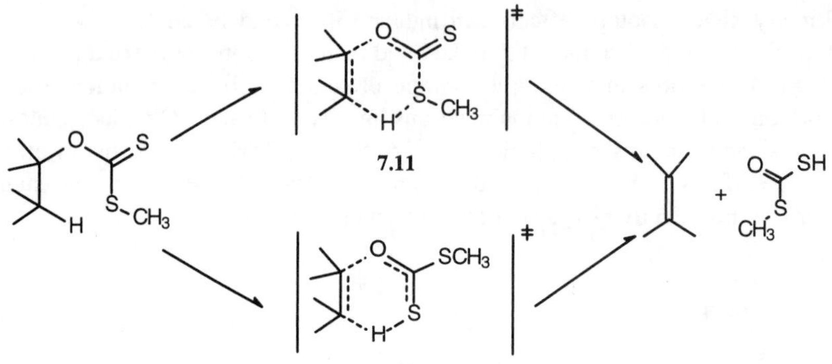

Scheme 7.15. Hydrogen transfers to the thion sulphur in the Tschugaeff reaction.

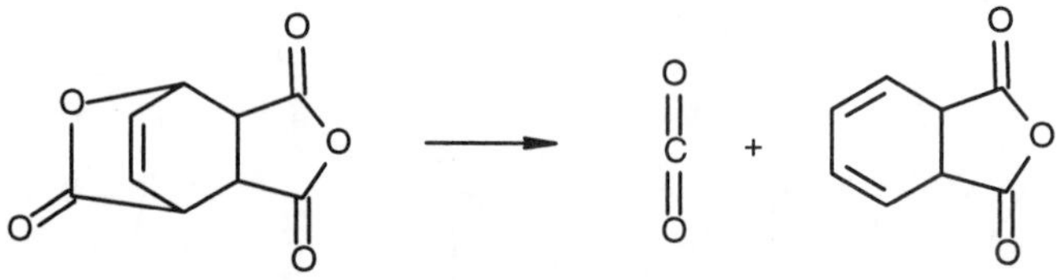

Scheme 7.16. Concerted non-synchronous *retro*-Diels-Alder decarboxylation.

The primary heavy atom isotope effect technique has been applied to the *retro*-Diels-Alder reaction (Scheme 7.16).[44] The primary isotope effects for the bridge oxygen ($^{16}k^{18}/k = 1.014$) and the bridge carbon ($^{12}k^{13}/k = 1.013$) indicate that both C-O and C-C bonds are breaking in the transition state. Heavy atom primary isotope effects are direct measures of bond orders in a bond undergoing change in the transition state[45] and in this case indicate a concerted process with C-C fission well advanced over C-O fission.

The *retro*-Diels-Alder reaction involving the thermal fragmentation of the maleic anhydride 1-methylfuran adduct[46] exhibits secondary deuterium kinetic isotope effects for the atoms shown in Scheme 7.17 of $k_{7.12}/k_{7.13} = 1.16$ and $k_{7.12}/k_{7.14} = 1.08$. The value of $k_{7.12}/k_{7.14}$ indicates that the bond *b* is breaking in a rate-limiting step. If the isotope effect of $k_{7.12}/k_{7.13}$ is due to deuteration at R_1 (ie *b* breaks in the rate-limiting step and *a* breaks in a fast step) then, after a substantial progression of reaction, the remaining reactant, starting from **7.12**, would be richer in **7.15** than in **7.16**. Neither **7.15** nor **7.16** become enriched, indicating that the depletion rates are the same and that the isotopic effect for fission of *b* is the same as that for *a*. Thus the isotopic effect of 1.16 is the sum of equal contributions from *a* and *b* and these bonds therefore break simultaneously in the transition state of the rate-limiting step.

Secondary α-deuterium isotope effects have been determined for the symmetrical *retro*-Diels-Alder reaction to give anthracene and ethylene (Scheme 7.18)[47] consistent with a concerted mechanism.

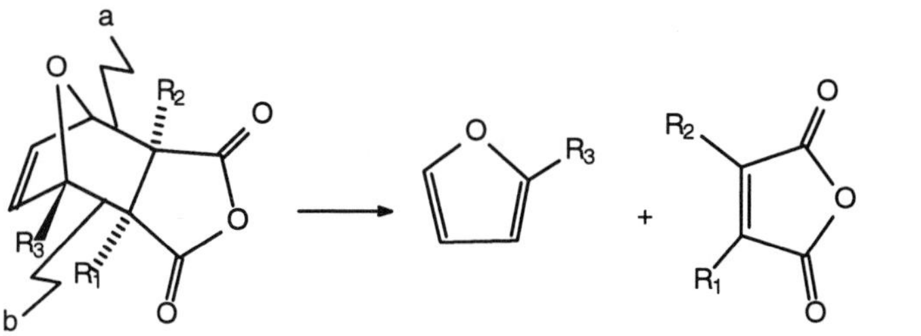

7.12 - 7.16

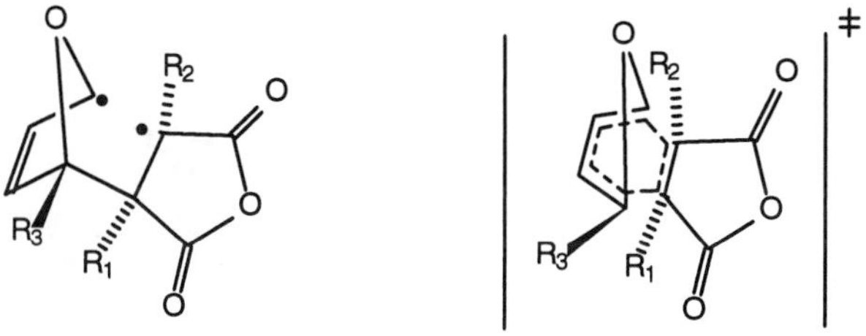

7.12 7.13 7.14 7.15 7.16

	7.12	7.13	7.14	7.15	7.16
R_1	H	D	H	D	H
R_2	H	D	H	H	D
R_3	H	H	D	H	H

Scheme 7.17. Secondary α-deuterium isotope effect on the *retro*-Diels-Alder reaction.

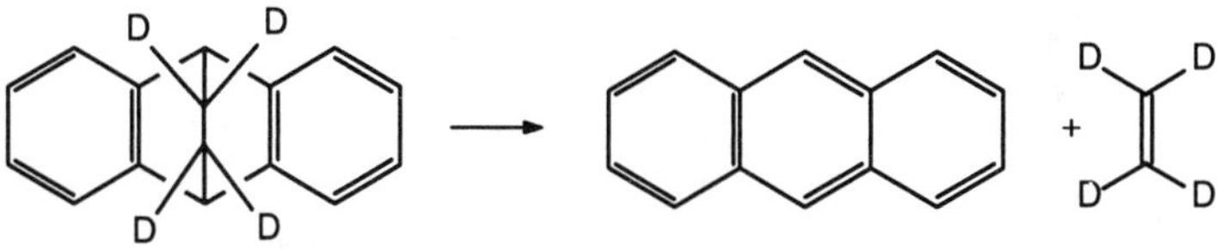

Scheme 7.18. A symmetrical *retro*-Diels-Alder reaction.

The temperature dependence of the primary isotope effect[48-50] was studied for the ene reaction between diethyl mesoxalate (**7.17**) and 3-phenylpropene (**7.18**) (Scheme 7.19). The primary hydrogen isotope effect of 2.56 indicates that the H-C and O-H bonds are being broken and formed respectively in the transition structure and the independence of the isotope effect on temperature was thought to be consistent with a "bent" structure for the hydrogen "in-flight" as indicated in Scheme 7.19 (**7.19** and **7.20**). The observation of inverse α-deuterium secondary isotope effects at both ends of the double bond (Scheme 7.19) indicates that both carbons are involved in bonding changes in the transition structure as in a (2+2) cyclic system (**7.19**). In allylic hydrogen abstractions where it is known that there is a bridged radical mechanism the inverse secondary deuterium isotope effects are different for both ends. There has been criticism of the application of the influence of temperature on kinetic isotope effects as a criterion of the geometry of the hydrogen transfer transition structure.[51] The main conclusion that the mechanism is cyclic still holds but it is certainly a moot point as to whether the hydrogen transfers in a straight line or *via* a "bent" geometry.

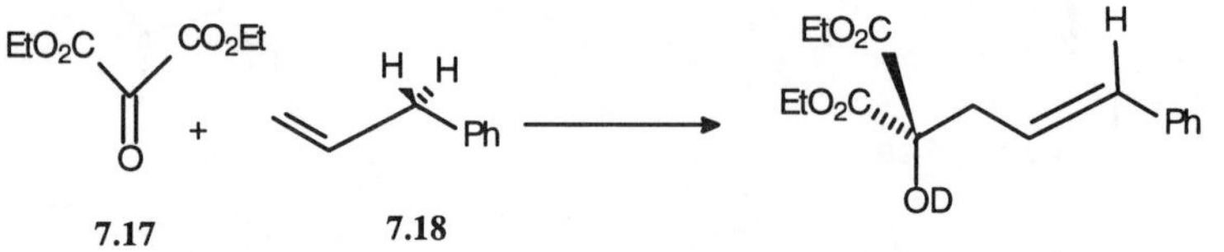

Scheme 7.19. Non-linear hydrogen transfer in an ene reaction.

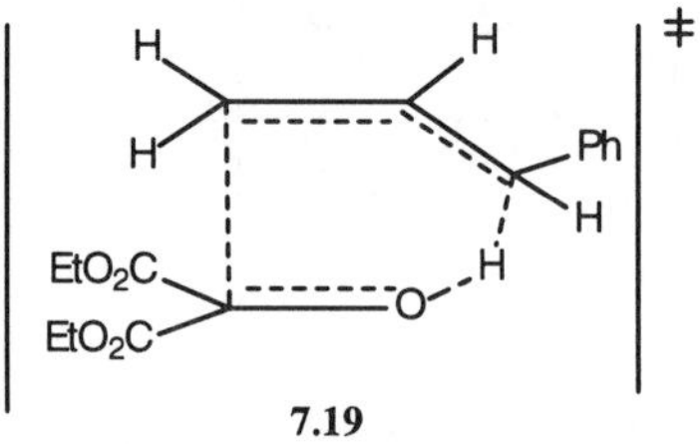

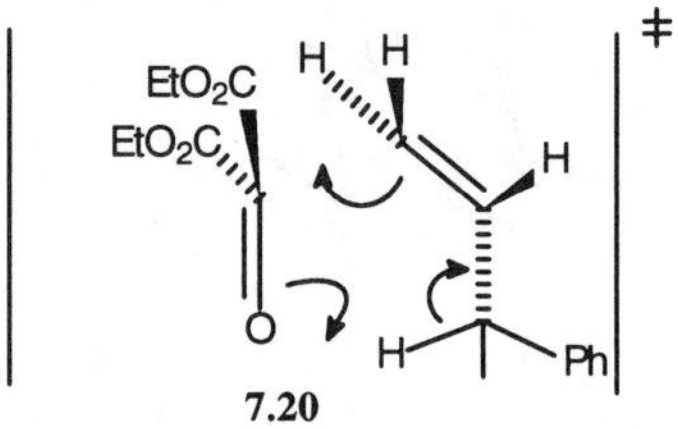

Multiple isotopically labelled substrates can in principle give transition state structures by comparison of the observed effects with a matrix of calculated effects where the isotopically sensitive bonds are varied within a range of strengths. The calculations are usually based on BEBO programs (Chapter 2) with force constants derived from spectroscopic studies or from *ab initio* calculations and standard single bond lengths adjusted according to bond orders of the particular model. The models follow the Wolfsberg-Stern "cut-off" procedure (Chapter 2) which omits calculations of vibrations associated with atoms more than two bonds from the isotopically substituted sites. The Claisen rearrangement of phenyl allyl ether was studied by Subotskowska, Saunders and Shine[52,53](Scheme 7.20). The isotope effects are calculated for models assuming a variation in C-O bond fission and in C-C bond formation from 10 to 90%. Not all the elements of the matrix are calculated but good agreement between the experimental and calculated isotope effects when the C-C bond formation and C-O bond fission were in the ranges 10 - 20% and 40 - 50% respectively for the Claisen rearrangement of the phenyl ether (Scheme 7.20).[53] Table 7.1 summarises the calculated isotope effects for the part of the bond order matrix which fits the data. The results qualitatively indicate that bond formation and bond fission have substantial heavy atom primary isotope effects consistent with their both occurring in the transition state of the rate-limiting step. The assumptions inherent in comparisons of isotope effects with values calculated for particular models are discussed in Chapter 2.

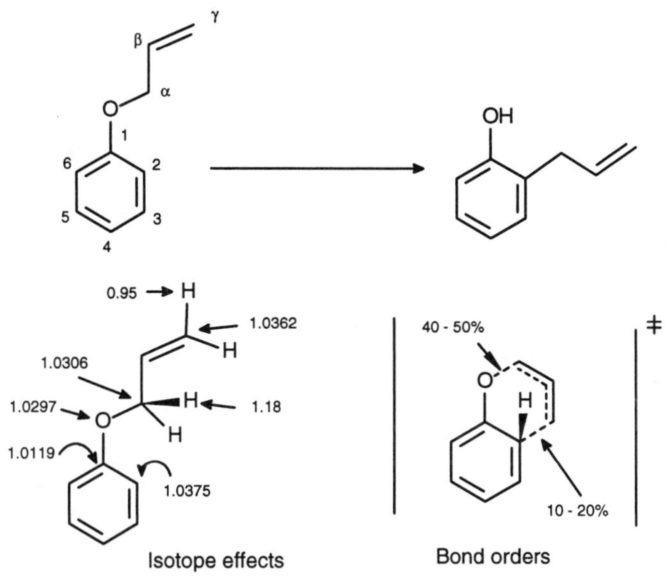

Scheme 7.20. Bond orders (as percentage) and observed isotope effects (see Table 7.1) for the Claisen rearrangement of phenyl allyl ether; Data are from *Refs. 52* and *53*.

The observed values of the C-O and C-C bond orders sum to less than unity and this is consistent with substantial bonding changes elsewhere in the transition structure presumably at the allyl trio of carbon atoms.

Table 7.1.
Grid of isotope effects predicted for various models of the phenylallyl ether Claisen rearrangement.[c, d, e]

C-O /%[b]	C-C /%[b]	1,8O	α-H$_2$	α-C	β-C	γ-C	γ-D$_2$	2-C	1-C	Dev$_{Heavy}$	Dev$_{H/D}$
60	20	1.0365	1.1306	1.0306	1.0200	1.0330	0.9159	1.0243	1.0154	0.0071	0.0552
50	10	*1.0345*	*1.1620*	*1.0321*	*1.0169*	*1.0394*	*0.9658*	*1.0294*	*1.0137*	0.0045	0.0226
exp		*1.0297*[a]	*1.18*[a]	*1.0306*[a]	*1.0148*[a]	*1.0362*[a]	*0.95*[a]	*1.0375*[a]	*1.0119*[a]		
40	20	*1.0323*	*1.1949*	*1.0036*	*1.0194*	*1.0410*	*0.9678*	*1.0320*	*1.0154*	0.0044	0.0226
60	10	1.0365	1.1306	1.0306	1.0171	1.0354	0.9397	1.0216	1.0138	0.0061	0.0432
50	20	1.0345	1.1620	1.0321	1.0197	1.0370	0.9412	1.0281	1.0154	0.0053	0.0178
40	10	1.0323	1.1949	1.0336	1.0168	1.0436	0.9931	1.0332	1.0138	0.0043	0.0471

a) Experimental values; b) Bond orders; c) Data are from Ref. 53; d) Identification of isotopically substituted atoms given in Scheme 7.20; e) See also Table 2.2.

Bond fission and bond formation in the aliphatic Claisen rearrangement, at C4-O and C6-C1 respectively, were monitored by α-deuterium secondary isotope effects[54,55] (Scheme 7.21). The comparison of kinetic isotope effect and the equilibrium isotope effect indicated bond fission (C4-O) of about 70% and bond formation (C6-C1) of about 16%.

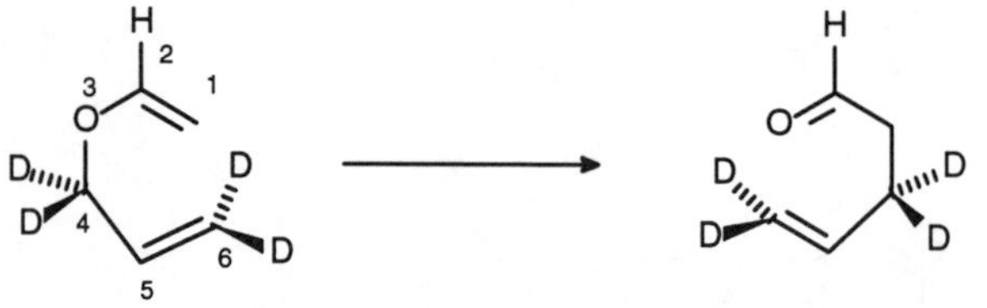

Scheme 7.21. Secondary α-deuterium isotope effect studies on the Aliphatic Claisen rearrangement.

Table 7.2.
Matrix of isotope effects calculated for the model of the Claisen rearrangement of vinyl allyl ether.[c, d, e]

C-O $I\%^b$	C-C $I\%^b$	^{18}O	4-D$_2$	4^{14}C	6^{14}C	6-D$_2$	2^{14}C	Dev$_{Heavy}$	Dev$_{H/D}$
60	20	1.0468	1.0785	1.0620	1.0202	0.9907	1.0267	0.0060	0.0194
40	20	1.0448	1.1105	1.0693	1.0192	0.9896	1.0258	0.0037	0.0216
exp		1.0506^a	1.092^a	1.072^a	1.0178^a	0.976^a	1.0271^a		
50	20	1.0459	1.1128	1.0674	1.0197	0.9902	1.0261	0.0038	0.0236
90	10	1.0399	1.0049	1.0489	1.0268	1.0791	1.0117	0.0173	0.1290
10	90	1.0322	1.2730	1.0575	1.0096	0.8668	1.0220	0.0141	0.1900
50	50	1.0391	1.1315	1.0580	1.0155	0.9317	1.0155	0.0120	0.0590

a) Experimental values; b) Bond orders; c) Data are from Ref 56; d) Identification of substituted atoms given in Scheme 7.21; e) See also Table 2.1.

Multiple kinetic isotope effects including heavy atom substitution were also studied for the aliphatic Claisen rearrangement[56,57] and Table 7.2 indicates part of the matrix of isotope effects calculated for models similar to those employed for the Claisen rearrangement of phenyl allyl ether. There is broad agreement with the results for the approach utilising simply the α-deuterium secondary isotope technique with the experimental isotope effects fitting those for the model with C-O bond fission advanced to 40-50% and C-C bond formation only 20% advanced.

7.6   THEORETICAL METHODS

Cyclic reactions have been the subject of many theoretical calculations with increasing levels of sophistication to predict transition structures (Chapter 2). In the case of the parent Diels-Alder reaction of ethylene with butadiene the results have progressed from CNDO/2 calculations which indicate the absence of an energy barrier to those which argue for diradical intermediates, concerted synchronous and concerted asynchronous pathways. The answers depend on the various approximations made to enable the calculations to be completed in a reasonable time.

The application of *ab initio* methods to cyclic reactions, even though they involve approximations, has centred on the prediction of transition structures which can then be utilised for calculating kinetic isotope effects for multiple isotopic substitution. Increasingly, the theoretical methods have been used to predict enthalpies of activation which can be used to calculate Gibbs' free energies. Although the comparisons of theory and experiment are usually made to test the validity of the theories, nevertheless the calculation of transition structures and verification by comparison with physical properties are important in addressing the question of the relative timing of bonding changes and also whether the reaction is concerted or not.

The aliphatic Claisen rearrangement of vinyl allyl ether (Scheme 7.21) was studied by calculating the matrix of isotope effects for the putative models using *restricted Hartree-Fock* (RHF)[58,59] and *complete active space self consistent field* (CASSCF)[58,59] methods* and the 6-31G* basis set.[60] Transition state structures derived from the *ab initio* calculations were employed together with the "QUIVER" program (see Chapter 2) to yield theoretical isotope effects to compare with the experimental values (Table 7.3). The CASSCF calculation for the Claisen rearrangement gives a sum

Table 7.3.
Matrix of isotope effects predicted for the model of the Claisen rearrangement of vinyl allyl ether from *ab initio* techniques.[b]

	^{18}O	$4-H_2$	$4-^{14}C$	$6-^{14}C$	$6-D_2$	$2-^{14}C$
	1.0506^a	1.092^a	1.072^a	1.0178^a	0.976^a	1.0271^a
RHF[c]	1.032	1.051	1.053	1.028	0.932	1.005
$\Delta/\%^e$	1.9	4.1	1.9	1	4.4	2.21
CASSCF[d]	1.043	1.186	1.069	1.015	0.964	1.011
$\Delta/\%^e$	0.8	9.4	0.3	0.3	1.2	1.61

a) Experimental values; b)Data are from Ref. 60; c) Values calculated with RHF; d) Values calculated with CASSF; e) Percentage difference between predicted and experimental values.

of the differences (Δ) of 4.21% compared with 11.41% for the RHF calculation if the value of the isotope effect for the $4-H_2$ position is omitted. The experimental enthalpy of activation (30.6 kcal/mole) is closer to the value calculated by CASSCF (42.5 kcal/mole) compared with that calculated using the RHF protocol (48.8 kcal/mole). The results as calculated by the CASSCF procedure indicate that C4-O bond fission is more advanced than C1-C6 bond formation in the transition structure and fission of the C-O bond extends to 70% and C-C bond formation is to the extent of 20% and are in agreement with those obtained by the "bracketting" technique using the BEBO calculations on the more primitive model.

Kinetic isotope effects for the Diels-Alder reaction provide strong evidence for synchroneity in agreement with the structures derived from *ab initio* theory[61-64]. The reaction of isoprene with maleic anhydride was studied in benzene solution and the experimental value of $\Delta H^{\ddagger}$ of 11.8 kcal/mole is about 4.1 kcal/mole less than that calculated by the *ab initio* method for the gas phase using a hybrid Hartree-Fock density function method (DFT). The

---

* The reader is referred to *Refs. 54* and *55* for detailed explanations of the acronyms used in theoretical papers.

RHF calculation gives 37.6 kcal/mole for $\Delta H^{\ddagger}$. The experimental isotope effects for ^{13}C and secondary deuterium isotope effects are proportional to the values (slope = 1) calculated from the DFT method. Both RHF and DFT calculations predict the *endo* structure and a small amount of asynchronous character. The DFT theory gives a bond order of 32% for the bond forming between maleic anhydride and the 1-C atom of isoprene and 29% for the bond forming to the 4-C atom.

The comparison of calculated rate constants with experimental values has been applied to the disproportionation reaction whereby hydrogen is transferred from diazene (N_2H_2) to another diazene molecule.[65-67] *Ab initio* calculations indicate that the free energy of activation ($\Delta G^{\ddagger}$) is dominated by entropy contributions. This indicates a potential energy surface which has almost no barrier and the optimised transition structure represents concerted bond formation and bond fission to the transferring hydrogens.[67] The disproportionation of diazene to nitrogen and hydrazine has a gas-phase rate constant equivalent to 9.7 kcal/mole of free energy[65] which compares with an *ab initio* prediction of 9.4 kcal/mole at 298 K.[67] Similar agreement is observed between *ab initio* results (for a concerted mechanism) and experimental values for the reaction of diazene with olefines. A stepwise process for the disproportionation reaction involving an initial hydrogen transfer step ($2trans N_2H_2 \rightarrow N_2H + N_2H_3$) may be excluded because it possesses a ΔH° of 26 kcal/mole well in excess of the activation enthalpy.[68]

FURTHER READING

Baldwin, J. E., Freedman, T. B., Yamaguchi, Y., and Schaefer, H. F., Secondary deuterium isotope effects on the isomerization of the trimethylene diradical to cyclopropane, *J. Amer. Chem. Soc., 118,* 10934, 1996.

Baldwin, J. E., and Reddy, P., Primary deuterium kinetic isotope effects for the thermal [1,7] sigmatropic rearrangement of 7-methylocta-1,3(Z),5(Z)-triene, *J. Amer. Chem. Soc., 110,* 8223, 1988.

Carpenter, B. K., Bimodal distribution of lifetimes for an intermediate from a quasi-classical dynamics simulation, *J. Amer. Chem. Soc., 118,* 10329, 1996.

Chen, J. S., Houk, K. N., and Foote, C. S., The nature of the transition structures of triazolinedione ene reactions, *J. Amer. Chem. Soc., 119,* 9852, 1997.

Dewar, M. J. S., and Wade, L. E., A study of the mechanism of the Cope rearrangement, *J. Amer. Chem. Soc., 99,* 4417, 1977.

Dewar, M. J. S., and Wade, L. E., The possible role of 1,4-cyclohexylene intermediates in Cope rearrangements, *J. Amer. Chem. Soc., 95,* 290, 1973.

Dewar, M. J. S., and Ford, G. P., Thermal decarboxylation of but-3-enoic acid. MINDO/3 calculations of activation parameters and primary kinetic isotope effects, *J. Amer. Chem. Soc., 99,* 8343, 1977.

Dunn, G. E., Carbon-13 kinetic isotope effects in decarboxylation, in *Isotopes in Organic Chemistry,* Buncel, E. and Lee, C. C., eds., Elsevier, Amsterdam, 1977, Chapter 1.

Engel, P. S., Nalepa, C. J., Horsey, D. W., Keys, D. E., and Grow, R. T., Thermolysis of azoalkanes containing the 2,3-diazabicyclo[2.2.2]oct-2-ene (DBO) skeleton, *J. Amer. Chem. Soc., 105,* 7102, 1983.

Firestone, R. A., Stereospecificity in Diels-Alder and 1,3-dipolar cyclo-additions does not prove the concerted mechanism, *Heterocycles, 25,* 61, 1987.

Ganem, B., The mechanism of the Claisen condensation: déjà vu all over again, *Angew. Chem. Int. Ed. Engl., 35,* 936, 1996.

Getty, S. J., and Berson, J. A., Stereochemistry and mechanism of the reverse ene reaction of *cis*-2-alkyl-1-alkenylcyclobutanes. Stereoelectronic control in a system showing marginal energetic benefit of concert, *J. Amer. Chem. Soc., 113,* 4607, 1991.

Houk, K. N., Gonzalez, J., and Li, Y., Pericyclic reaction transition states: passions and punctilios, 1935 - 1995, *Acc. Chem. Res., 28,* 81, 1995.

Houk, K. N., Li, Y., and Evanseck, J. D., Transition structures of hydrocarbon pericyclic reactions, *Angew. Chem. Int. Ed. Engl., 31,* 682, 1992.

Houk, K. N., The frontier molecular orbital theory of cycloaddition reactions, *Accs. Chem. Res., 8,* 361, 1975.

Houk, K. N., Lin, Y-T., and Brown, F. K., Evidence for the concerted mechanism of the Diels-Alder reaction of butadiene with ethylene, *J. Amer. Chem. Soc., 108,* 554, 1986.

Rhoads, S. J., Rearrangements proceeding through "no mechanism" pathways: the Claisen, Cope, and related rearrangements, in *Molecular Rearrangements,* de Mayo, P., ed., Interscience, New York, 1963, Chapter 11.

Roth, W. R., and Unger, C., Concerted and non-concerted reaction paths: thermolysis of dispiro[2.2.2.2]deca-4,9-diene, *Liebig's Ann. Chem.,* 1361, 1995.

Singleton, D. A., Merrigan, S. R., Liu, J., and Houk, K. N., Experimental geometry of the epoxidation transition state, *J. Amer. Chem. Soc., 119,* 3385, 1997.

Skancke, P. N., How can quantum chemical calculations contribute to the elucidation of chemical reactions? Some theory and examples, *Acta Chem. Scand., 47,* 629, 1993.

Townshend, R. E., Ramunni, G., Segal, G., Hehre, W. J., and Salem, L., Organic transition states. V. The Diels-Alder reaction, *J. Amer. Chem. Soc., 98,* 2190, 1976.

Vance, R. L., Rondon, N. G., Houk, K. N., Jensen, H. F., Borden, W. K., Kormornicki, A., and Wimmer, E., Transition structures for the Claisen rearrangement, *J. Amer. Chem. Soc., 110,* 2314, 1988.

Wessel, T. E., and Berson, J., Stabilization of a putative cyclohexane-1,4-diyl intermediate elicits an antarafacial Cope rearrangement via a stepwise mechanism. Pyrolysis of (*R,E*)-5-methyl-1,2,6-octatriene to 4-methyl-3-methylene-1,5-heptadiene, *J. Amer. Chem. Soc., 116,* 495, 1994.

Wiest, O., Houk, K. N., Black, K. A., and Thomas, B., Secondary kinetic isotope effects of diastereotopic protons in pericyclic reactions: a new mechanistic probe, *J. Amer. Chem. Soc., 117,* 8594, 1995.

Willi, A. V., Kinetic and other isotope effects in cleavage and formation of bonds to carbon, in *Isotopes in Organic Chemistry,* Buncel, E. and Lee, C. C., eds., Elsevier, Amsterdam, 1977, Chapter 5.

Woodward, R. B., and Hoffmann, R., *The Conservation of Orbital Symmetry,* Academic Press, New York, 1970.

REFERENCES

1. Dell, C. P., Cycloaddition in synthesis, *J. Chem. Soc. Perkin Trans. 1,* 3873, 1998.

2. Mann, J., *Chemical Aspects of Biosynthesis,* Oxford University Press, Oxford, 1994.

3. Rhoads, S. J., Rearrangements proceeding through "no mechanism" pathways: the Claisen, Cope, and related rearrangements, in *Molecular Rearrangements,* de Mayo, P., ed., Interscience, New York, 1963, Chapter 11.

4. Rhoads, S. J., and Raulins, N. R., Claisen and Cope rearrangements, *Organic Reactions, 22,* 1, 1975.

5. Woodward, R. B., and Hoffmann, R., *The Conservation of Orbital Symmetry,* Academic Press, New York, 1970.

6. Baldwin, J. E., Andrist, A. H., and Pinschmidt, R. K., Orbital-symmetry-disallowed energetically concerted reactions, *Accs. Chem. Res., 5,* 402, 1972.

7. Kassel, L. S., *Kinetics of Homogeneous Gas Reactions,* Reinhold Publishing Corporation, New York, 1932, p. 148.

8. Pearson, R. G., Symmetry rules for chemical reactions, *Accs. Chem. Res., 4,* 152, 1971.

9. Benson, S. W., and Srinivasan, R., Some complexities in the reaction system: $H_2 + I_2 = 2HI$, *J. Chem. Phys., 23,* 200, 1955.

10. **Sullivan, J. H.,** Rates of reaction of hydrogen with iodine, *J. Chem. Phys., 30,* 1292, 1959.

11. **Sullivan, J. H.,** Mechanism of the "bimolecular" hydrogen-iodine reaction, *J. Chem. Phys., 46,* 73, 1967.

12. **Perkins, M. J.,** *Radical Chemistry,* Ellis Horwood, New York and London, 1994.

13. **Fossey, J., Lefort, D., and Sorba, J.,** *Free Radicals in Organic Chemistry,* John Wiley & Sons, Chichester and New York, 1995.

14. **Horn, B. A., Herek, J. L., and Zewail, A. H.,** *Retro*-Diels-Alder femtosecond reaction dynamics, *J. Amer. Chem. Soc., 118,* 8755, 1996.

15. **Bauer, S. H.,** Operational criteria for concerted bond breaking in gas-phase molecular elimination reactions, *J. Amer. Chem. Soc., 91,* 3688, 1969.

16. **Sustmann, R., Rogge, M., Nüchter, U., and Harvey, J.,** A stopped-flow kinetic study of the reaction of *trans*-1-(dimethylamino)-1,3-butadiene with dimethyl dicyanofumarate, *Chemische Berichte, 125,* 1665, 1992.

17. **Breslow, R., and Dowd, P.,** The dimerization of triphenylcyclopropene, *J. Amer. Chem. Soc., 85,* 2729, 1963.

18. **Creary, X., Casingal, V. P., and Leahy, C. E.,** Solvolytic elimination reactions. Stepwise or concerted? *J. Amer. Chem. Soc., 115,* 1734, 1993.

19. **Carpenter, B. H.,** Anatomy of a bond cleavage. The dynamics of rearrangement of a strained-ring hydrocarbon, *J. Org. Chem., 57,* 4645, 1992.

20. **Carpenter, B. H.,** Intramolecular dynamics for the organic chemist, *Accs. Chem. Res., 25,* 520, 1992.

21. **Borden, W. T.,** Diradicals as reactive intermediates, in *Reactive Intermediates,* Jones, M., Jr., and Moss, R. A., eds., Wiley & Sons, New York, Vol. 2, 1981, p. 175.

22. **Berson, J. A.,** Hypothetical biradical pathways in thermal unimolecular rearrangements, in *Rearrangements in Ground and Excited States,* de Mayo, P., ed., Academic Press, New York, 1980, p. 311.

23. **Newman-Evans, R. H., and Carpenter, B. K.,** Temperature dependence of selectivity as a criterion for mechanism. Rearrangement of bicyclo[2.1.1]hexene-5-d and two phenyl derivatives, *J. Amer. Chem. Soc., 106,* 7794, 1984.

24. **Owens, K. A., and Berson, J. A.,** Exclusion of a 1,4-cyclohexenediyl as a metastable intermediate in the [3,3] sigmatropic rearrangement of a 1-hexen-5-yne, *J. Amer. Chem. Soc., 110,* 627, 1988.

25. **Owens, K. A., and Berson, J. A.,** Stereochemistry of the thermal acetylenic Cope rearrangement. Experimental test for a 1,4-cyclohexenediyl as a mechanistic intermediate, *J. Amer. Chem. Soc., 112,* 5973, 1990.

26. **Houk, K. N., Firestone, R. A., Munchausen, L. L., Mueller, P. H., Arison, B. H., and Garcia, L. A.,** Stereospecificity of 1,3-dipolar cycloadditions of *p*-nitrobenzonitrile oxide to *cis*- and *trans*-dideuterioethylene, *J. Amer. Chem. Soc., 107,* 7227, 1985.

27. **Amyes, T. L., and Richard, J. P.,** How do reaction mechanisms change? Appearance of concerted pericyclic eliminations for the reaction of cumyl derivatives, *J. Amer. Chem. Soc., 113,* 8960, 1991.

28. **Doering, W. von E., Toscanos, V. G., and Beasley, G. H.,** Kinetics of the Cope rearrangement of 1,1-dideuteriohexa-1,5-diene, *Tetrahedron, 27,* 5299, 1971.

29. **Doering, W. von E.,** Mechanism in the system of dimers of butadiene, in *23rd International Congress of Pure and Applied Chemistry (Butterworths, London), 1,* 237, 1971.

30. **Doering, W. von E.,** Impact of upwardly revised ΔH^o_f of primary, secondary, and tertiary radicals on mechanistic constructs in thermal reorganisations, *Proc. Nat. Acad. Sci. U. S. A., 78,* 5279, 1981.

31. **Gajewski, J. J., and Conrad, N. D.,** On the mechanism of the Cope rearrangement, *J. Amer. Chem. Soc., 100,* 6268, 1978.

32. **Gajewski, J. J., and Conrad, N. D.,** Variable transition state structure in the Cope rearrangement as deduced from secondary deuterium isotope effects, *J. Amer. Chem. Soc., 100,* 6270, 1978.

33. **Goldstein, M. J., and Benzon, M. S.,** Inversion and stereospecific cleavage of bicyclo[2.2.0]hexane, *J. Amer. Chem. Soc., 94,* 5119, 1972.

34. **Goldstein, M. J., and Benzon, M. S.,** Boat and chair transition state configuration of 1,5-

hexadiene, *J. Amer. Chem. Soc., 94,* 7147, 1972.

35. **Goldstein, M. J., and Benzon, M. S.,** The linear analysis of labeling experiments. An application to sigmatropic rearrangements of 1,5-hexadiene, *J. Amer. Chem. Soc., 94,* 7149, 1972.

36. **Chang, M. H., Jain, R., and Dougherty, D. A.,** Chemical activation as a probe of reaction mechanism. Synthesis and thermal decomposition of 2,3-diazabicyclo[2.1.1]hex-2-enes, *J. Amer. Chem. Soc., 106,* 4211, 1984.

37. **Shen, K. K., and Bergman, R. G.,** An azo compound route to spiropentene thermolysis intermediates. Formation of vibrationally excited organic molecules in the thermal decomposition of pyrazolines, and evidence concerning the distribution of excess energy in reacytion products, *J. Amer. Chem. Soc., 99,* 1655, 1977.

38. **Bigeleisen, J., and Friedman, L.,** C^{13} Isotope effect in the decarboxylation of malonic acid, *J. Chem. Phys., 17,* 998, 1949.

39. **Yankwich, P. E., Promislow, A. L., and Nystrom, R. F.,** C^{14} and C^{13} intramolecular isotope effects in the decarboxylation of liquid malonic acid at 144.5°, *J. Amer. Chem. Soc., 76,* 5893, 1954.

40. **Bigley, D. B., and Thurman, J. C.,** Studies in decarboxylation. Part V. Kinetic isotope effects in the gas-phase thermal decarboxylation of 2,2-dimethyl-4-phenylbut-3-enoic acid, *J. Chem. Soc. (B),* 941, 1967.

41. **Bigley, D. B., and May, R. W.,** Studies in decarboxylation. Part IV. The effect of alkyl substituents on the rate of gas-phase decarboxylation of some βγ-unsaturated acids, *J. Chem. Soc. (B),* 557, 1967.

42. **Bigley, D. B.,** Studies in decarboxylation Part V. Kinetic isotope effects in the gas-phase thermal decarboxylation of 2,2-dimethyl-4-phenylbut-3-enoic acid, *J. Chem. Soc. (B),* 941, 1967.

43. **Bader, R. F., W., and Bourns, A. N.,** A kinetic isotope study of the Tschugaeff reaction, *Can. J. Chem., 39,* 348, 1961.

44. **Goldstein, M. J., and Thayer, G. L.,** Mechanism of a Diels-Alder reaction. II. The structure of the transition state, *J. Amer. Chem. Soc., 87,* 1933, 1965.

45. **Page, M. I., and Williams, A.,** *Organic and Bio-Organic Mechanisms,* Addison Wesley Longman, Harlow, Essex, 1997, Chapter 4.

46. **Seltzer, S.,** The mechanism of a Diels-Alder reaction, *J. Amer. Chem. Soc., 85,* 1360, 1963.

47. **Taageppera, M., and Thornton, E. R.,** Secondary deuterium isotope effects. The transition state in the reverse Diels-Alder reaction of 9,10-dihydro-9,10-ethanoanthracene. A potentially general method for experimentally determining transition state symmetry and distinguishing concerted from stepwise mechanisms, *J. Amer. Chem. Soc., 94,* 1168, 1972.

48. **Kwart, H., and Brechbiel, M. W.,** Transition state geometry in the ene reactions of mesoxalic esters, *J. Org. Chem., 47,* 3353, 1982.

49. **Kwart, H., Brechbiel, M. W., Acheson, R. M., and Ward, D. C.,** Observations on the geometry of hydrogen transfer in [1,5] sigmatropic rearrangements, *J. Amer. Chem. Soc., 104,* 4671, 1982.

50. **Kwart, H.,** Temperature dependence of the primary kinetic hydrogen isotope effect as a mechanistic criterion, *Accs. Chem. Res., 15,* 401, 1982.

51. **Anhede, B., and Bergman, N-Å.,** Transition state structure and the temperature dependence of the kinetic isotope effect, *J. Amer. Chem. Soc., 106,* 7634, 1984.

52. **Kupczyk-Subotkowska, L., Saunders, W. H., and Shine, H. J.,** The Claisen rearrangement of allyl phenyl ether: heavy-atom kinetic isotope effects and bond orders in the transition structure, *J. Amer. Chem. Soc., 110,* 7153, 1988.

53. **Kupczyk-Subotkowska, L., Subotkowski, W., Saunders, W. H., and Shine, H. J.,** Claisen rearrangement of allyl phenyl ether. 1-^{14}C and β-^{14}C kinetic isotope effects. A clearer view of the transition structure, *J. Amer. Chem. Soc., 114,* 3441, 1992.

54. **Gajewski, J. J., and Conrad, N. D.,** Aliphatic Claisen rearrangement transition state structure from secondary α-deuterium isotope effects, *J. Amer. Chem. Soc., 101,* 2747, 1979.

55. **Gajewski, J. J., and Conrad, N. D.,** Variable transition state structure in 3,3-sigmatropic shifts from α-secondary deuterium isotope effects, *J. Amer. Chem. Soc., 101,* 6693, 1979.

56. **Kupczyk-Subotkowska, L., Saunders, W. H., Shine, H. J., and Subotkowski, W.,** Thermal

rearrangement of allyl vinyl ether: heavy-atom kinetic isotope effects and the transition structure, *J. Amer. Chem. Soc., 115,* 5957, 1993.

57. **Kupczyk-Subotkowska, L., Saunders, W. H., and Shine, H. J.,** Carbon kinetic isotope effects and transition structures in the rearrangements of allyl vinyl ethers, *J. Amer. Chem. Soc., 116,* 7088, 1994.

58. **Hirst, D. M.,** *A Computational Approach to Chemistry,* Blackwell Scientific Publications, Oxford, 1990, pp. 29, 46.

59. **Hehre, W. J., Radom, L., Schleyer, P. v. R., and Pople, J. A.,** *Ab Initio Molecular Orbital Theory,* Wiley & Sons Inc., New York, 1986.

60. **Yoo, H. Y., and Houk, K. N.,** Transition structures and kinetic isotope effects for the Claisen rearrangement, *J. Amer. Chem. Soc., 116,* 12047, 1994.

61. **Beno, B. R., Houk, K. N., and Singleton, D. A.,** Synchronous or asynchronous? An "experimental" transition state from a direct comparison of experimental and theoretical kinetic isotope effects for a Diels-Alder reaction, *J. Amer. Chem. Soc., 118,* 9984, 1996.

62. **Gajewski, J. J., Peterson, K. B., Kagel, J. R., and Huang, Y. C. J.,** Transition state structure variation in the Diels-Alder reaction from secondary deuterium kinetic isotope effects: the reaction of nearly symmetrical dienes and dienophiles is nearly synchronous, *J. Amer. Chem. Soc., 111,* 9078, 1989.

63. **Gajewski, J. J., Peterson, K. B., and Kagel, J. R.,** Transition state structure variation in the Diels-Alder reaction from secondary deuterium kinetic isotope effects: the reaction of a nearly symmetrical diene and dienophie is nearly synchronous, *J. Amer. Chem. Soc., 109,* 5545, 1987.

64. **Van Sickle, D. E., and Rodin, J. O.,** The secondary deuterium isotope effect on the Diels-Alder reaction, *J. Amer. Chem. Soc., 86,* 3091, 1964.

65. **Tang, H. R., McKee, M. L., and Stanbury, D. M.,** Absolute rate constants in the concerted reduction of olefins by diazene, *J. Amer. Chem. Soc., 117,* 8967, 1995.

66. **Tang, H. R., and Stanbury, D. M.,** Direct detection of aqueous diazene: its UV spectrum and concerted dismutation, *Inorg. Chem., 33,* 1388, 1994.

67. **McKee, M. L., Squillacote, M. E., and Stanbury, D. M.,** *Ab initio* investigation of dihydrogen transfer from *cis*-1,2-diazene, *J. Phys. Chem., 96,* 3266, 1992.

68. **Stanbury, D. M.,** Kinetic-behavior of diazene in aqueous-solution, *Inorg. Chem., 30,* 1293, 1991.

Chapter 8

Enzyme Reactions

The concerted mechanism has long been associated with theories of enzyme action, and as far back as the 1950's Swain and Brown[1] proposed a concerted chemical model which has remained as a talisman in studies aimed at understanding enzyme mechanisms. *However*, there is no *a priori* reason why a concerted mechanism should be faster than its corresponding stepwise one[2] and the resultant energy barrier in a concerted process may be the partial or complete sum of the individual barriers. It is only when there is coupling between the processes that concertedness may become advantageous. Enzymatic catalysis is distinguished by the involvement of non-covalent binding interactions between complexed substrate molecules and the enzyme active site. These binding interactions have the potential to act concertedly with each other during the act of complexation[3-14] and with covalent bonding changes.

Diagnosis between concerted and stepwise mechanisms in enzyme catalysis has proved difficult simply because of the complexity of the possible enzyme-substrate interactions. This chapter draws examples from enzyme chemistry where concertedness has been considered and indicates the bonding changes involved. It is not possible at present to provide answers as definite as those for the simpler chemical systems and it is important to reiterate that concerted mechanisms may arise as an indirect consequence of, say, the *snug fit* of substrate in active site and that there is no general catalytic advantage to be gained from a concerted over a stepwise enzyme mechanism.

The general theory of enzyme catalysis refers to complexation of the substrate at the active site by a combination of forces. All catalytic and specificity advantages arise from the *snug fit* in the active site of the substrate reactants which are aligned so that the centres undergoing bonding or binding changes are precisely located against the centres on the enzyme which promote reaction.[15] Enzymes create sites of catalysis (the active sites) which have structures, when filled with substrates, similar to those of the transition structure of the catalytic processes.[16] This provides a relatively sound knowledge base from which to understand the catalytic functions because the substrate's surroundings are characterised by x-ray crystallography much more precisely than is possible for reactions in free solution. The possibility arises that the mechanistic definition of an enzyme reaction could be made at a higher resolution than that for a reaction not involving complexation[17] and structural studies therefore play a very important part in discussing enzyme mechanisms.

The *snug fit* theory requires that all the reactive centres of the substrate be held against the reacting groups on the enzyme at the *same time* or *during* the period between exit or entry of the substrate from and to the active site. Entering or exiting of substrate or product could involve stepwise interactions but during this period ($t_{1/2} \sim 10^{-9}$sec) the prerequisite for a concerted reaction may be satisfied by the conversion of the energy of binding interactions into chemical transformations.

8.1   PROTON TRANSFER

Proton transfer concerted with heavy atom reorganisation in simple systems is well documented in Chapter 3. An enzyme, with its catalytic site filled with substrate, can exploit the proton more successfully than solution reactions because the electrophilicity of the transferring proton is not diluted by solvent interactions and because the active site-substrate interactions can locate the catalytic hydrogen directly into its optimal position for reaction.

8.1.1 Proton transfer coupled with acyl group migration

Acyl group transfer to a pro-nucleophile bearing a proton (Nu-H) involves three processes: formation of a tetrahedral intermediate, proton transfer from the pro-nucleophile and collapse of the intermediate by departure of the leaving group possibly assisted by proton transfer (Scheme 8.1). The proton is usually transferred in the catalytic mechanism of proteases by the intervention of a general base, such as the imidazolyl function, and often the overall process involves an acyl enzyme (RCO-enzyme) intermediate which reacts with the pro-nucleophile (Nu-H) to give products.

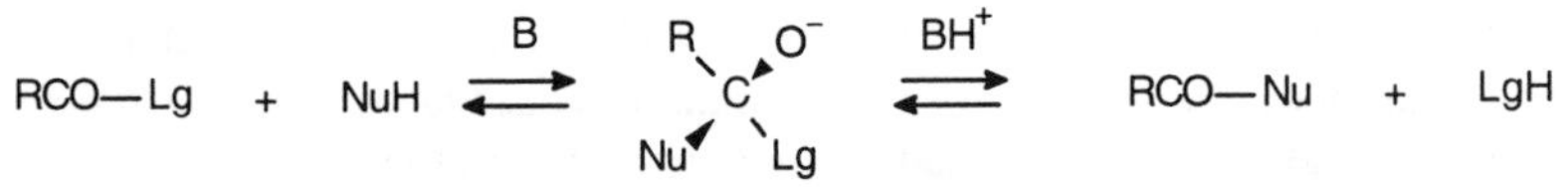

$$\text{RCO—Lg} \;+\; \text{NuH} \;\underset{}{\overset{B}{\rightleftharpoons}}\; \begin{array}{c} R \quad O^- \\ \diagdown \diagup \\ C \\ \diagup \diagdown \\ Nu \quad Lg \end{array} \;\overset{BH^+}{\underset{}{\rightleftharpoons}}\; \text{RCO—Nu} \;+\; \text{LgH}$$

Scheme 8.1. Proton transfer coupled with acyl group migration.

A typical mechanism for the hydrolysis of an amide catalysed by a serine protease such as α-chymotrypsin is displayed in Scheme 8.2. The aminolysis of acyl-protease (essentially the microscopic reverse of the hydrolysis in Scheme 8.2) exhibits a very small positive β_{nuc} (0.1) indicating very little charge development in the transition state which is consistent with a concerted process. The concerted mechanism for decomposition of T_- is untenable as a major pathway because the basicity of nitrogen in T_- is expected to be similar to that of the parent amine; proton transfer from the imidazolium cation (ImH^+) to this nitrogen will therefore be thermodynamically favourable for most amide substrates and the proton will jump to the nitrogen more rapidly than the intermediate can break down. The expulsion of the fully protonated amine will therefore occur through a stepwise path (through $T_\pm$) rather than through the concerted mechanism.[18]

Concerted general acid-base catalysis of serine attack to *form* T_- is permitted because the stepwise process involves a thermodynamically unfavourable proton transfer from serine hydroxyl to imidazolyl (step A in Scheme 8.3) and an unfavourable step involving protonation of serine oxygen ether (B in Scheme 8.3). The thermodynamically unfavourable proton transfer (ImH$^+$ to Ser-O-C) becomes favourable when the (C-O-C) bond is cleaved.[19]

The Charge Relay System. A structural feature common to most of the serine proteases[20] is the juxtaposition of an aspartic acid and histidine residue. This feature gave rise to the concept of the charge relay system in proteases whereby the aspartate anion assists proton transfer at the remote serine centre by a concerted relay of charge through an intervening imidazolyl function (Scheme 8.4). A similar juxtaposition of aspartic acid or asparagine and

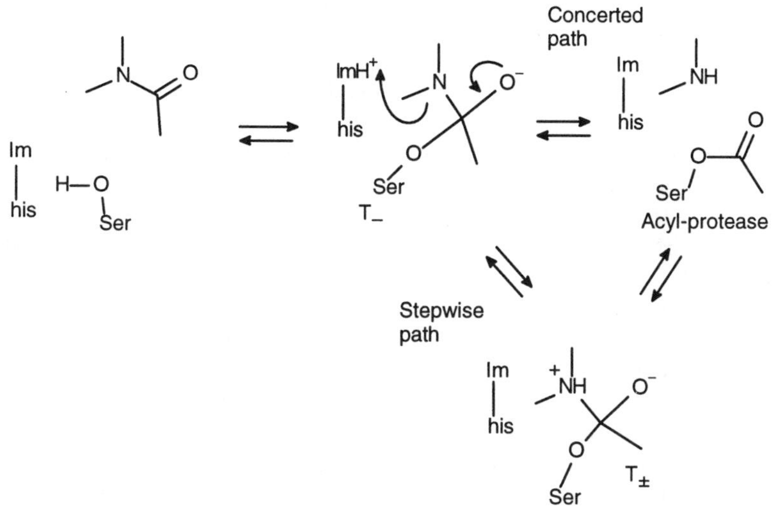

Scheme 8.2. Formation of an acyl-enzyme in hydrolysis of an amide catalysed by a serine protease.

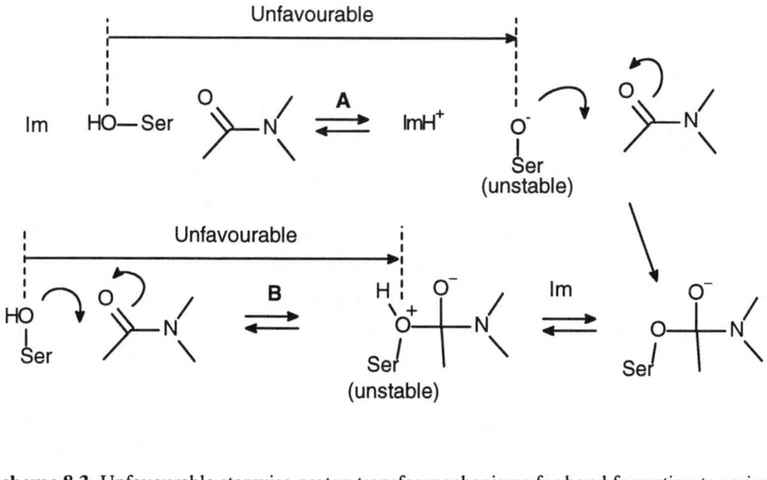

Scheme 8.3. Unfavourable stepwise proton transfer mechanisms for bond formation to serine.

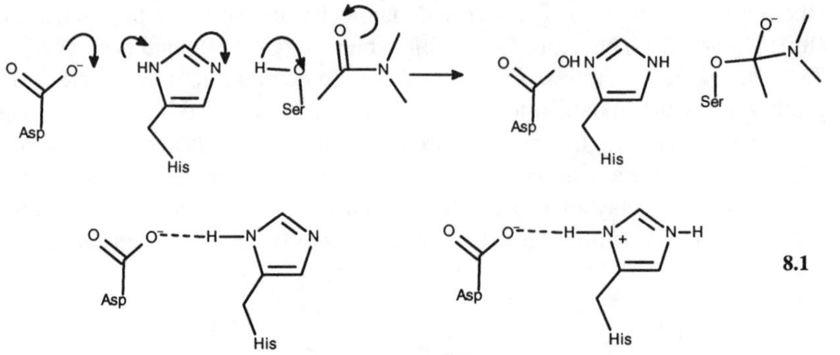

Scheme 8.4. The charge relay system in serine proteases.

histidine residues in cysteine proteases has been noted.[21] The concerted process illustrated in Scheme 8.4 involves two protons "in-flight" and effectively spreads the developing positive charge on the imidazolyl moiety over the carboxylate group of the aspartate function. Proton inventory studies indicate that about 50% of the systems show two protons in-flight and 50% show one proton in-flight;[22-28] this supports the notion that the charge-relay system operates for some substrates but not for others and that a two-proton transfer is not a *general* prerequisite for catalysis in proteases. Studies on chymotrypsin[29] show that the nature of the acid changes relative to the acyl component of the substrate. Those acyl enzymes with high values of pK_a have substantial zwitterionic character (**8.1**, in Scheme 8.4). Those acyl enzymes with low pK_a's have a weak or non-existent electrostatic interaction which therefore serves no useful purpose in catalysis.

Whereas the mechanisms of serine proteases undoubtedly involve proton transfer from the serine hydroxyl to the imidazolyl base this is not the case with the cysteine proteases. Papain has been shown to possess a cysteine-25 and histidine-159 which are essentially present as the zwitterionic species over the pH-range of catalytic activity.[30-32] No general base is required to remove the thiol proton and it is possible that the imidazolyl group is responsible in its conjugate acid form for protonating the leaving group rather than in its base form for accepting a proton from the nucleophilic thiol group.

The charge relay system probably does not make the serine more intrinsically nucleophilic but rather derives from the following:[33] to provide the correct tautomeric form of the histidyl residue, to orientate the imidazolyl ring optimally for accepting a proton from the serine hydroxyl, and to provide electrostatic stabilisation of the developing imidazolium cation.

Molecular orbital calculations have been made[34-37] on model systems utilising the atomic coordinate data from x-ray crystallographic results[38] for proteases. The observations that the aspartate assists the catalysis were to be expected on simple chemical intuition. Most of the theoretical studies omitted to employ factors relating to entropy and to stabilisation of the developing oxyanion.

8.1.2 1,2-Proton transfer

Isotopes of hydrogen have proved very useful in determining the overall mechanism of triosephosphate isomerase (TIM) where proton transfer is the key step. The enzyme catalyses the migration of a proton between C2 and C3 positions in the interconversion of G3P and the more stable DHAP. When the isomerase was first studied by hydrogen isotope labelling techniques no exchange with solvent protons was observed. This observation is consistent with a concerted, hydride ion transfer process (Scheme 8.5). Improved techniques demonstrated that while most of the proton is delivered back to the substrate in the enzyme complex a small percentage goes to the solvent resulting in exchange. Scheme 8.6 illustrates the overall mechanism of the catalytic reaction for the enzyme. The 1-pro-R proton of DHAP is excised in the reaction yielding G3P[39] and the fact that about 6% of the label in 1[(R)-^{3}H]-DHAP is retained[40] indicates that the proton shuttles between a single face of the enediol intermediate and a single base.

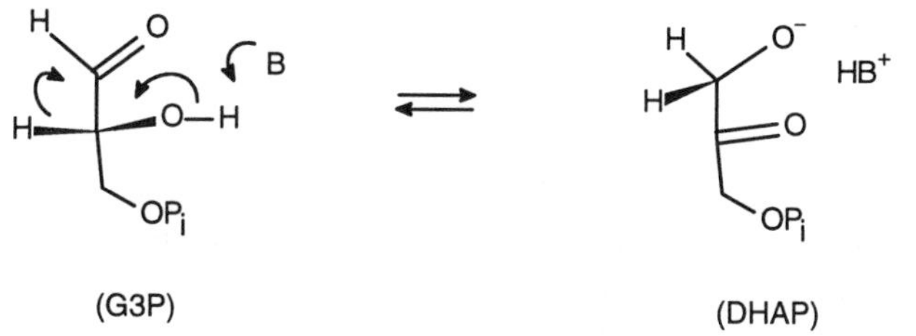

(G3P)   (DHAP)

Scheme 8.5. Concerted hydride ion transfer in the originally proposed mechanism for the triose phosphate isomerase catalysis; DHAP = 1,3-dihydroxyacetone phosphate, G3P = glyceraldehyde-3-phosphate.

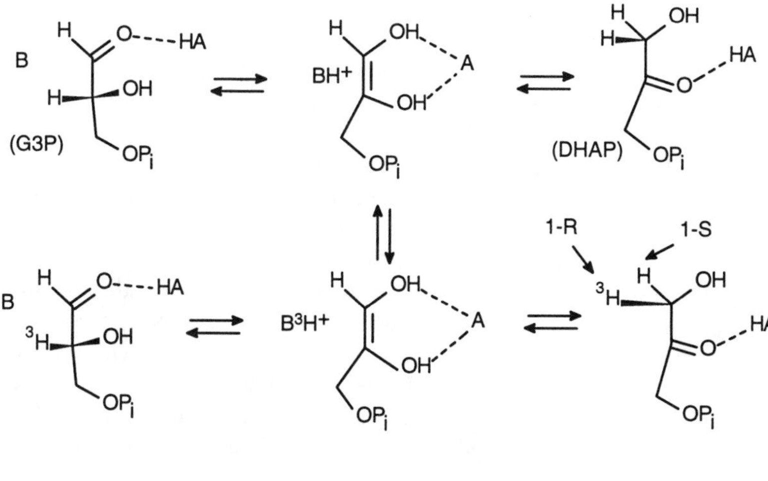

Scheme 8.6. Schematic mechanism for the triose phosphate isomerase reaction; A = histidyl-95 and, possibly, lysyl-13, B = glutamyl-165.

Chemical studies of the action of bromoketone and epoxide inhibitors[41,42] and results from the x-ray crystallographic studies indicate that the base (B) is likely to be glu-165. The identity of (A) in Scheme 8.6 may be the imidazolyl group of his-95 because activity as well as the polarisation of the carbonyl group is lost on point mutation of this for glutamine.[43,44] The results do not indicate whether or not the electrophilic interaction between (HA), the carbonyl oxygen, and the 2-hydroxyl is concerted with proton transfer to the base.

Albery and Knowles[45,46] utilised triose phosphate isomerase as a vehicle to test the concept of catalytic perfection whereby evolutionary selective pressures lead to the enzyme's activity being improved to the point where further improvement is not necessary because the rate becomes limited by diffusion. The isotopic exchange between the ene-diol complexes in Scheme 8.6 "clocks" the lifetimes of these intermediates.

8.1.3 1,3-Proton transfer

Aspartate aminotransferase catalyses the transfer of the NH_2 group between aspartic acid and 2-oxoacids *via* the intervention of a pyridoxal coenzyme (Scheme 8.7). The kinetically significant step, a 1,3-proton transfer is thought to involve lysine-258 as a proton relay system (Scheme 8.8); replacement of the lysine-258 with an alanine residue by site-directed mutation[47] yields a mutant enzyme which is only weakly active but which can have activity restored by exogenous amines (B, in Scheme 8.8). The rate constant obeys a two parameter free energy relationship (Equation 8.1) which includes a steric parameter (namely the molecular volume).

$$\log k_B = 0.39 \, pK_a - 0.055 \, \text{Molecular Volume} - 0.7 \qquad (8.1)$$

The molecular volume effect is relatively large and consistent with the requirement that the amine must be accommodated in a restricted cavity vacated when lysine-258 is exchanged for the alanine residue. The ß value indicates that some 40% of the full charge is developed on the nitrogen base in agreement with that found by Auld and Bruice[48] in studies of a model system.

Partial retention of label in the aspartate aminotransferase reaction has been demonstrated for the 1,3-hydrogen transfer which must therefore involve a syn elimination process consistent with a single base catalytic group.[49] The rate of the 1,3-shift is less than that of proton exchange with the medium indicating that the catalytic group is exposed to the bulk solvent in the ternary complex between pyridoxal phosphate coenzyme, substrate and enzyme. The primary kinetic isotope effect for the hydrogen on the C_a position is small and moreover is essentially independent of the isotopic composition of the solvent.[50] This result is consistent with concerted transfer of the hydrogen between 1 and 3 positions (Scheme 8.8).

The rearrangement catalysed by Δ^5-3-ketosteroid isomerase involves

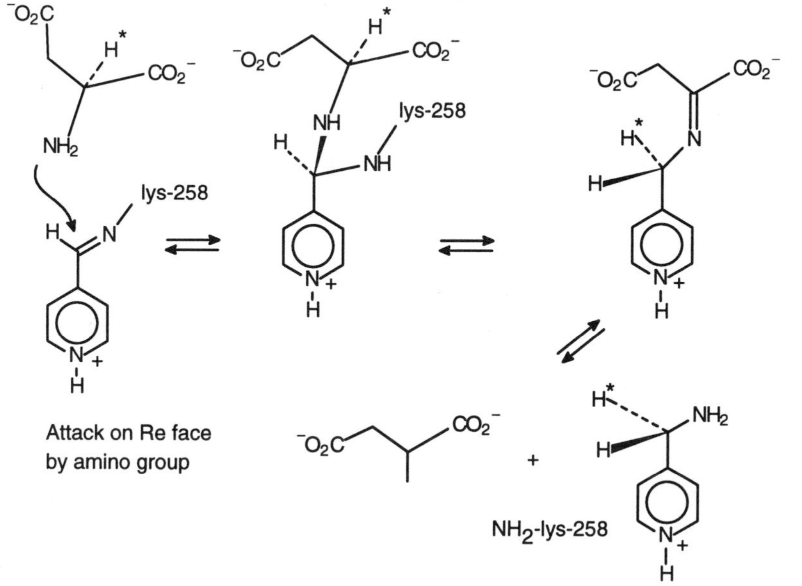

Scheme 8.7. The first step in transamination from aspartate catalysed by aspartate aminotransferase.

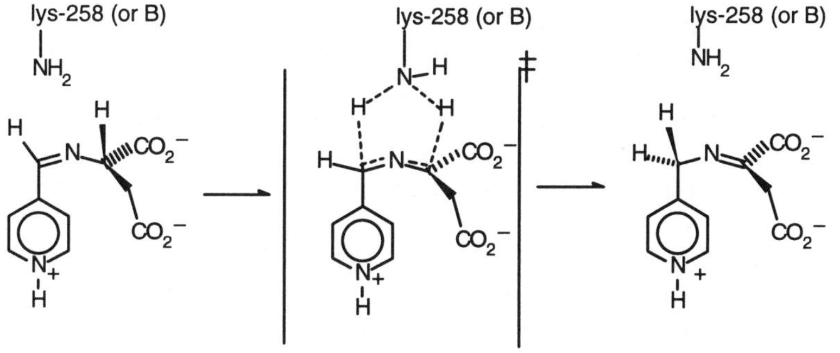

Scheme 8.8. The concerted transfer of proton between the 1 and 3 positions in mutant and native (lys-258) aspartate aminotransferase.

transfer of the hydrogen from the 4β-position to the 6β-position (Scheme 8.9). The hydrogen is not exchanged with the solvent but it is unlikely that this involves direct transfer since spectroscopic evidence has been obtained for a dienol intermediate.[51,52] The reaction is probably in two parts with transfer of the 4β-hydrogen to a base (the asp-38 residue) and this is the rate-limiting step. The deuterium isotope effect of k^H/k^D is 6.1 in H_2O and studies with the mutant (tyr-14-phe) indicate a rate reduced by some $10^{4.7}$-fold.[53,54] The data are consistent with a concerted mechanism (Scheme 8.9) where tyrosine-14 acts as a general acid by protonating the enolic oxygen as it is formed. The intracomplex proton transfers are much faster than diffusion of the proton to and from solvent.

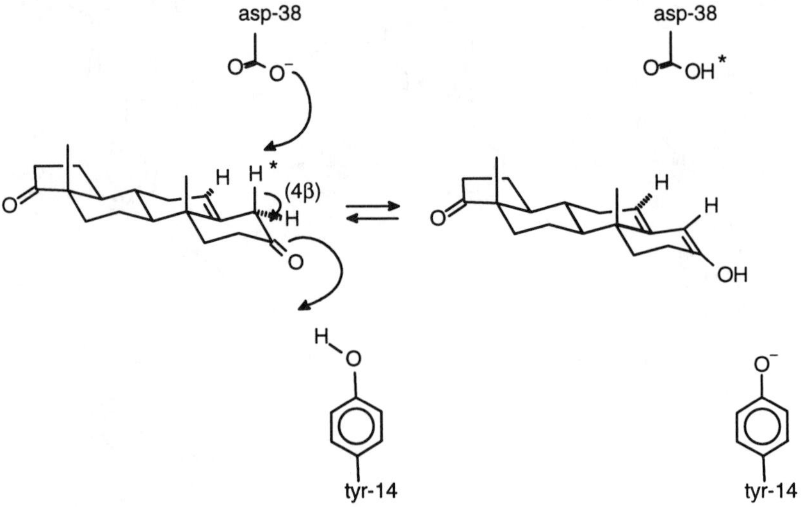

Scheme 8.9. Δ^5-3-Ketosteroid isomerase; formation of dienol intermediate.

8.1.4 Hydrogen-bonds and enzymic catalysis

The manifest importance of hydrogen-bonds[55-58] in the structure of enzymes and the observation of general acid-base catalysis in model reactions beg the question of the importance of hydrogen-bonding in catalysis.[59-64] Proton transfer in ice can take place 70-fold faster than in water because of the favourable direction of the hydrogen-bonds and conversely proton transfer between the NH of peptide α-helices and water is slow because of the unfavourable hydrogen-bonding. This explanation was developed into the concept of *facilitated proton transfer* where the hydrogen-bonds in an enzyme system are favourable for proton transfer in a particular catalytic scheme.[65-68] Salicylate derivatives are known to have a highly effective catalytic propensity which is thought to derive from the strong hydrogen-bond between carboxylate and phenolic oxygen in the products of reaction as illustrated for the catalysed hydrolysis of an acetal (Scheme 8.10).[69] This hypothesis was tested by the use of models where similar strong hydrogen-bonding could be observed in the product species (Scheme 8.10); simple provision of hydrogen-bonding is not sufficient to ensure enhanced catalysis.

Strong hydrogen-bonding has been advanced in discussions of enzyme catalysed processes where a carbanion cannot readily be stabilised for example in mandelate racemase (Scheme 8.11).[70-75] In the absence of enzyme the proton transfer is too slow to measure because the dianion (Scheme 8.11) has very high energy (although such species are known) and a concerted acid-base mechanism (Scheme 8.12) could circumvent this problem.

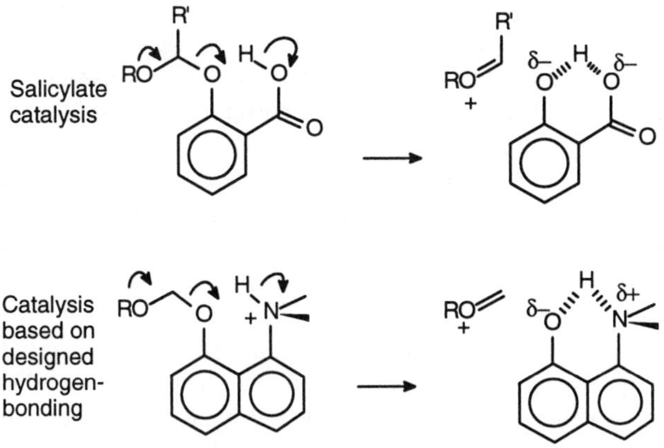

Scheme 8.10. Strong hydrogen-bonding interactions in the product assist catalysis.

Scheme 8.11. Mandelate racemase reaction.

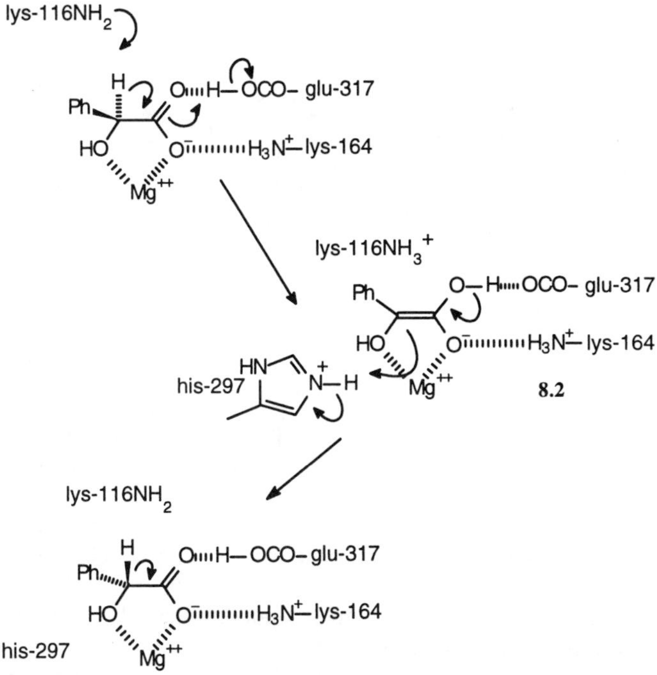

Scheme 8.12. Concerted acid-base catalysis in mandelate racemase.

The mechanism was modified to account for the fact that the enolate species derived from a carboxylate anion (**8.2** in the Scheme 8.12) is still thermodynamically very unfavourable (pK_a >19). The involvement of a short, strong hydrogen-bond from glu-317 to the carboxylate in (**8.2**) was proposed to allow the system to mimic a neutral acid.

Short, strong hydrogen-bonds. There is currently fruitful controversy about the importance of short, strong hydrogen-bonds in enzyme catalysis as opposed to the alternative electrostatic stabilisation.[76,77] Such bonds would have only a low barrier between the two minima and are sometimes called "low barrier hydrogen-bonds." The formation of the strong hydrogen-bond (or electrostatic interaction) in the intermediate could be concerted with the other bonding changes. The nature of the increased stability of the intermediate, whether it be short, strong hydrogen-bonding or electrostatic interactions, is still an open question,[78-81] and in the cases to be described the involvement of the one has not been proven over the other.

In the case of serine proteases it was proposed that the function of the oxyanion pocket in catalysis is not only to stabilise the oxyanion by hydrogen-bonding but also to partially transfer a proton. The pK_a of the NH donor amide and that of the neutral hydroxyl of the tetrahedral intermediate are close and stabilisation of this intermediate could occur through two short, strong hydrogen-bonds in which the protons of the peptide NH's of gly-193 and ser-195 are partially transferred to the basic oxygen (Scheme 8.13).

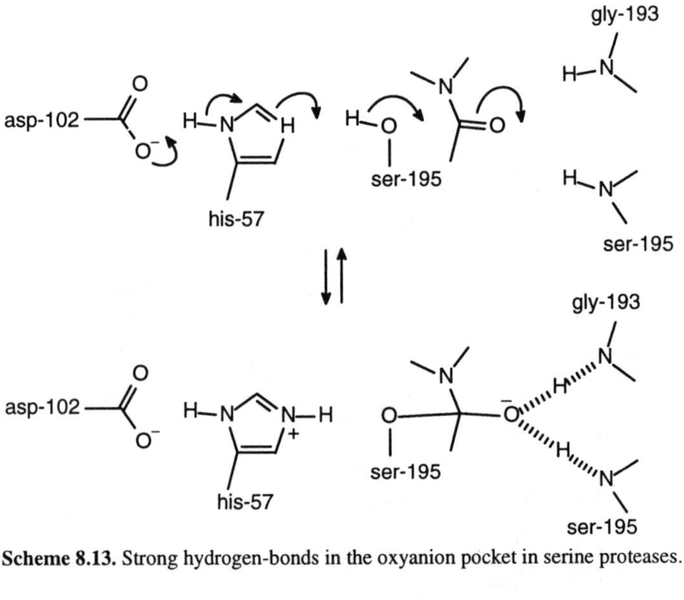

Scheme 8.13. Strong hydrogen-bonds in the oxyanion pocket in serine proteases.

The concept of short, strong hydrogen-bonding has also been applied to other hydrolases and ribonuclease has been cited where a proposed pentacoordinated intermediate is stabilised by two short, strong hydrogen-bonds between lys-41, gln-11 and/or the peptide NH of phe-120 and the two oxygens bearing the unit negative charge (Scheme 8.14).

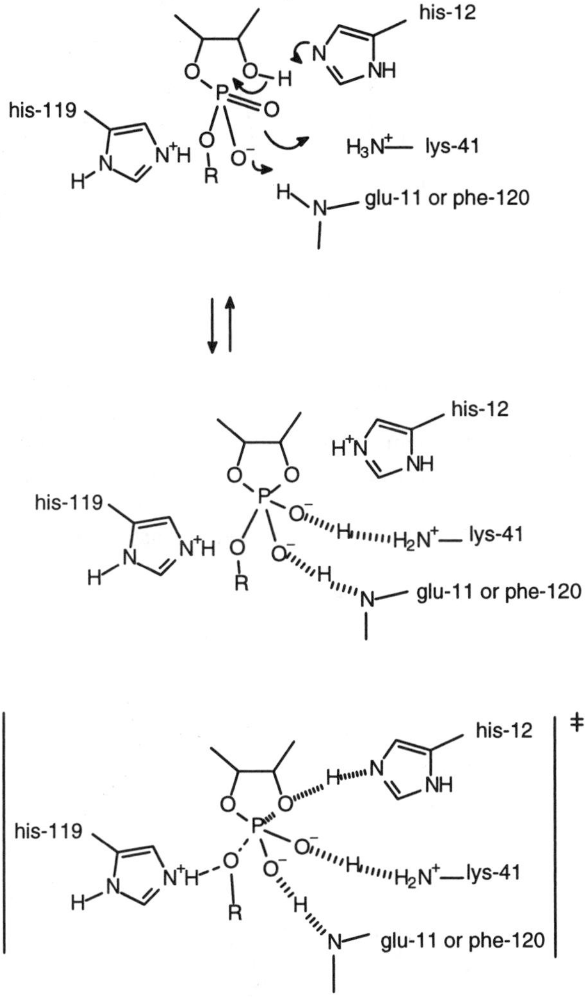

Scheme 8.14. Short, strong hydrogen-bonding and ribonuclease catalysis.

Both imidazolyl groups (of his-119 and of his-12) in the ribonuclease catalysed reaction are thought to be involved in a concerted mechanism and, moreover, that P-O bond forming is in concert with fission of the P-O bond in the leaving group (OR). The mechanism would involve a strong hydrogen-bond or electrostatic interaction to stabilise the transition structure (see Scheme 8.14). If the serine proteases have concerted C-O formation and C-Lg fission, as indeed may be the case for activated esters, the mechanism in Scheme 8.13 would also need to be modified so that it is the *transition structure* which is stabilised (Scheme 8.15).

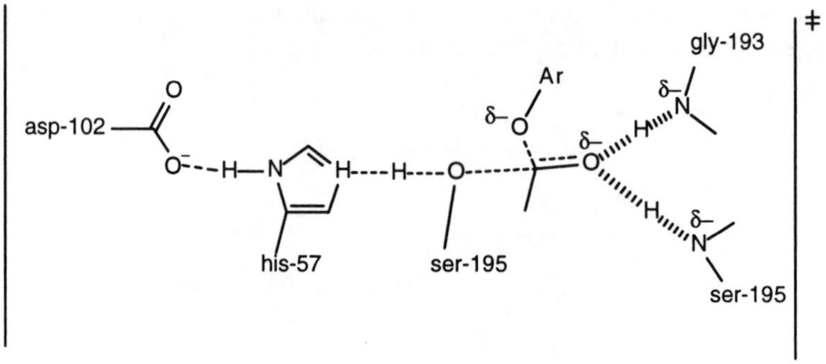

Scheme 8.15. Short, strong hydrogen-bonding stabilising the oxyanion component in the transition structure in serine protease catalysis.

The ionisation of the α-proton in the first step of citrate synthase is assisted by aspartate-375 acting as a general base and by the NH of the neutral imidazolyl side chain of his-274 acting as an electrophile (Scheme 8.16) and a strongly hydrogen-bonded enolate ion intermediate may be postulated. Histidine-320 acts in its protonated form in the second step and "genuine" general acid catalysis with full proton transfer can occur.

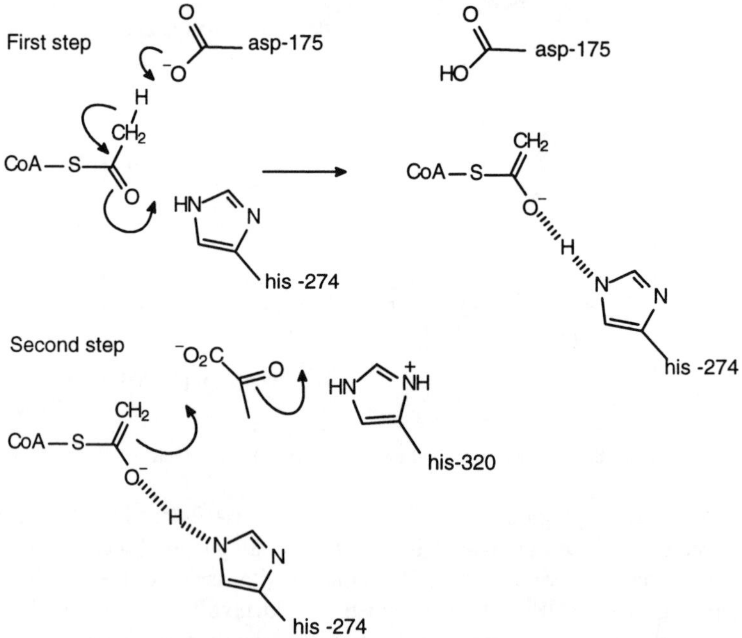

Scheme 8.16. Citrate synthase.

It is noted that triose phosphate isomerase (TIM) also involves a neutral imidazolyl group (his-95) as an "acid" in the ionisation step and this could also be rationalised on the basis of a short, strong hydrogen-bond or electrostatic interaction (Scheme 8.17).

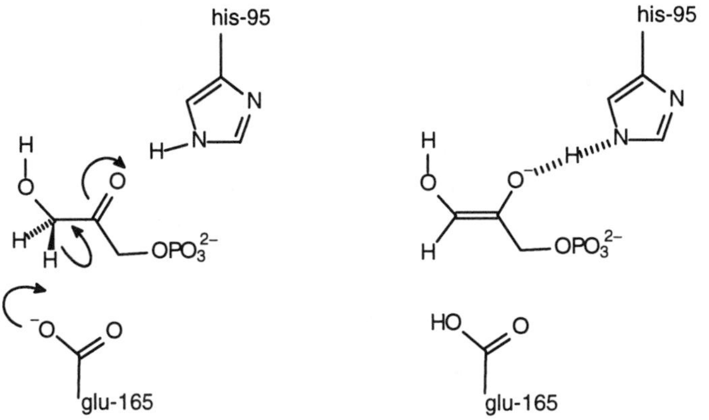

Scheme 8.17. The ionisation mechanism of triose phosphate isomerase (TIM).

Jencks' *libido rule*[19] indicates that general acid catalysis occurs in those cases where the liberated group has a higher pK_a than that of the postulated proton donor. Short, strong hydrogen-bonds or electrostatic interactions do not need to be postulated, for example, in leaving group departure in glycosidases or proteases and the action of his-320 as a general acid in the coupling step of citrate synthase (Scheme 8.16) will therefore be as a regular proton donor.

Short, strong hydrogen-bonds were proposed to counter the thermodynamic problem which the concerted proton transfer proposal does not resolve. Although there have been well-documented cases of such bonds in general[82] there has been little evidence for them (only a few *bona-fide* strong hydrogen-bonds have been demonstrated in model systems) and their existence in enzyme mechanisms is still under active discussion.

Deuterium fractionation factors. There appears to be a connection between low barrier hydrogen-bonds and bases with low isotopic fractionation factors in enzyme reactions. The deuterium fractionation factor of a hydrogen is the tendency for the deuterium to enrich the position relative to its content in water. Fractionation factors close to unity are observed for CO_2H, imidazolyl NH, NH_2 and hydroxyl groups where the proton readily exchanges with solvent water. The thiol group is the only amino acid side chain which has a fractionation factor in the range < 0.5.

The concept of low barrier hydrogen-bonds can account for fractionation factors of less than 0.5 even though they possess no thiol group.[83,84] A double potential well hydrogen-bond has a fractionation factor near unity when the hydrogen is attached to oxygens because this is similar to the structural features possessed by water. As the barrier decreases the proton moves more freely between the potential minima and hydrogen becomes weakly covalently bonded to both oxygens and the fractionation factor decreases. The use of fractionation factors to diagnose the involvement of groups such as thiols in proline racemase is therefore suspect.

The chemical shift of the proton between $N^{\delta 1}$ of his-57 and the β-carboxyl group of asp-102 and the deuterium isotope effect on the chemical shift indicate the existence of a low barrier hydrogen-bond (Scheme 8.18) in the charge relay system in the ground state structure of serine proteases.[85,86] It should be noted that these results do not imply that the charge relay system has any great effect on the catalytic reactivity of the serine-195 but complement the orientation and other factors described in Section 8.1.1.

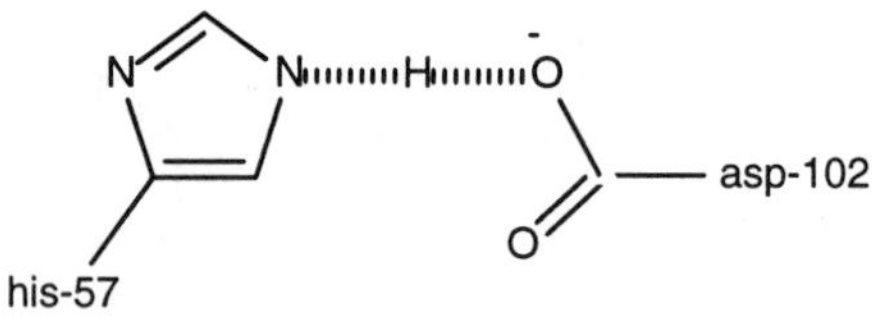

Scheme 8.18. A low barrier hydrogen-bond between asp-his in α-chymotrypsin.

8.2   THEORETICAL CALCULATIONS

Since the advent of high speed computers the advances in *ab initio* and semi-empirical molecular orbital theory and in molecular mechanics calculations have led to a steady flow of increasingly reliable predictions of enzyme mechanisms.[87,88] It could be said that the calculations for enzymes are more predictive than those for model systems in solution where there is a difficulty in accounting for solvation. In the enzyme case the current precise knowledge of atomic coordinates enables the location, in principle, of all the importance interactors. In this case the enzyme becomes, to a certain extent, more like a gas phase reaction because most of the bulk solvent does not interact with the enzyme substrate complex.

The task in enzymology is almost impossible if the calculations are carried out without restraint on the atomic coordinates because it is very difficult to search all the possible coordinates for energy minima. When a saddle point is detected there is nothing to say that it is the *correct* one and moreover there is an entropy requirement in order to convert potential energy determined directly from the calculations into free energy which governs the reaction. The possession of high precision x-ray crystallographic data for enzymes has simplified the calculation problem by excluding particular space coordinates for the minima search. Recent work has centred on hybrid methods to model enzyme catalysis; quantum mechanics is employed for the active site-substrate region, and molecular mechanics models the much larger part of the system which is not directly involved in the chemical reaction.[89-91]

Early work utilised either the active site fragments of the enzyme-substrate interaction or simply models of the supposed interaction. It is doubtful if any of the early work had any diagnostic impact for enzyme mechanisms but it should not be dismissed completely because it represents the gradual development of a predictive protocol. It has to be said that the development is not yet complete but when coupled with experimental results such as isotope effects molecular orbital theory provides a powerful force in achieving an understanding of the mechanisms of enzymes.

The catalytic mechanism of papain has been studied using a hybrid quantum mechanical - molecular mechanical potential employing AM1 for the active site and the AMBER force field for the bulk of the enzyme not directly involved in catalysis. The interesting result of this work is that the formation of the C-S bond is predicted to be concerted with hydrogen-bonding type interactions at the carbonyl oxygen and the leaving nitrogen and with C-N bond fission (Scheme 8.19). It is clear from the data that the C-N bond has not extended very much further than a single bond length in the transition state on the way to the thiol ester which ressembles a tetrahedral intermediate with a long C-S bond.

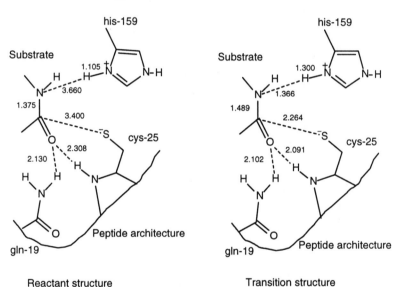

Scheme 8.19. Quantum mechanical-molecular mechanics calculation of the transition state for papain. Bond lengths (Å) are from *Refs. 89* and *90*.

Molecular orbital studies on the charge relay system indicate that it does not contribute substantially to catalysis;[37,92] in the case of the oxyanion pocket computer simulation of structures where the constituents of this motif are deleted predict that it is essential for catalytic activity.[93,94]

Warshell applied his empirical valence bond approach[88] to the problem of guanosine triphosphate (GTP) hydrolysis catalysed by the *ras p21* proteins. The γ-phosphate oxygen was considered to act as a general base to assist water attack at the phosphorus (Scheme 8.20). The overall free energy calculated for both steps is 24 kcal/mole which is in good agreement with the observed activation free energy. However, Herschlag[95] has pointed out that model studies indicate that GTP hydrolysis in solution is dissociative and that since there would be very little bond formation to the incoming water nucleophile there would be little advantage from removal of a proton in the transition state due to the small amount of charge development (Scheme 8.21).

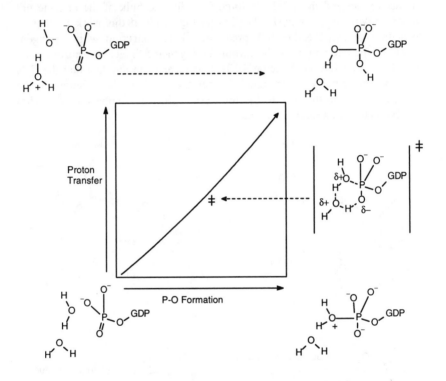

Scheme 8.20. More O'Ferrall-Jencks type diagram for the formation of a pentacoordinate intermediate in the hydrolysis of GTP catalysed by *ras p21* (*Ref. 88*). GDP = guanosine diphosphate. The dotted line (north) refers to a combined proton transfer and P-O bond formation coordinate, whereas the southern line is simply the P-O bond formation coordinate.

Molecular orbital and molecular mechanics techniques can be used to predict, for example, transition structures of enzyme reactions which can be determined with increasing confidence by comparing physical properties such as multiple isotope effects with those calculated for a grid of putative structures. An aspect of the simulation approach which is very useful in enzyme studies includes the power to judge the fit of a substrate or inhibitor into a potential receptor cavity and judging the timing of bonding changes to estimate charges in the transition structure in the enzyme active site.

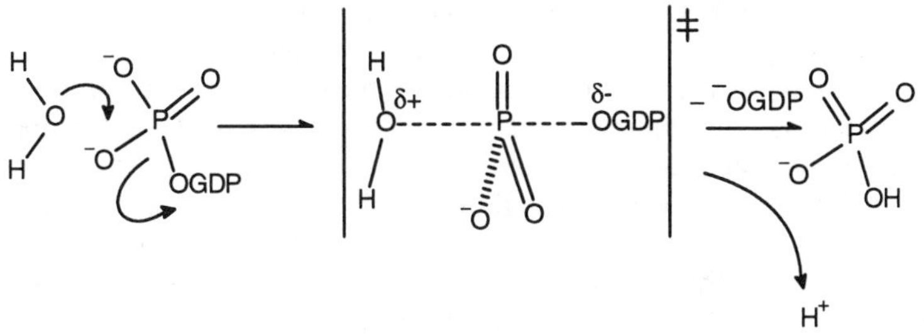

Scheme 8.21. Dissociative mechanism for the model hydrolysis of GTP.

8.3 ISOTOPES

8.3.1 Acyl group transfer

Bender proposed that the chymotrypsin-catalysed hydrolysis of amide involves concerted displacement of the leaving group to yield acyl enzyme.[96] Heavy atom isotope effects were measured for the reaction of 4-nitrophenyl acetate with chymotrypsin, carbonic anhydrase, papain and *Aspergillus* acid protease;[97] the resultant isotope effects are closely similar to those which are known to be concerted for attack of model oxygen and sulphur nucleophiles in aqueous solution. The smaller inverse β-deuterium isotope effects (Scheme 8.22) are interpreted to indicate greater H-bonding or electrostatic interaction between enzyme and ester carbonyl in the transition state compared with the reactant state.

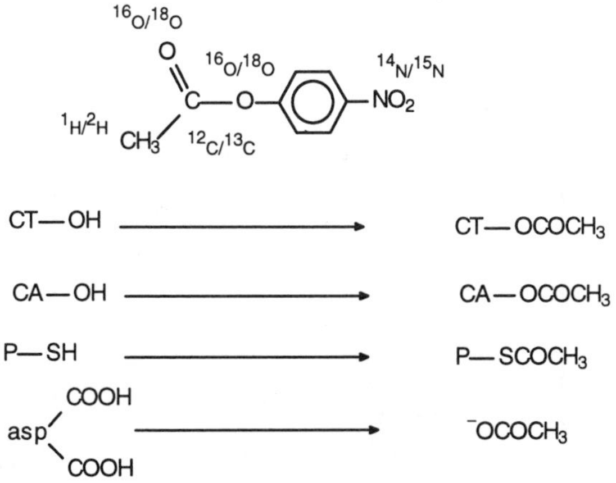

Scheme 8.22. Heavy atom and β-deuterium isotope effects in the hydrolysis of 4-nitrophenyl acetate catalysed by enzymes (*Ref. 97*).

Although the formation of the pentacoordinate intermediate has been the predominant mechanism in discussions of ribonuclease catalysis, studies with model systems have indicated that it could simply be a trigonal bipyramidal transition structure. Heavy atom isotope effects for the reaction are illustrated in Scheme 8.23.[98] The normal isotope effect of 1.6% for the leaving oxygen indicates that the bond order to the leaving group is lost in the transition structure. This is consistent with either a concerted displacement at phosphorus or a stepwise process where breakdown of the phosphorane intermediate is rate-limiting. A normal isotope effect of 0.5% for the non-bridge oxygens indicates that there is *loss* of bond order consistent with a concerted mechanism which is slightly associative. The enzyme reaction has been studied by use of substituent effects and the transition state can be defined regarding the fission of the P-O bond to the leaving group (Scheme 8.24). There is substantial bond fission to the leaving oxygen as judged by the value of α_{lg} ($-0.59/1.74 = -0.34$).[99]

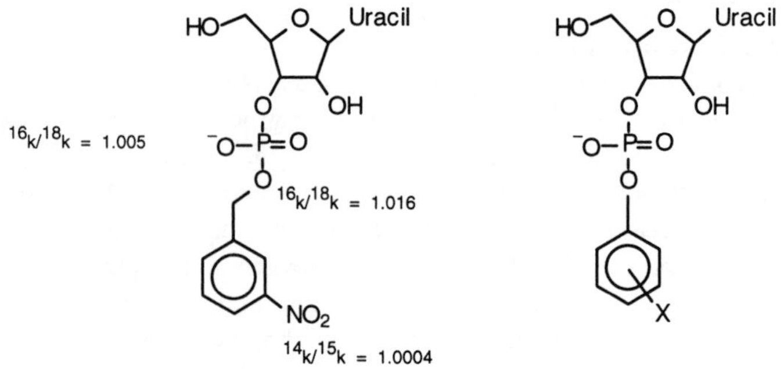

Scheme 8.23. Heavy atom isotope effects for the ribonuclease reaction. Isotope effect data are from *Ref. 98*.

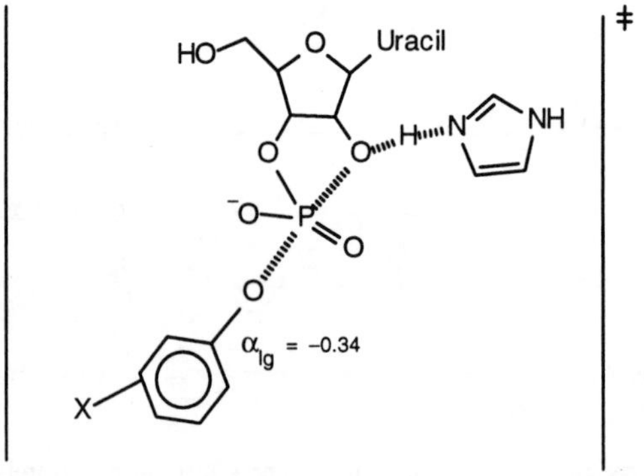

Scheme 8.24. Substituent effects on the leaving group for the ribonuclease reaction. The value of α_{lg} is from *Ref. 99*.

Inversion of configuration at the phosphorus in the ribonuclease reaction is consistent with a concerted mechanism. In the case of alkaline phosphatase retention of configuration in a two-step mechanism involving a phosphoryl enzyme intermediate is also consistent with two concerted displacements involving inversion and this is consistent with the ^{18}O-isotope effects in the alkaline phosphatase catalysed hydrolysis of phospho-monoesters.[100]

8.3.2 Double isotope fractionation

Unlike ordinary chemical reactions enzymatic ones often have kinetically significant complexation steps which can modify the kinetic isotope effect[101,102] even though they have isotopically insensitive transition structures. In Scheme 8.25 deuteriation will increase the significance of the second step in case 1 thus *increasing* the observed ^{13}C isotope effect. In the case of the stepwise process (case 2) deuteriation will decrease the significance of the second step causing the ^{13}C isotope effect to *decrease*. We have not included mechanisms where C-C bond change precedes proton transfer but these, although unlikely, can be studied by the double isotope fractionation technique.

Case 1   Concerted

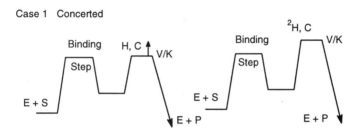

Deuteriation increases ^{13}C-isotope effect

Case 2   Stepwise

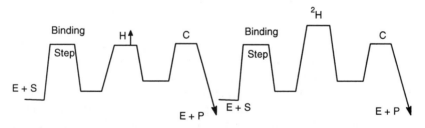

Deuteriation decreases ^{13}C-isotope effect

Scheme 8.25. Effect of isotopically insensitive steps on enzyme kinetic isotope effects.

The double isotope fractionation technique has been applied to the malate synthase reaction which is essentially a Claisen condensation (Scheme 8.26).[103] The stereochemistry at the C2 carbon of the acetylcoenzyme-A

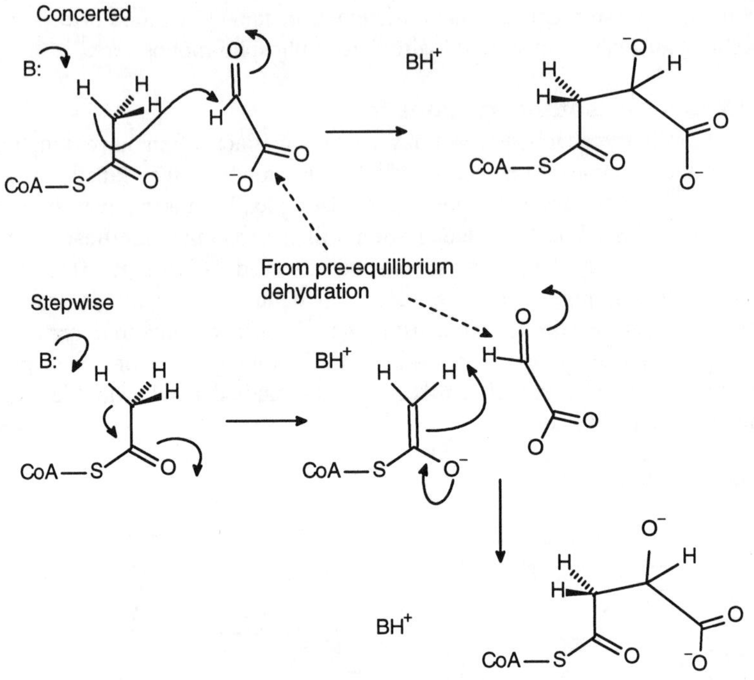

Scheme 8.26. Stepwise and concerted mechanisms for malate synthase.

(acetyl-S-CoA) is inverted during condensation.[103,104] In the case of malate synthase (and Claisen enzymes in general) there is a lack of exchange of the α-proton. While both results are suggestive of a concerted mechanism they are not completely diagnostic and double isotope fractionation indicates a stepwise process. The ^{13}C isotope effect at the aldehydic carbon of the glyoxylate ion when [^{2}H$_3$]-acetylSCoA is substituted for [^{1}H$_3$]-acetylSCoA provides a useful distinction between concerted and stepwise enzymatic processes. If only one isotope effect is observed the mechanism is stepwise but if both ^{2}H and ^{13}C isotope effects are expressed the mechanism may be either concerted or balanced stepwise where both transition states are kinetically significant. A distinction can be made between these cases by observing the consequence of isotopic substitution at one site on the kinetic isotope effect at the other. If the isotopically sensitive step is cleanly rate-limiting deuterium substitution will have no effect on the ^{13}C isotope effect. If the step is *not* cleanly rate-limiting deuterium isotope substitution will increase the kinetic significance of that transition state and the ^{13}C-isotope effect will rise toward the intrinsic value. Deuterium substitution will slow down the first step in a balanced stepwise process and thus decrease the kinetic significance of the second ^{13}C sensitive step. Thus deuterium substitution will result in a smaller observed ^{13}C effect. The observation that ^{13}(V/K)$_H$ = ^{13}(V/K)$_D$ = 1.0037 (essentially no isotope effect) requires that the malate synthase follows a stepwise path. The lack of isotope exchange with solvent and the stereochemical results probably arise from the *snug fit* of the substrate at the active site which could inhibit rotational motion and also exclude solvent from the transferring proton.

The double isotope fractionation method was applied originally to the decarboxylation of malic acid catalysed by malic enzyme (Scheme 8.27). The

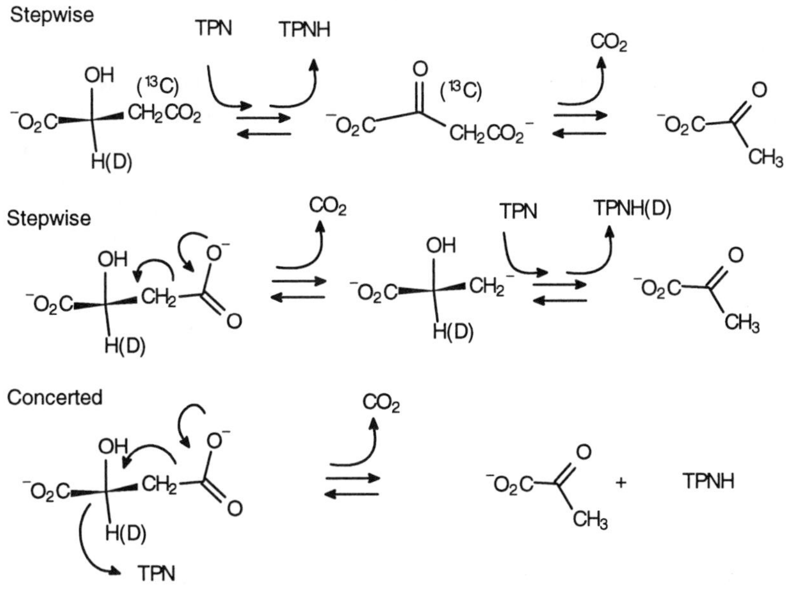

Scheme 8.27. Decarboxylation of malic acid catalysed by malic enzyme.

deuterium isotope effect of 1.47 and the ^{13}C kinetic isotope effects of 1.031 and 1.025 with malate and malate-2-d respectively immediately exclude the concerted pathway and are consistent with hydride transfer preceding the decarboxylation. In this case the deuteriation slows down the first step thus making the second step less kinetically significant and decreasing the ^{13}C kinetic isotope effect (case 2 in Scheme 8.25).[102]

Proline racemase catalyses the inversion of configuration of D- and L-proline; two bases are involved (Scheme 8.28). Let us first assume that the

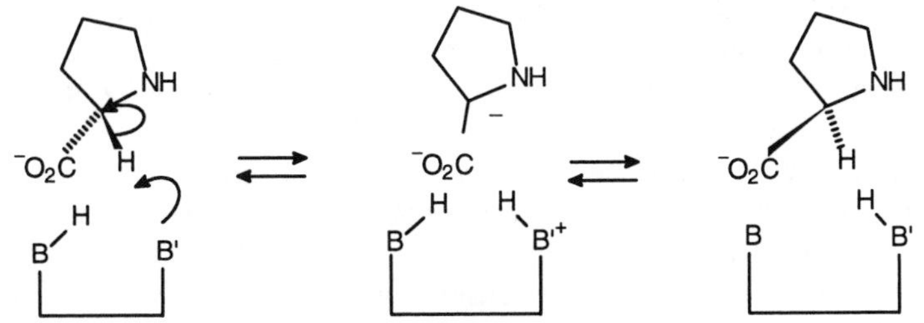

Scheme 8.28. Proline racemase.

mechanism is stepwise; in the first step the C2 proton of D-proline is abstracted to give a carbanion and the second step involves delivery of a solvent-derived proton to this carbanion. In mixed solvent H_2O/D_2O the L-proline product will have a lower deuterium content than the solvent. If the experiment is repeated in the same solvent with D-[2^2H]proline the abstraction is slowed by deuterium substitution but the proton delivery step will be *less* rate-limiting and product deuterium content increases towards that of solvent. If proton transfers are concerted the observed product content will be the same for both isotopes. It is therefore evident that the proline racemase reaction is either stepwise with catalytic groups having fractionation factors close to 0.5 *or* concerted. Thiols have fractionation factors within this range and these are thought to be the bases involved[105] although the fractionation factor is not a particularly good diagnostic tool for the identity of acidic groups (Section 8.1.5).

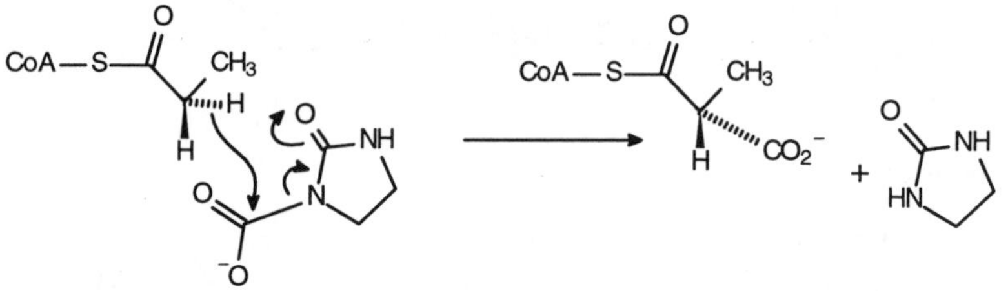

Scheme 8.29. Carboxylation of propionyl-SCoA.

Carboxylation of propionyl-SCoA by N-carboxybiotin to yield S-methylmalonyl-SCoA involves retention of configuration at C2 and this was explained by a concerted process (Rétey and Lynen)(Scheme 8.29).[106,107] The concerted mechanism involves breaking of the C-H bond and making a C-C bond leading to both a ^{13}C and a ^{2}H primary kinetic isotope effect. A stepwise path requires at least one of these effects but may show *both* if the two isotope-sensitive transition states are partly rate-determining. The ^{13}C effect is 1.0227 and for pyruvate-d$_3$ the ^{13}C-isotope effect is 1.0141; thus deuteriation reduces the isotope effect from 2.3% to 1.4% indicating that proton removal and C-C formation are in different steps thus *excluding* the concerted pathway. The retention observed by Rétey and Lynen would then be due to the two substrates being held together after proton transfer during which time the carboxyl group transfers. This process is a D_E+A_E mechanism and, whereas in a non-enzyme process the propionyl-SCoA anion would rotate in the ion-pair complex, this is prevented in the enzyme reaction by the *snug fit* of substrate in active site.

The crotonase catalysed dehydration of 3-hydroxybutyrylpantetheine (Scheme 8.30) has substantial primary ^{2}H and ^{18}O isotope effects for the indicated atoms (^{2}H(V/K) = 1.60 and ^{18}O(V/K) = 1.053). In this case both bonds may be broken in the same transition state of a *concerted* mechanism; it is possible however that the isotope effects reflect two transition states of nearly identical energy.[108]

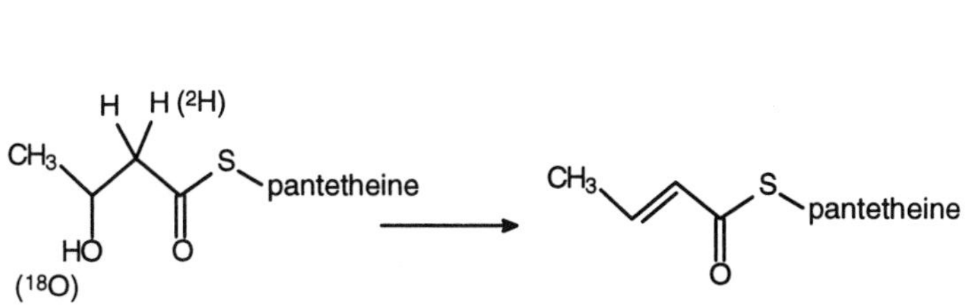

Scheme 8.30. Crotonase catalysed dehydration of 3-hydroxybutyrylpantetheine.

The α-hydrogen of phenylalanine does not exchange with the solvent in the presence of phenylalanine ammonia lyase (Scheme 8.31). The value of the deuterium isotope effect of 2.0 (dideuteriation at C3) and ^{15}N isotope effects of 0.9921 and 1.0047 for the deuteriated and unlabelled substrates excludes stepwise processes with carbanion formation and a concerted pathway but is consistent with an E1cB process. The *pro*-3S-hydrogen is eliminated stereoselectively.[109]

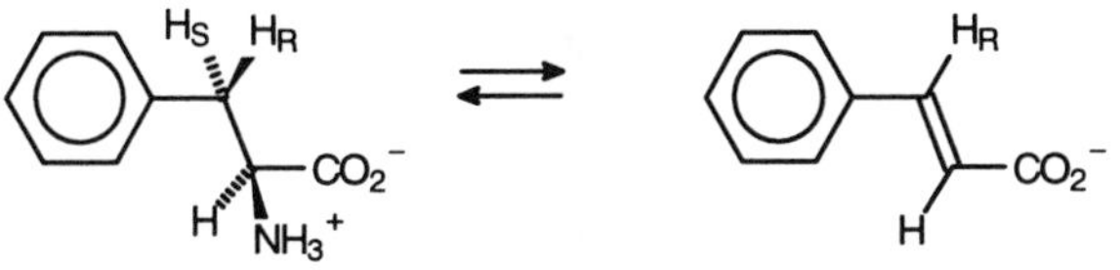

Scheme 8.31. Phenylalanine ammonia lyase.

6-Phosphogluconate dehydrogenase catalyses the oxidative decarboxylation of 6-phosphogluconate to give ribulose-5-phosphate (Scheme 8.32). The ^{13}C kinetic isotope effects and the deuterium isotope effect show that the chemical reaction in the enzyme-substrate complex is *stepwise* with hydride ion transfer preceding decarboxylation.[110]

Enzyme catalysed decarboxylation of prephenate involves hydride transfer to the B-face of NAD. The increase in ^{13}C isotope effect on V/K as the deoxoprephenate is deuteriated (1.0033 and 1.0103) and the deuterium isotope effect of 2.34 indicate that both effects are on the same (concerted) step. If the mechanism were stepwise the ^{13}C-isotope effect would have decreased because deuteriation would slow down the ^{13}C-sensitive step (Scheme 8.33). If the intrinsic isotope effect for ^{13}C and deuterium is 1.0155 the transition state would have little C-C bond fission. The incorporation of ^{18}O into the substrate during acid catalysed decarboxylation of prephenate and deoxoprephenate indicates a stepwise mechanism (Scheme 8.33) in contrast to that of the enzyme catalysed one.[111]

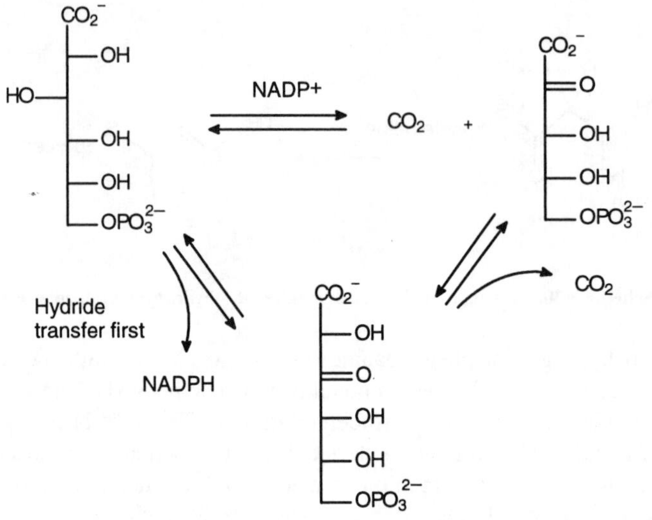

Scheme 8.32. 6-Phosphogluconate dehydrogenase.

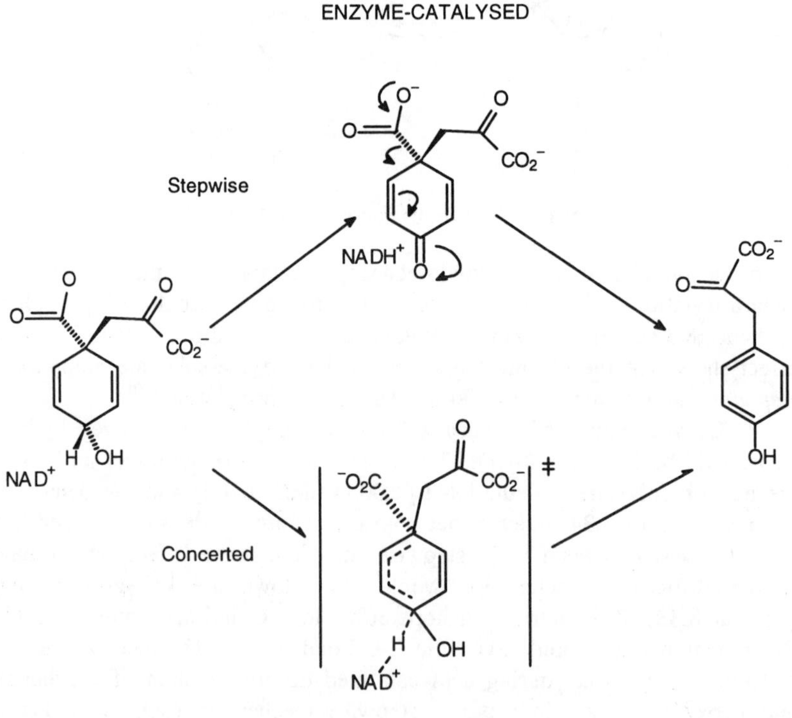

ENZYME-CATALYSED

Scheme 8.33. Decarboxylation of prephenate.

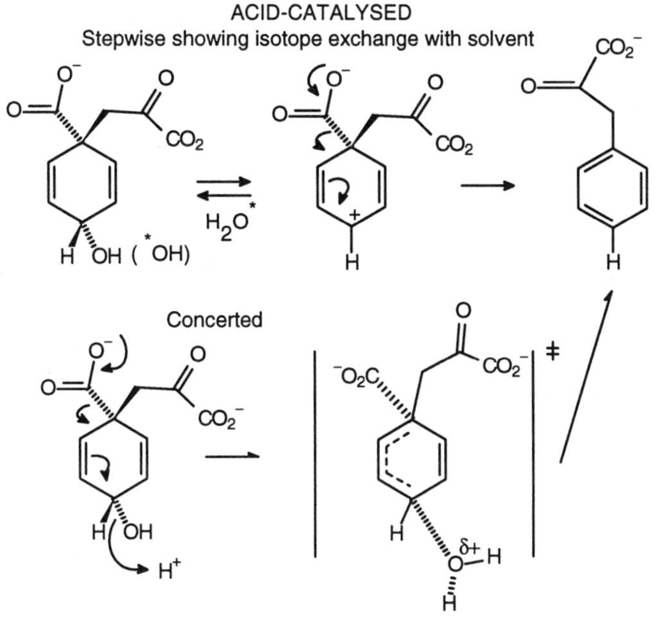

ACID-CATALYSED
Stepwise showing isotope exchange with solvent

Concerted

Scheme 8.33 *(continued).* Decarboxylation of prephenate.

8.4   CARBOHYDRASES

We shall consider in this section lysozymes, chitinases and ß-galactosidases which are members of the carbohydrase family of enzymes. Lysozyme from hen's egg white was the first enzyme to have its structure solved by x-ray crystallography, and today the structures of many lysozymes have been determined including mutant species and enzyme inhibitor complexes. The extraordinarily long primary sequence of β-galactosidase[112] and its only recently determined three-dimensional structure[113] has meant that the structural aspects of the enzyme have not kept pace with the kinetic studies which are well developed for this enzyme.

In recent years the emphasis in mechanism has been on the relationship between the lifetime of a carbenium ion and the concerted process whereby the carbenium ion is not explicitly formed. Indeed the methoxycarbenium ion ($CH_3OCH_2^+$), which is related to the action of carbohydrases, has been estimated to have a lifetime of about 10^{-15} seconds in water, and its components are therefore involved in transfer reactions in an open or "exploded" A_ND_N transition state (**8.3** in Scheme 8.34).[114,115]

Lysozyme catalyses the fission of the glycoside link in oligosaccharides of N-acetylglucosamine and its derivatives. The mechanism proposed as a result of the initial x-ray crystallographic results involves general acid catalysis by the glutamate-35 residue and stabilisation of an *oxocarbenium* ion intermediate by the anion of the aspartic acid-52 residue (Scheme 8.34). An acylal intermediate is not observed, but the observation of strict retention of configuration is evidence for the carbenium ion model *or* a concerted process involving an acylal intermediate; neither mechanism of Scheme 8.34 has been convincingly demonstrated yet.

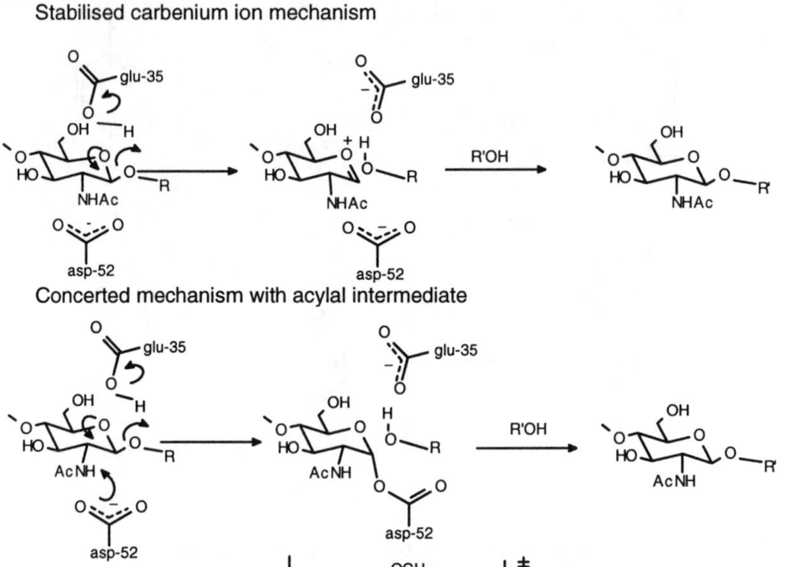

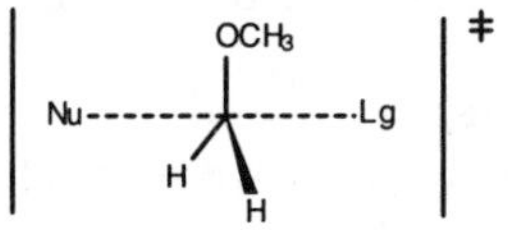

Stabilised carbenium ion mechanism

Concerted mechanism with acylal intermediate

Scheme 8.34. Mechanisms for lysozyme and its analogues.

Lysozyme also catalyses transfer of the glycosyl group to acceptors other than water with retention of configuration at C1. A front-side A_ND_N process has not been excluded for this but it is generally considered that the retention at C1 arises from two inversions (Scheme 8.34). Although the site of the catalytic process is in little doubt it is chastening to note that model-building studies provide the only evidence that the carboxylic residues 35 and 52 are indeed involved. The glutamic acid is proposed to act in the acid form and the aspartate in its base form, but the problem of kinetic ambiguity (whereby the aspartate residue acts as acid and the glutamate as anion) has not been absolutely resolved. In the models of the enzyme substrate complex the glutamate carboxylic acid is close to the ring oxygen and the aspartate carboxylate anion to the C1 carbon.

The putative acylal intermediate in the lysozyme mechanism would involve C1 attached covalently to the oxygen of the aspartate-52 group. Studies of the effect of the leaving group structure on the reactivity are consistent with substantial bond fission. The ^{18}O-isotope effect on the leaving oxygen $(1.047)^{116}$ compares with the value (1.025) for model reactions where the oxygen receives a proton. The α-deuterium secondary isotope effect of between 1.11 and 1.19 from various substrates indicates a reduced bond order at C1 consistent with sp³ to sp^2 rehybridisation.[117,118] Thus the mechanism would have an open transition structure for formation of the acylal intermediate (Scheme 8.34).

The nature of an enzyme intermediate could be difficult to identify if its formation (k_2) were slower than its decomposition (k_3); such is the case with the glycosyl lysozyme where all the kinetic data refer to its formation. In lysozyme it is therefore difficult to distinguish between a transition state on the way to either an sp^2 or an sp^3 intermediate. For ß-galactosidase $k_2>k_3$ and the α-deuterium secondary isotope effects at C1 on k_3 of 1.1 and 1.21 for the enzyme and a number of isozymes are consistent with an sp^3 to sp^2 rehybridisation as expected for a covalent intermediate. An ionic intermediate requires an inverse isotope effect.[116,119,120] However, reference to Chapter 5 indicates that α-deuterium secondary isotope effects are not compellingly diagnostic between S_N2 and S_N1 processes.

Early structural studies considered that the putative oxocarbenium ion would only interact covalently with aspartate-52 to form an acylal with difficulty due to changes elsewhere in the oligosaccharide. In oligosaccharide-enzyme complexes the distances between C1 of the putative fission site and the nearest oxygen of the carboxyate ion are of the order of 3 Ångstroms. It seems inconceivable that the known very high reactivity of such species does not force it to collapse by conjugation with any weak adjacent nucleophile. It is well known that oxocarbenium ions are very highly reactive in water.[115] In the lysozyme case, where there is retention of configuration at C1, one face of C1 in the glycosyl cation must be protected so that the lifetime of the cation would have to exceed that of a diffusion process enabling the leaving group to depart and be replaced by an acceptor nucleophile. The demonstration of acylal intermediates in other glycosidases involving retention of configuration[121-125] is difficult to ignore and together with the absence of data to the contrary one is led to accept the simplest explanation, namely the formation of an acylal intermediate in the lysozyme mechanism.

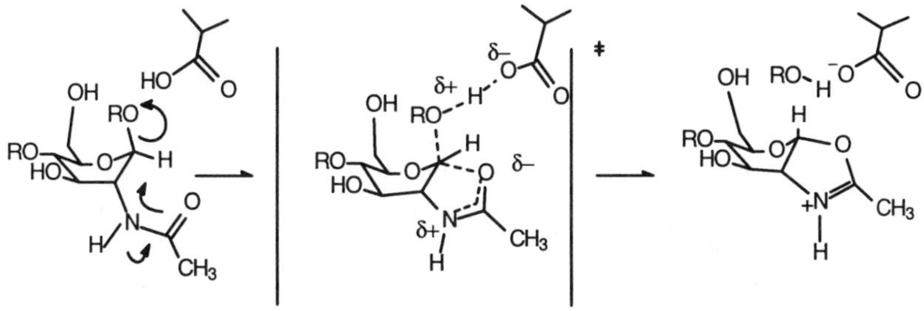

Scheme 8.35. Family 18 chitinase retains configuration at C_1 by intervention of an oxazoline ion.

Carbohydrases can be classified into families according to their sequence structure and chitinases belong to families 18 and 19.[126] The family 18 chitinases, of which lysozyme is an example, show retention of configuration at C_1 during hydrolysis. Family 19 chitinases exhibit inversion of configuration.

The family 18 chitinase from *Serrata marcescens* has been examined by *ab initio* quantum mechanical calculations at the HF/6-31G** level which predict that bond fission at C_1 leads to an oxazoline ion intermediate; only the oxazoline ion is able to orient in the enzyme active site to allow stereoselective attack by water (Scheme 8.35).[127-129] The x-ray crystallographic structure from the chitinase from barley (*Hordeum vulare* L., a member of the family 19 chitinases) has two acidic residues (glu 67 and glu 89) separated by about 9.2Å in the active site. The formation of an oxazoline ion is prevented by hydrogen-bonding of the C2' N-acetyl group of the glycosyl moiety (see Scheme 8.36) and a general base catalysed attack of water on the C1 position assisted by general acid assistance of leaving group departure is predicted by the simulation method.

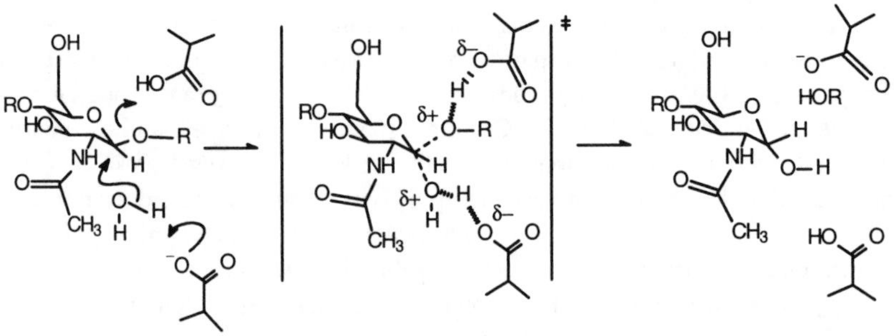

Scheme 8.36. Mechanism for displacement in family 19 chitinases showing inversion at C1.

8.5   ASPARTYL PROTEASES

Pepsin was one of the first proteases to be studied as a pure protein[130] but its assay technique for substrate hydrolysis has proved to be a large barrier to progress. The realisation that the HIV-1 protease has a mechanistic similarity enabled work in this area to be extensively funded and many aspartyl proteases now have fully documented atomic coordinates.[38,131,132] The mechanism of action of aspartyl proteases is not clear at present although it seems to require the action of two carboxylic acid groups probably as a carboxylate ion-carboxylic acid pair (Scheme 8.37). Although there is a similarity between residue requirements at the active site between aspartyl proteases and glycosidases there appear to be no other similarities except that ester-forming reagents inhibit both types of enzyme.

A few non-natural substrates form relatively stable covalently-bound intermediates with aspartyl proteases (Scheme 8.37). The intermediate in the sulphite reaction is unstable and degrades to yield the free enzyme whereas the products from the other reactions are relatively stable and the modified amino acid can be recovered by hydrolysis of the protein. On the basis of these reactions, with the knowledge that the reagents have good leaving groups, it is not surprising that if there were an acyl enzyme intermediate in the aspartyl protease catalysed peptide hydrolysis it would be formed slowly because of its poor leaving group and would decompose quickly. Despite this obvious chemical knowledge and precedent the literature favours mechanisms with non-acyl enzyme species. There are two main mechanisms (Scheme 8.38) each of which involves concerted processes:[140] one type involves an acyl enyme (located on an aspartate residue as an anhydride link) and the other involves a tetrahedral intermediate formed between water and the peptide link.

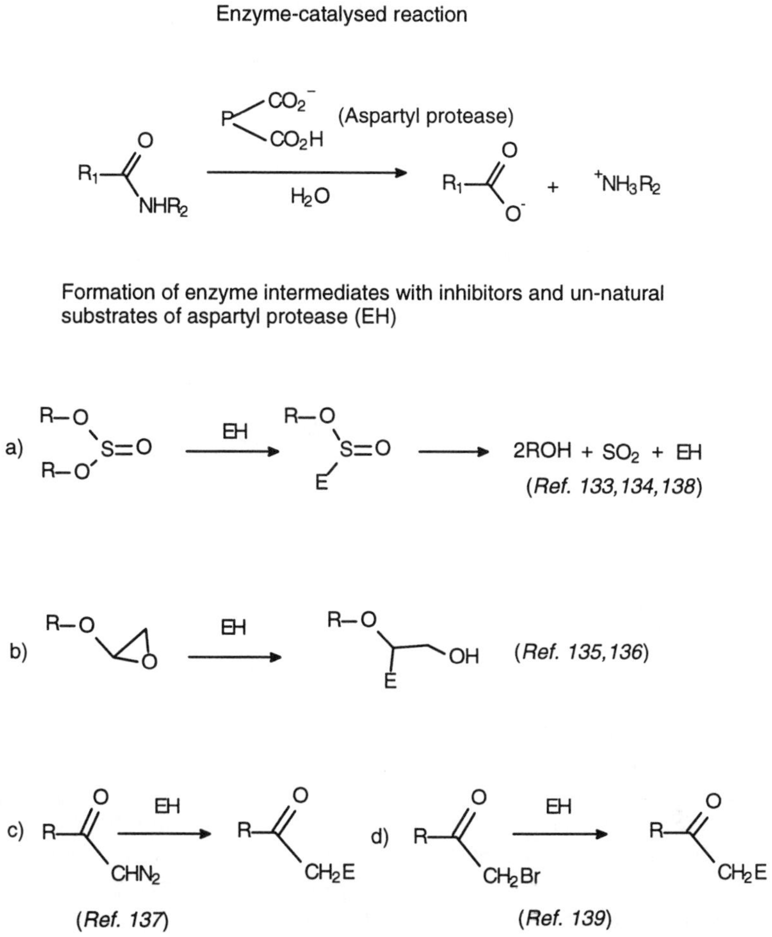

Enzyme-catalysed reaction

Formation of enzyme intermediates with inhibitors and un-natural substrates of aspartyl protease (EH)

a) (Ref. 133,134,138)

b) (Ref. 135,136)

c) (Ref. 137)   d) (Ref. 139)

Scheme 8.37. Some reactions of aspartyl proteases with substrates and inhibitors (Ref. 130).

X-Ray crystallographic studies have been made on enzyme-inhibitor complexes with aspartyl proteases where the central portion of the peptide inhibitor is proposed as a transition state analogue of the statine type (Scheme 8.39). [141-143] Although the transition state analogues prove to be excellent inhibitors it is not certain that much of the binding derives from the reaction centre component. The situation is further hampered by subjectivity in the choice of mechanisms from which to derive the transition state analogue. It would be very unwise to decide that since an inhibitor design based on a subjective mechanism is excellent that such a mechanism is correct, particularly when there is an alternative mechanism which is more chemically appropriate.

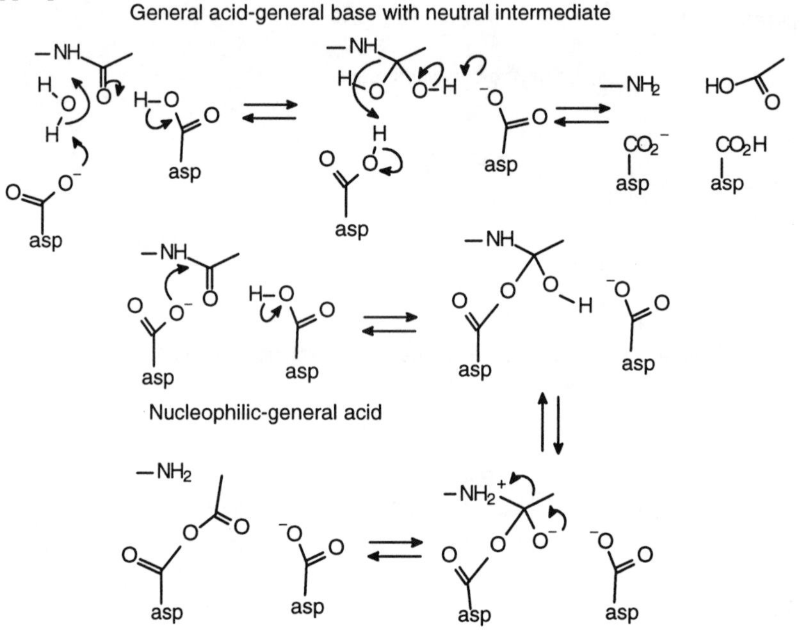

General acid-general base with neutral intermediate

Nucleophilic-general acid

Scheme 8.38. Some catalytic mechanisms for the aspartyl proteases.

The mechanism involving a tetrahedral adduct formed by general base catalysed attack of water on the peptide bond of the substrate fits the molecular orbital calculations (Scheme 8.40);[140] this structure is consistent with the observation of [18]O exchange from enriched solvent into the peptide carbonyl group of the substrate remaining after substantial hydrolysis has occurred (Scheme 8.41). Labelled oxygen is incorporated into the tetrahedral intermediate in the first step by attack of asp-125. Donation of a proton to the unlabelled, original oxygen occurs from asp-25 which becomes protonated by transfer of a proton from asp-125(Scheme 8.41).[144,145] However, exchange is also consistent with the anhydride mechanism (Scheme 8.42) if the covalently bound tetrahedral intermediate were to partition to an acyl imidate species by release of hydroxyl (as water) followed by attack of labelled water on the imidate. Protonation of the nitrogen in the anhydride mechanism (Scheme 8.42) after formation of the neutral tetrahedral intermediate is diffusion controlled at the pH optimum of the enzymic catalysis (pH 4) because the amine has a pK_a in the region of 9 - 10.

Substrate:

Angiotensinogen ...his-pro-phe-his-leu-CO-NH-val-ile-his...

Inhibitors:

L-363,564 (*statine type*)

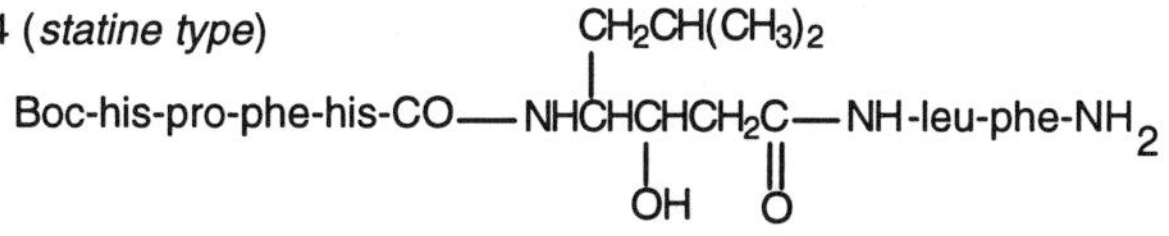

$$CH_2CH(CH_3)_2$$
$$|$$
Boc-his-pro-phe-his-CO——NHCHCHCH_2C——NH-leu-phe-NH_2
$$\underset{OH}{|} \quad \underset{O}{\|}$$

H-142: pro-his-pro-phe-his-CH_2-NH-val-ile-his-lys

Aspartyl protease endothiapepsin - L-363,564 complex:

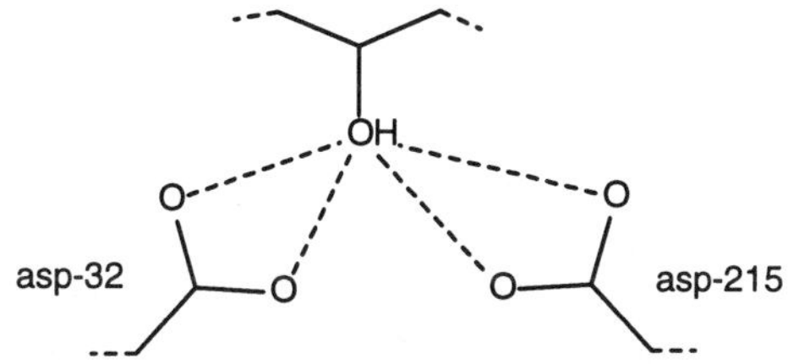

Scheme 8.39. Transition state analogues for aspartyl proteases; the arrangement shown for the L-363,564-endothiapepsin complex is absent in the H-142 complex. The numbering scheme is from *Ref. 141.*

Substrate $CH_3CONHser$-gln-asn-tyr-pro-ile-val$COOCH_3$

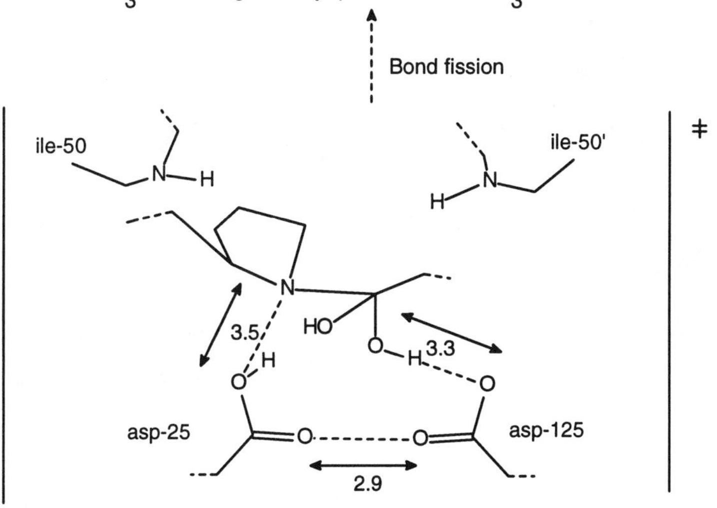

Bond fission

Scheme 8.40. Transition structure predicted for the aspartyl protease. Bond distances in Ångstroms are from *Ref. 140.*

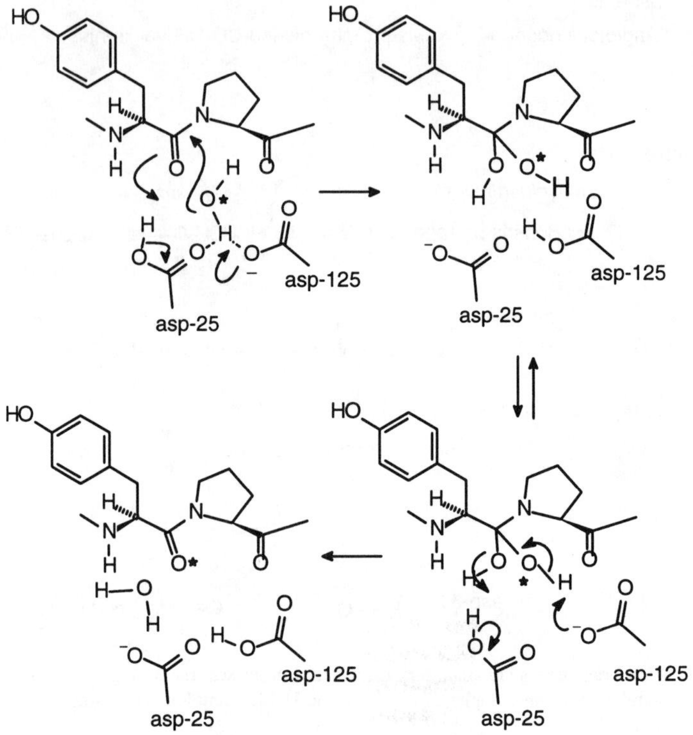

Scheme 8.41. A mechanism of ^{18}O exchange from solvent into the carbonyl oxygen of the substrate catalysed by aspartyl proteases.

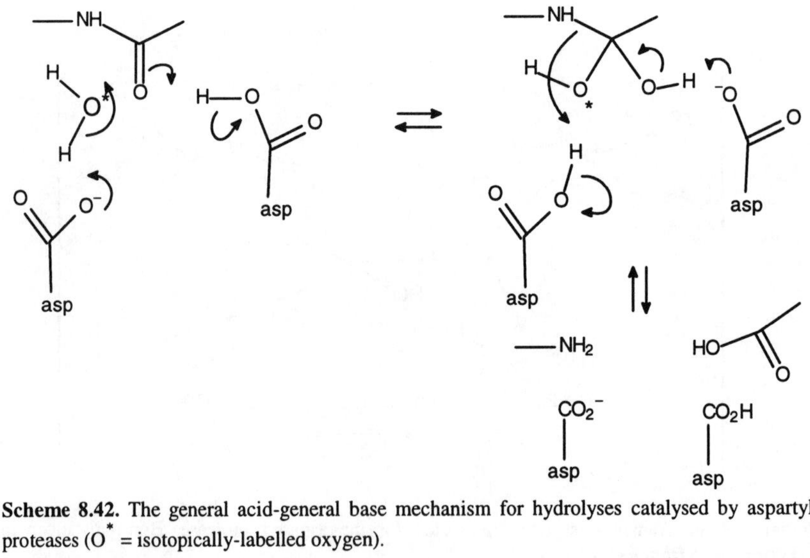

Scheme 8.42. The general acid-general base mechanism for hydrolyses catalysed by aspartyl proteases ($O^* =$ isotopically-labelled oxygen).

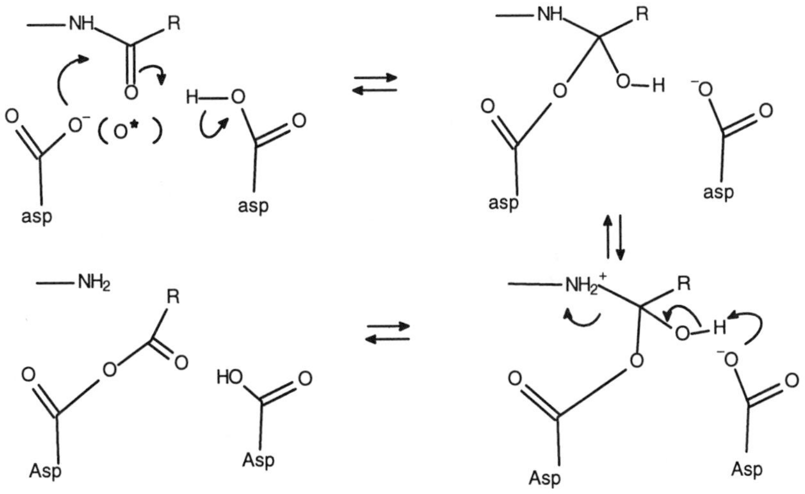

Scheme 8.42 (*continued*). The anhydride mechanism for hydrolyses catalysed by aspartyl proteases (O* = isotopically-labelled oxygen).

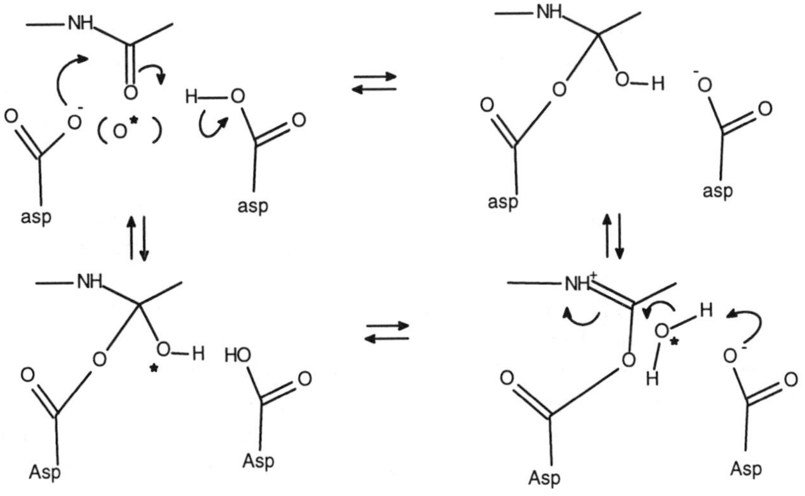

Scheme 8.42 (*continued*). Exchange mechanism *via* the anhydride for hydrolyses catalysed by aspartyl proteases (O* = isotopically-labelled oxygen).

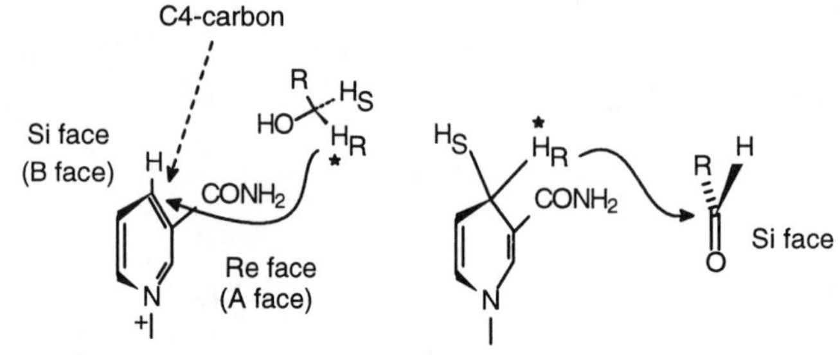

Scheme 8.43. Stereochemistry of hydrogen transfer between pyridinium ion nucleotide and the primary alcohol in alcohol dehydrogenases.

8.6 HYDRIDE TRANSFER

Liver alcohol dehydrogenase (LADH) is an example of redox enzymes which catalyse the interconversion of alcohol and aldehyde or aldehyde and carboxylic acid by hydrogen transfer to the pyridine ring of a nicotinamide coenzyme residue (such as NAD$^+$P) to form a reduced pyridine nucleus.[146] *In vivo*, the reduced pyridine is re-oxidised to the nicotinamide group *via* an electron-transport pathway linked to oxygen.

The stereochemistry of the LADH enzyme system requires that the coenzyme and substrate occupy the active site of the enzyme in such a way that only the H$_R$ hydrogen is transferred between the Re face of the pyridinium ion and the Re face of the aldehyde (Scheme 8.43). The relatively large primary isotope effect for the transfer of the hydrogen indicates that this step is rate-limiting.[147]

Residues implicated in the active site include histidine-67, cysteine-46 and cysteine-174 which ligate one of the zinc(II) ions and in the free enzyme there is a water molecule which completes a distorted tetrahedral complex of the zinc. Two mechanisms have been proposed both of which invoke substantial arrays of concerted bond changes. In the first, the zinc(II)-coordinated water, with a reduced pK$_a$, acts as an acid catalyst (Scheme 8.44). The second mechanism involves displacement of the coordinated water by the alcohol or carbonyl and the electrophilic component of the catalytic step includes both zinc(II) and the hydroxyl proton of serine-48 (Scheme 8.44). There is no substantial evidence in favour of either stepwise or concerted bond changes in the proposed mechanism.

A radical mechanism is excluded for the enzyme because the product of oxidation of bicyclo[4.1.0]heptan-2-ol is the ketone wherein the cyclopropyl ring remains intact.[148]

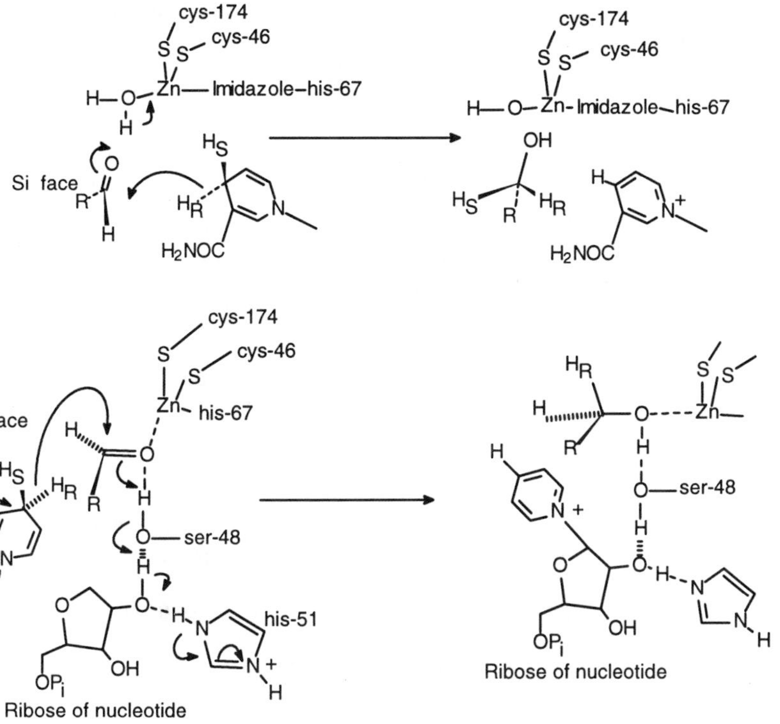

Scheme 8.44. Mechanisms for the reduction of the in liver alcohol dehydrogenase: a) water coordinated to the zinc(II) acting as a general acid; b) zinc(II) as a Lewis acid by direct coordination with the carbonyl of the aldehyde.

Intrinsic ^{2}H and ^{13}C isotope effects for oxidation of benzyl alcohol catalysed by the liver alcohol dehydrogenase enzyme indicate an aldehyde-like transition state;[149] the absence of normal solvent isotope effects[150] is consistent with proton transfer steps which are not rate-limiting and not concerted with the hydride ion transfer. X-Ray crystallographic results indicate that an imidazolium ion (his-195) and a guanidinium ion (arg-109) are located in lactate dehydrogenase close to the oxygen of the lactate moiety and could act as electrophilic acceptors of the ketonic oxygen of the pyruvate product; an aspartate-histidine diad (asp-168/his-195) probably functions to position the imidazolium ion for interaction with the oxygen of the lactate moiety (Scheme 8.45).

The glyceraldehyde phosphate dehydrogenase system (Scheme 8.46) involves reduction from acid to aldehyde level and this includes the imidazolium ion of histidine-176 acting as the electrophilic activator of the carbonyl function in the substrate.

The hydrogen isotope effect for the hydrogen being transferred during the formate dehydrogenase reaction is $^Hk/^Dk = 2.2$ and the carbon isotope effect

($^{12}k/^{13}k = 1.042$) is the same for both protio or deuterio formates.[151,152] This indicates that the reaction is concerted and that substrate binding and release are not rate-limiting (Scheme 8.47).

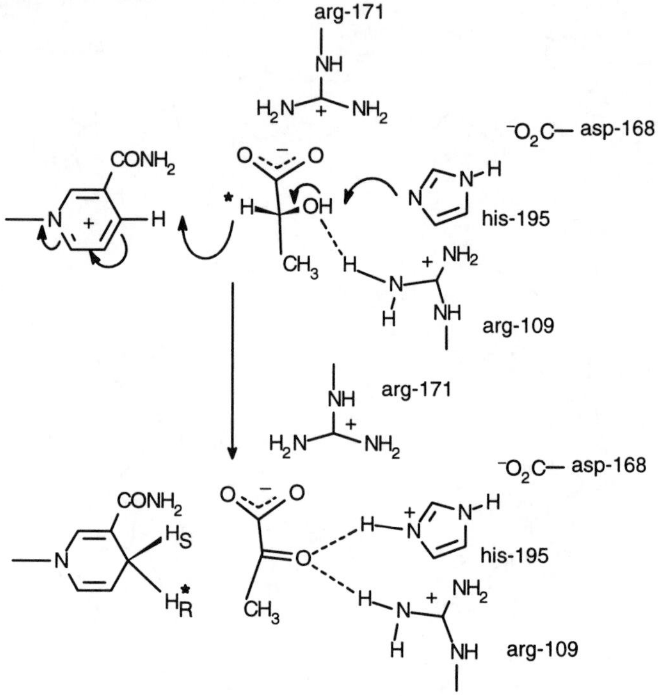

Scheme 8.45. Schematic diagram of the mechanism for lactate dehydrogenase.

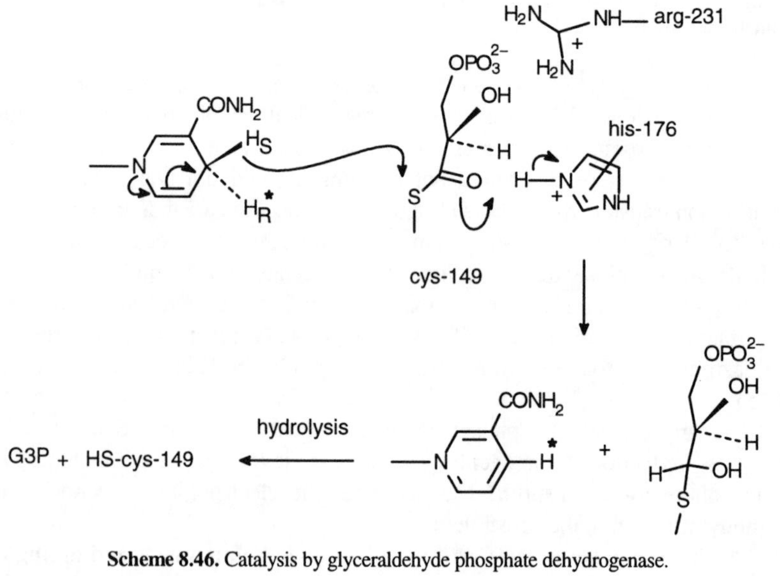

Scheme 8.46. Catalysis by glyceraldehyde phosphate dehydrogenase.

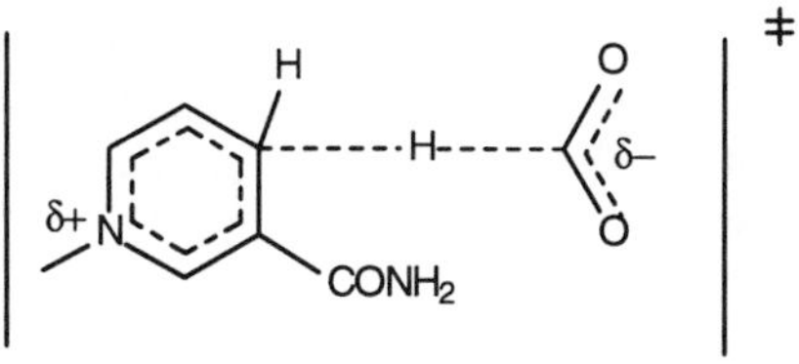

Scheme 8.47. Concerted hydride transfer in the formate dehydrogenase reaction.

There appears to be reasonably good evidence that bond formation and fission in hydrogen transfer is concerted in the pyridine-dihydropyridine systems. The coupling of this two-bond process with proton transfer and other electrophilic bonding changes in Schemes 8.44 - 8.47 is still only reasonable speculation but it is likely that concert obtains in view of the alignment of the reactants afforded by the complexation of substrate with coenzyme and active site.

8.7 AMINE OXIDASES

D-Amino acid oxidase, a flavoenzyme, catalyses the oxidation of a D-amino acid to the corresponding imino species which then hydrolyses to ammonia and 2-oxoacid (Scheme 8.48).[153] The reaction could involve prior carbanion

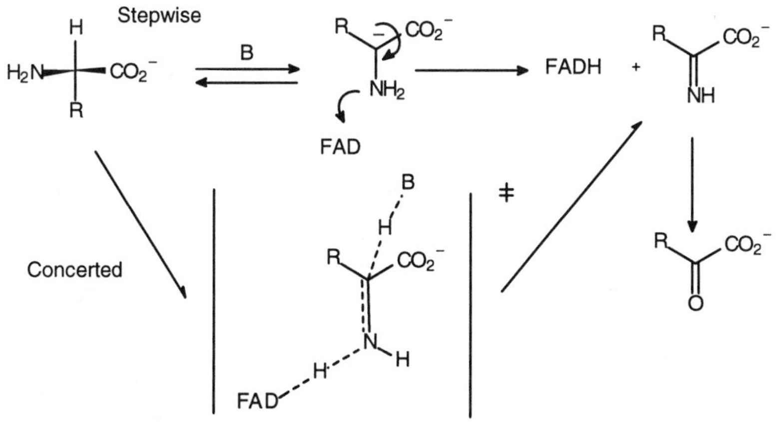

Scheme 8.48. Oxidation of a D-amino acid.

formation by proton abstraction followed by electron transfer to the oxidised flavin. However, Miura and Miyake[154] consider that the formation of the carbanion is very unfavourable indeed and the activation energy should be greater than 29.9 kcal/mole whereas the reported value for D-amino acid oxidase/D-alanine is 17.6 kcal/mole. If the enzyme were to proceed via the carbanion intermediate the activation energy of carbanion formation would need to be lowered by 11.9 kcal/mole (equivalent to lowering the pK_a of the 2-proton by more than 9 pK_a units). The reaction of monoamine oxidase with β-chloroalanine gives no elimination product (Scheme 8.49).[154] If the reaction were to involve the carbanion (Scheme 8.49) then an elimination product would be expected and the absence of such a product is therefore interpreted to support the concerted mechanism (although it cannot be said that the results are absolutely diagnostic).

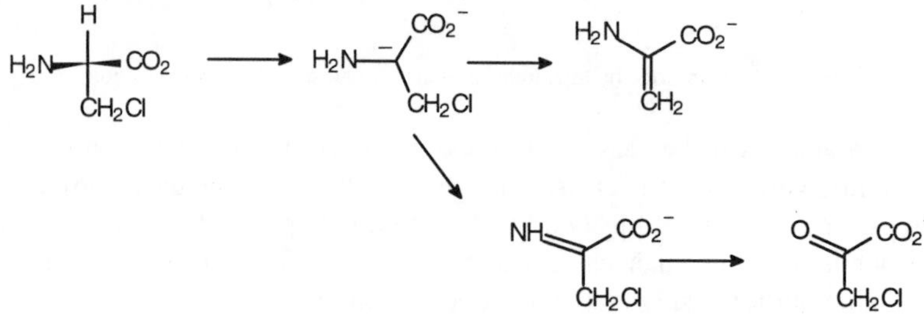

Scheme 8.49. Reaction of D-amino acid oxidase on β-chloroalanine.

8.8   CARBON-CARBON BOND FORMATION

Carbon carbon bond formation and fission reactions most often occur *via* attack of a nucleophilic carbon (usually carbanionic) at an electrophilic carbon centre *via* aldol-type condensation reactions. These reactions are central to both primary and secondary metabolic processes.

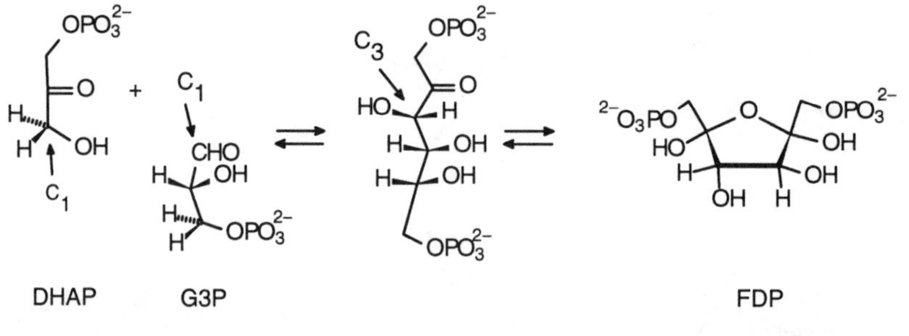

Scheme 8.50. The overall aldolase reaction; FDP = fructose-1,6-diphosphate.

Class I aldolases require lysine as an essential component and catalyse the aldol condensation reaction between DHAP and G3P to form FDP (fructose-1,6-diphosphate) (Scheme 8.50), and only the pro-S hydrogen of DHAP is replaced by the C1-carbon of G3P; no label is incorporated at C3 of FDP when the synthesis is carried out in D_2O solvent. The reaction occurs in two stages involving removal of the proton from C1 of DHAP followed by addition of the putative carbanion to the G3P. Aldolase also catalyses the stereospecific exchange of the 1-[(S)-^{3}H] proton of DHAP in the absence of G3P and the exchange rate can be reduced by adding G3P; the pH-dependence of the kinetics is similar for both exchange and condensation reactions. The stereochemistry of the exchange is opposite to that catalysed by triose phosphate isomerase where the (1-[(R)-^{3}H-) is displaced but is the same as that in the full aldolase reaction. The DHAP-enzyme complex exhibits increased absorbance between 240 and 270nm consistent with formation of an imino function[155] and isolation of the enzyme-DHAP complex inhibited by labelled $NaBH_4$ reagent, followed by amino acid sequence analysis, indicates that the link is between the DHAP and lys-229 of the enzyme.

Scheme 8.51 illustrates the simplest mechanism for catalysis of hydrogen exchange based on the results given above. Reduction of the imine intermediate by sodium borohydride is stereospecific and the observation of the stereochemical configuration of the product indicates that the Si face of the DHAP attached to the enzyme by the imino function faces out into bulk solvent.[156,157]

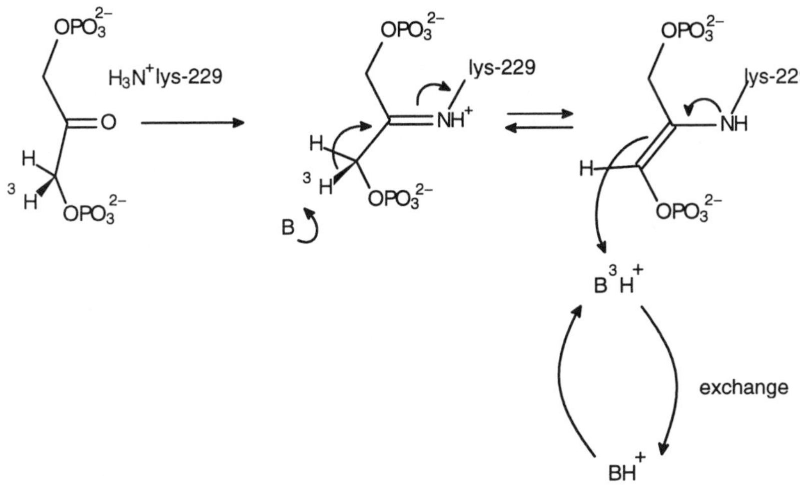

Scheme 8.51. The mechanism of proton exchange at dihydroxyacetone phosphate catalysed by aldolase.

Scheme 8.52 shows the imine-forming step where the hydroxyl lost from the carbinolamine intermediate (as water) presumably departs from the Si face (which is attacked by $NaBH_4$). The side chain amino group of lysine-229 points up from the floor of a cavity in the active site. The model, which presents the Si face of the imino group to the solvent, has been interpreted to involve the groups (lys-146 and arg-148) and (lys-41, arg-42 and arg-303) respectively in interaction with the C1 and C6 phosphate ions. Lysine-146 appears to be close enough to be involved in proton transfer at the carbinolamine nitrogen in lysine-229 and at the oxygen in the dehydration step (Scheme 8.52). The hydroxyl group of tyrosine-363 appears to be involved in addition of a proton to C3.[157] While the proposed concerted mechanisms (Schemes 8.51 and 8.52) appear attractive there is no substantial evidence for concertedness as opposed to the corresponding stepwise mechanisms.

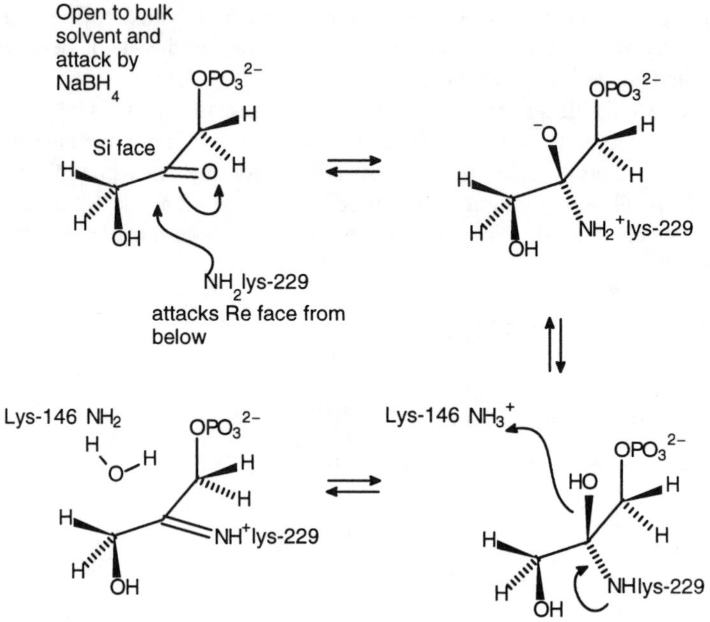

Open to bulk solvent and attack by NaBH$_4$

$OPO_3{}^{2-}$

Si face

H

H

H

OH

O

H

NH$_2$lys-229 attacks Re face from below

$\rightleftharpoons$

$OPO_3{}^{2-}$

H

H

O

H

H

OH

NH$_2{}^+$lys-229

$\updownarrow$

Lys-146 NH$_2$

H

O—H

$OPO_3{}^{2-}$

H

H

H

H

OH

NH$^+$lys-229

$\rightleftharpoons$

Lys-146 NH$_3{}^+$

HO

$OPO_3{}^{2-}$

H

H

H

H

OH

NHlys-229

Scheme 8.52. Stereochemistry of the imine forming step in the aldolase catalysed reaction (the lysine-146 appears to have a role in proton transfer at both nitrogen and oxygen of the carbinolamine). The Re face of the dihydroxyacetone phosphate is the same face as the Si face of fructose 1,6-diphosphate shown[156] to be the site of attack by the terminal amino group of lysine-229.

The carboxylation of ribulose-1,5-biphosphate is the conduit for most of the carbon occurring in the living world.[158] The reaction involves addition of carbon dioxide to the ribulose molecule followed by carbon-carbon bond fission to yield two molecules of glycerophosphate (Scheme 8.53) catalysed

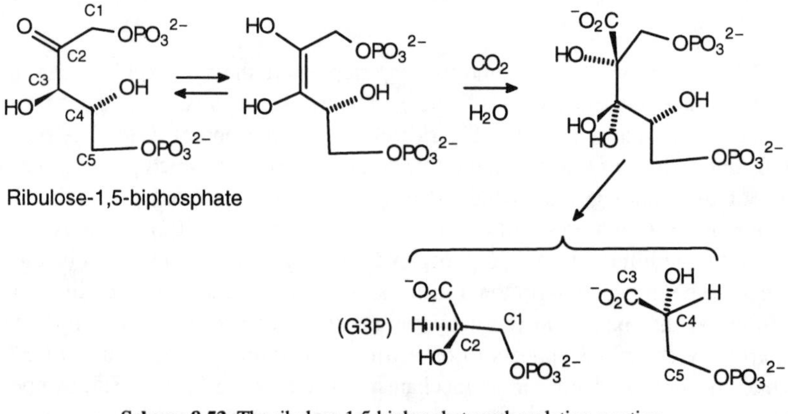

C1

O

C2

$OPO_3{}^{2-}$

C3

HO

C4

OH

C5

$OPO_3{}^{2-}$

Ribulose-1,5-biphosphate

$\rightleftharpoons$

HO

$OPO_3{}^{2-}$

HO

OH

$OPO_3{}^{2-}$

CO_2

H_2O

^-O_2C

HO

$OPO_3{}^{2-}$

HO

OH

HO

$OPO_3{}^{2-}$

(G3P)

^-O_2C

H

C1

HO

C2

$OPO_3{}^{2-}$

C3

OH

^-O_2C

H

C4

C5

$OPO_3{}^{2-}$

Scheme 8.53. The ribulose-1,5-biphosphate carboxylation reaction.

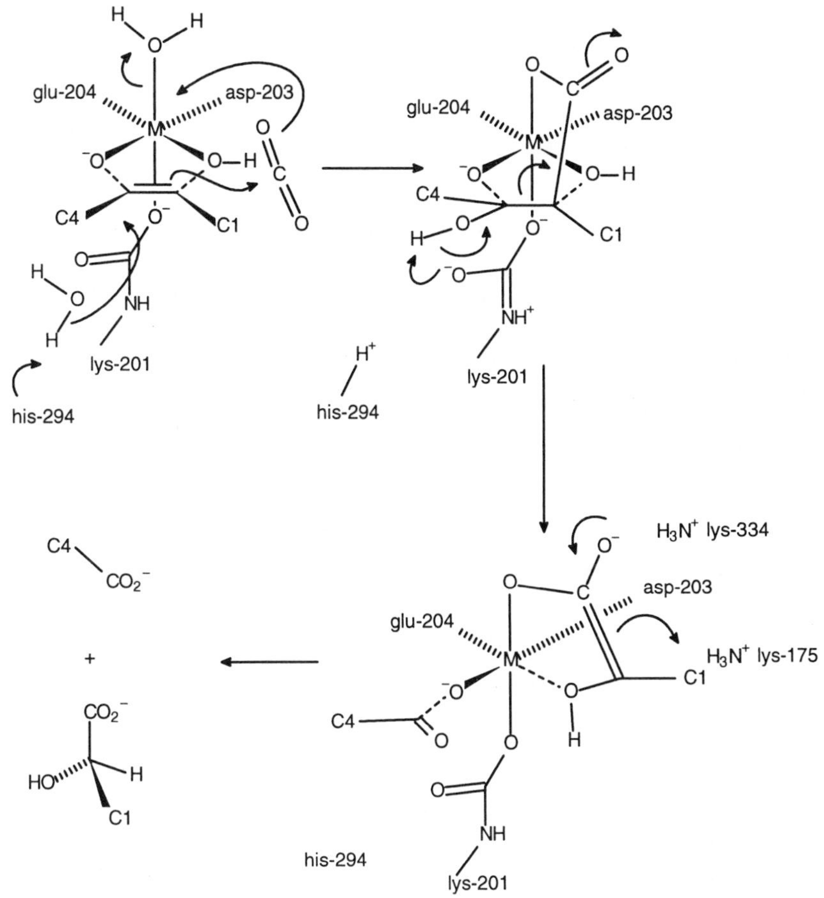

Scheme 8.54. The carboxylation and carbon-carbon bond fission mechanisms in the ribulose-1,5-biphosphate carboxylation reaction.[158] The C4 and C1 carbons are those identified in Scheme 8.53.

by ribulose bisphosphate carboxylase (rubisco). The carboxylation step involves bond formation to CO_2 at C2 in advance of bond formation with water at C3, and the mechanism is illustrated in Scheme 8.54.

8.9 CLOCKS

The alkane mono-oxygenase from *Methylococcus capsulatis* (*Bath*) hydroxylates (R)-$[1-^2H_1, {}^3H_1]$ ethane to yield ethanol with retention of configuration.[159] Provided the carbon-carbon bond rotation of an ethyl (Scheme 8.55) radical or cation is not inhibited when close-packed in the active site then retention of configuration would indicate a concerted mechanism. The primary isotope effect ($^Hk/^Dk$) of 5.2 indicates that the hydrogen is in flight in the transition state which must therefore involve attack at the carbon concerted with hydrogen abstraction.[159]

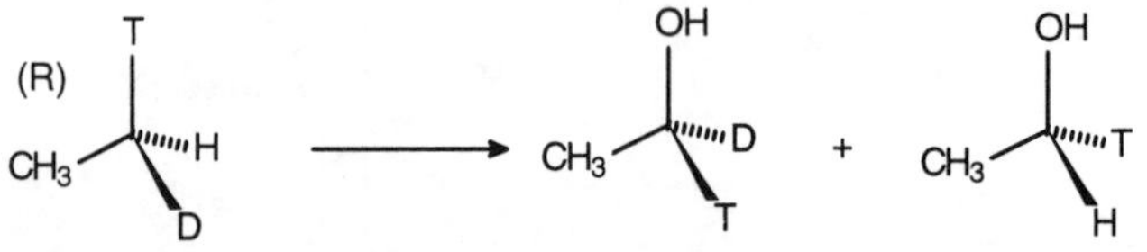

Scheme 8.55. Ethane hydroxylation.

The hydroxylation catalysed by cytochrome P450 has been studied by use of an alkane (**8.4** → **8.6**) which according to model studies gives a radical which would rearrange as shown (Scheme 8.56) with a rate constant of 6 10^{11} sec^{-1}.[160] The initial hydroxylation product, **8.6**, undergoes further reaction to give **8.7** so that the total flux of the hydroxylation is measured by the sum of **8.6** and **8.7**.

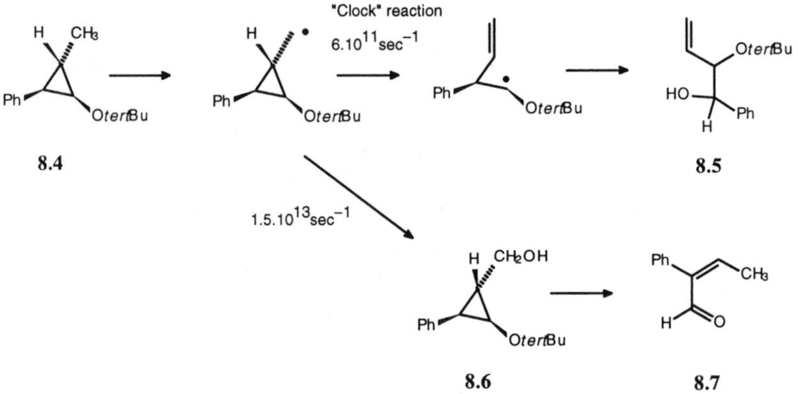

Scheme 8.56. Rearrangement of cyclopropylmethyl radicals acts as a clock for its lifetime in hydroxylation catalysed by cytochrome P450.

The ratio of the products (**8.6** + **8.7**)/**8.5** in the hydroxylation catalysed by cytochrome P450 together with the rearrangement rate constant (6 10^{11} sec^{-1}) enables the capture rate of the radical to be estimated (1.5 10^{13} sec^{-1}). The capture rate is too high for the radical to be a true intermediate and the formation of the **8.6** and **8.7** from **8.4** must therefore involve radical abstraction (of H$^{\bullet}$) concerted with the capture by hydroxyl.

8.10 CYCLIC REACTIONS

The only pericyclic process so far observed in primary metabolism is the Claisen rearrangement of chorismate, a key intermediate in the shikimic acid pathway, to prephenate.[161,162] This reaction (Scheme 8.57) is catalysed by the chorismate mutase from *E.coli* that has been characterised to 2.2Å resolution by x-ray crystallography.[163] Chorismate mutase from *bacillus subtilis* has also been characterised by x-ray crystallography.[164,165] Both native enzymes, when complexed with the competitive inhibitor (**8.8**), exhibit surprising similarities at the active sites although the overall primary sequences and secondary structures have few homologies. The active site interactions for both enzymes are illustrated in Scheme 8.58.

The Claisen rearrangement is itself catalysed by Lewis acids by coordination with the vinyl oxygen (Scheme 8.59),[161] and it is thought that the electrophilic interactions provided by the enzyme with the bound substrate hold the substrate in its boat conformation and assist in the sigmatropic rearrangement.

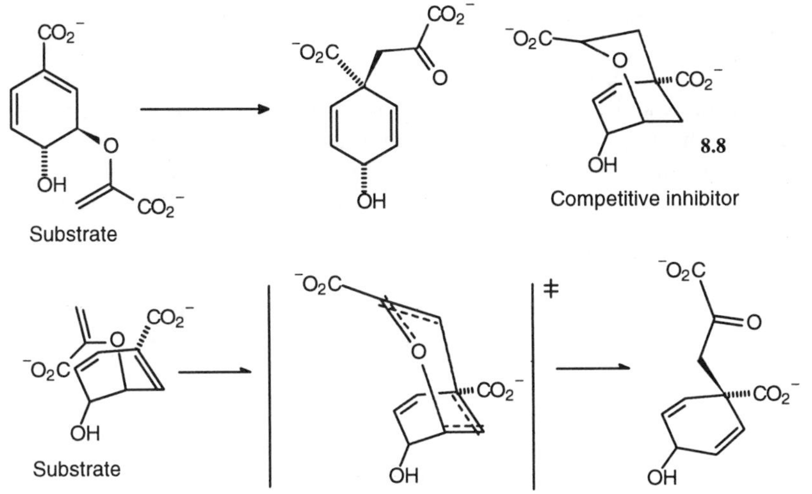

Scheme 8.57. Rearrangement of chorismate to prephenate.

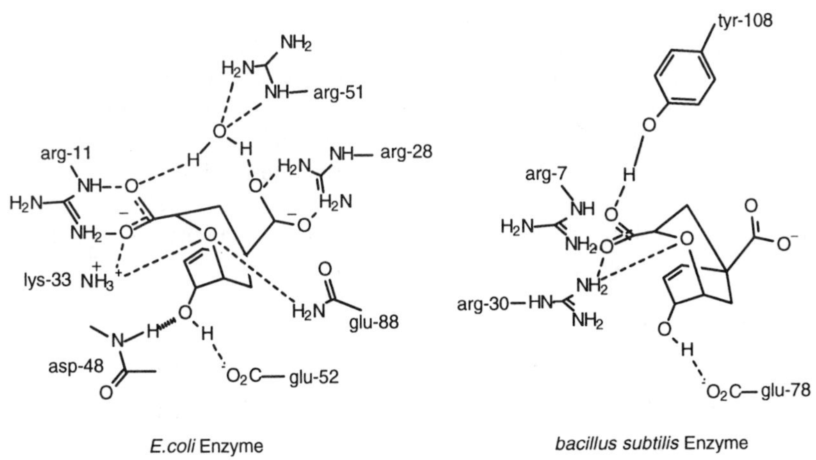

E.coli Enzyme   bacillus subtilis Enzyme

Scheme 8.58. Active site interactions between transition state analogue and chorismate mutase from various sources (*Ref. 163*).

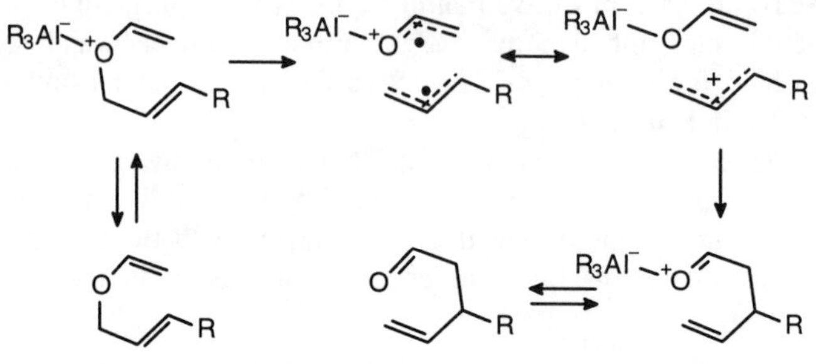

Scheme 8.59. Rearrangement of vinyl allyl ethers catalysed by Lewis acids.

Quantum mechanical - molecular mechanics and *ab initio* calculations for the bacillus subtilis enzyme predict that there is a significant electrostatic stabilisation of the transition structure through interaction between the arg-90[166] guanidinium group and the ether oxygen analogous to the Lewis acid interaction in the catalysed *model* Claisen rearrangement and the bonding is

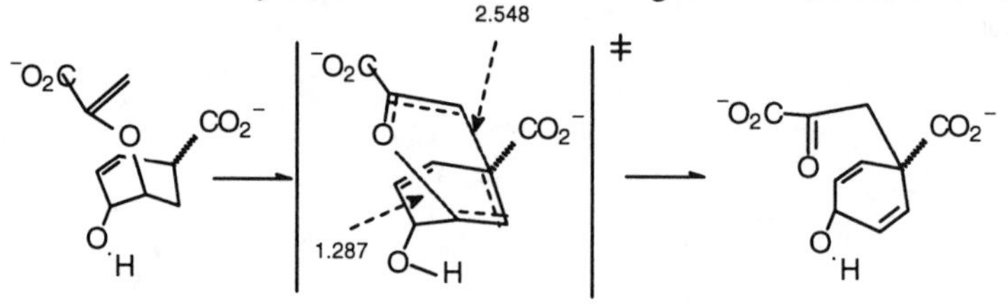

Scheme 8.60. Bonding in the model chorismate mutase transition structure calculated by quantum mechanics; the bond distances, in Ångstroms, are from *Ref. 167.*

illustrated in Scheme 8.60. *Ab initio* studies on model Claisen rearrangements indicate a concerted mechanism.[167,168] Squalene cyclase catalysis is characterised by precise stereoselective control of the formation of the four rings and in bio-mimetic cyclisations this is achieved by a stepwise process. It is established that the cyclisation in the enzyme catalysed reaction involves stepwise cyclisation *via* discrete carbenium ion intermediates followed by a series of 1,2-methyl and hydride shifts to yield lanosterol. Restricted Hartree Fock calculations using basis sets at 6-31G* and 6-31+G* levels were utilised to calculate transition structures for models of the enzyme which are illustrated in Scheme 8.61[169] and indicate concertedness.

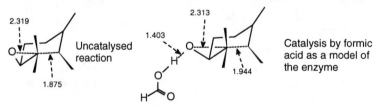

Scheme 8.61. Quantum mechanical model of the transition structure for squalene cyclase catalysis; bond lengths, given in Ångstroms, are from *Ref. 169.*

FURTHER READING

Adams, M. J., Oxido-reductases - pyridine nucleotide dependent enzymes, in *Enzyme Mechanisms*, Page, M. I., and Williams, A., eds., Royal Society of Chemistry, London, 1987, p. 477.

Allen, L. C., A model for the hydrogen-bond, *Proc. Natl. Acad. Sci. USA, 72,* 4701, 1975.

Allen, L. C., A simple model for the hydrogen-bonding, *J. Amer. Chem. Soc., 97,* 6921, 1975.

Babbitt, P. C., Mrachko, G. T., Hasson, M. S., Huisman, G. W., Kolter, R., Ringe, D., Petsko, G. A., Kenyon, G. L., and Gerlt, J. A., A functionally diverse enzyme superfamily that abstracts the α protons of carboxylic acids, *Science, 267,* 1159, 1995.

Botting, N. P., Isotope effects in the elucidation of enzyme mechanisms, *Natural Products Reports, RSC (London)* 337, 1994.

Cleland, W. W., Low-barrier hydrogen-bonds and low fractionation factor bases in enzymatic reactions, *Biochemistry, 31,* 317, 1992.

Cleland, W. W., What limits the rate of an enzyme-catalysed reaction? *Accs. Chem. Res., 8,* 145, 1975.

Cleland, W. W., Andrews, T. J., Gutteridge, S., Hartman, F. C., and Lorimer, G. H., Mechanism of rubisco: the carbamate as general base, *Chem. Rev., 98,* 549, 1998.

Cleland, W. W., and Kreevoy, M. M., Low-barrier hydrogen-bonds and enzymic catalysis, *Science, 264,* 1887, 1994.

Cook, P. F. (ed.), *Enzyme Mechanisms From Isotope Effects,* CRC Press, Boca Raton, 1991.

Dewar, M. J. S., and Storch, D. M., Alternative view of enzyme reactions, *Proc. Natl. Acad. Sci. USA, 82,* 2225, 1985.

Dewar, M. J. S., New ideas about enzyme reactions, *Enzyme, 36,* 8, 1986.

Eigen, M., Proton transfer, acid-base catalysis, and enzymatic hydrolysis. Part I: Elementary processes, *Angew. Chem. Int. Ed., 3,* 1, 1964.

Emsley, J., Very strong hydrogen-bonding, *Chemical Society Reviews, 9,* 91, 1980.

Fruton, J. S., The active site of pepsin, *Accs. Chem. Res., 7,* 241, 1974.

Ganem, B., The mechanism of the Claisen rearrangement: Déjà vu all over again, *Angew. Chem. Int. Ed. Engl., 35,* 936, 1996.

Gani, D., and Wilkie, J., Metal ions in the mechanism of enzyme-catalysed phosphate monoester hydrolyses, *Structure and Bonding, 89,* 133, 1997.

James, M. N. G., Hsu, I-N., and Delbaere, L. T. J., Mechanism of acid protease catalysis based on the crystal structure of penicillopepsin, *Nature, 267,* 808, 1977.

Jencks, W. P., Binding energy, specificity, and enzyme catalysis: the circe effect, *Adv. Enzymol., 43,* 219, 1975.

Jencks, W. P., *Catalysis in Chemistry and Enzymology,* McGraw-Hill, New York, 1969.

Kirby, A. J., Efficiency of proton transfer catalysis in models and enzymes, *Accs. Chem. Res., 30,* 290, 1997.

Kluger, R. Ionic intermediates in enzyme-catalyzed carbon-carbon bond formation: patterns, prototypes and proposals, *Chem. Revs., 90,* 1151, 1990.

Knowles, J. R., and Albery, W. J., Perfection in enzyme catalysis: the energetics of triosephosphate isomerase, *Accs. Chem. Res., 10,* 105, 1977.

Knowles, J. R., Tinkering with enzymes: what are we learning, *Science, 236,* 1252, 1987.

Kraut, J., How do enzymes work? *Science, 242,* 533, 1988.

Murakami, Y., Kikuchi, J-I., Hisaeda, Y., and Osamu, H., Artificial enzymes, *Chem. Rev., 96,* 721, 1996.

Richard, J. P., A consideration of the barrier for carbocation-nucleophile combination reactions, *Tetrahedron, 51,* 1535, 1995.

Scheiner, S., and Lipscomb, W. N., Molecular orbital studies of enzyme activity: catalytic mechanism of serine proteinases, *Proc. Natl. Acad. Sci. USA, 73,* 432, 1976.

Schowen, R. L., and Schowen, K. B., Reactive intermediates and the question of concertedness in the catalytic power of enzymes, *Croatica Chemica Acta, 65,* 779, 1992.

Silverman, R. B., Radical ideas about monoamine oxidase, *Accs. Chem. Res., 28,* 335, 1995.

Sinnott, M. L., Catalytic mechanisms of enzymic glycosyl transfer, *Chem. Revs., 90,* 1171, 1990.

Walsh, C. T., Liu, J., Rusnak, F., and Sakaitani, M., Molecular studies on enzymes in chorismate metabolism and the enterobactin biosynthesis pathway, *Chem. Rev., 90,* 1105, 1990.

Wang, J. H., Directional character of proton transfer in enzyme catalysis, *Proc. Natl. Acad. Sci. USA, 66,* 874, 1970.

Wang, J. H., Facilitated proton transfer in enzyme catalysis, *Science, 161,* 328, 1968.

Warshel, A., *Computer Modelling of Chemical Reactions in Enzymes and Solution,* John Wiley, New York, 1991.

REFERENCES

1. Swain, C. G., General acid-base concerted mechanism for enolisation of acetone, *J. Amer. Chem. Soc., 72,* 4578, 1950.

2. Jencks, W. P., Concertedness and enzyme catalysis, *Chemical approaches to understanding enzyme catalysis: Biomimetic chemnistry and transition state analogs, Proceedings of the 26th OHOLO conference, Israel, 1981,* Green, B. S., Ashani, Y., and Chipman, D. (eds.), *Studies in organic chemistry, 10,* 2, Elsevier, Amsterdam, 1981.

3. Jencks, W. P., On the attribution and additivity of binding energies, *Proc. Natl. Acad. Sci. USA, 78,* 4046, 1981.

4. Jencks, W. P., Binding energy, specificity, and enzyme catalysis: the circe effect, *Adv. Enzymol., 43,* 219, 1975.

5. Rappaport, H. P., An analysis of the binding of related ligands, *J. Theoret. Biol., 58,* 253, 1976.

6. Rappaport, H. P., Evaluation of group contribution to ligand binding, *J. Theoret. Biol., 79,* 157, 1979.

7. Williams, D. H., and Westwell, M. S., Aspects of weak interactions, *Chemical Society Reviews, 27,* 57, 1998.

8. Fersht, A. R., Leatherbarrow, R. J., and Wells, T. N. C., Binding energy and catalysis: a lesson from protein engineering of the tyrosyl-tRNA syntetase, *Trends in Biol. Sci., 11,* 321, 1986.

9. Fersht, A. R., Shi, J-P., Knill-Jones, J., Lowe, D. M., Wilkinson, A. J., Blow, D. M., Brick, P., Carter, P., Waye, M. M. Y., and Winter, G., Hydrogen-bonding and biological specificity analysed by protein engineering, *Nature, 314,* 235, 1985.

10. Wells, T. N. C., and Fersht, A. R., Hydrogen-bonding in enzymatic catalysis analysed by protein engineering, *Nature, 316,* 656, 1985.

11. Wells, T. N. C., and Fersht, A. R., Use of binding energy in catalysis analysed by mutagenesis of the tyrosyl-tRNA synthetase, *Biochemistry, 25,* 1881, 1986.

12. Jencks, W. P., Intrinsic binding energy, enzymic catalysis, and coupled vectorial processes, in *From Cyclotrons to Cytochromes. Essays in Molecular Biology and Chemistry,* Kaplan, N. O., ed., Academic Press, New York, 1982, p. 485.

13. Mader, M. M., and Bartlett, P. A., Binding energy and catalysis: the implications for transition state analogs and catalytic antibodies, *Chem. Revs., 97,* 1281, 1997.

14. Jencks, W. P., *Catalysis in Chemistry and Enzymology,* McGraw-Hill, New York, 1969.

15. Jencks, W. P., and Page, M. I., On the importance of togetherness in enzymic catalysis, *Fed. Eur. Biochem. Soc. (Enzymes: structure, function) Meet. Proc, 29,* 45, 1972.

16. Knowles, J. R., Enzyme catalysis: not different, just better, *Nature, 350,* 121, 1991.

17. Knowles, J. R., A feast for chemists, *Chem. Revs., 90,* 1078, 1990.

18. Satterthwait, A. C., and Jencks, W. P., The mechanism of the aminolysis of acetate esters, *J. Amer. Chem. Soc., 96,* 7018, 1974.

19. Jencks, W. P., Requirements for general acid-base catalysis of complex reactions, *J. Amer. Chem. Soc., 94,* 4731, 1972.

20. Blow, D. M., Structure and mechanism of chymotrypsin, *Accs. Chem. Res., 9,* 145, 1976.

21. Storer, A. C., and Ménard, R., Catalytic mechanism in papain family of cysteine peptidases, *Methods Enzymol., 244,* 486, 1994.

22. Hogg, J. L., Morris, R., and Durrant, N. A., Proton inventories of a serine protease charge-relay model in an aprotic solvent, *J. Amer. Chem. Soc., 100,* 1590, 1978.

23. Elrod, J. P., Hogg, J. L., Quinn, D. M., Venkatasubban, K. S., and Schowen, R. L.,

Protonic reorganization and substrate structure in catalysis by serine proteases, *J. Amer. Chem. Soc., 102,* 3917, 1980.

24. **Pollack, E., Hogg, J. L., and Schowen, R. L.,** One-proton catalysis in the deacetylation of acetyl-α-chymotrypsin, *J. Amer. Chem. Soc., 95,* 968, 1973.

25. **Quinn, D. M., Elrod, J. P., Ardis, R., Friesen, P., and Schowen, R. L.,** Protonic reorganisation in catalysis by serine proteases: acylation by small substrates, *J. Amer. Chem. Soc., 102,* 5358, 1980.

26. **Quinn, D. M., Venkatasubban, K. S., Kise, M., and Schowen, R. L.,** Protonic reorganization and substrate structure in catalysis by amidohydrolases, *J. Amer. Chem. Soc., 102,* 5365, 1980.

27. **Venkatasubban, K. S., and Schowen, R. L.,** The proton inventory technique, *Crit. Rev. Biochem., 17,* 1, 1984.

28. **Stein, R. L., Strimpler, A. M., Hori, H., and Powers, J. C.,** Catalysis by human leukocyte elastase: proton inventory as a mechanistic probe, *Biochemistry, 26,* 1305, 1987.

29. **Hamilton, S. E., and Zerner, B.,** Solvent effects on the deacylation of acyl-chymotrypsins: a critical comment on the charge-relay hypothesis, *J. Amer. Chem. Soc., 103,* 1827, 1981.

30. **Johnson, F. A., Lewis, S. D., and Shafer, J. A.,** Determination of a low pK_a for histidine-159 in the S-methylthio derivative of papain by nuclear magnetic resonance spectroscopy, *Biochemistry, 20,* 44, 1981.

31. **Lewis, S. D., Johnson, F. A., and Shafer, J. A.,** Effect of cysteine-25 on the ionisation of histidine-159 in papain as determined by proton magnetic resonance spectroscopy. Evidence for a his-159-cys-25 ion pair and its possible role in catalysis, *Biochemistry, 20,* 48, 1981.

32. **Ménard, R., Khouri, H. E., Plouffe, C., Laflamme, P., Dupras, R., Vernet, T., Tessier, D. C., Thomas, D. Y., and Storer, A. C.,** Importance of hydrogen-bonding interactions involving the side chain of asp-158 in the catalytic mechanism of papain, *Biochemistry, 30,* 5531, 1991.

33. **Fersht, A. R., and Sperling, J.,** The charge relay system in chymotrypsin and chymotrypsinogen, *J. Mol. Biol., 74,* 137, 1973.

34. **Rullmann, J. A. C., Bellido, M. N., and Duijnen, P. Th. van,** The active site of papain. All-atom study of interactions with protein matrix and solvent, *J. Mol. Biol., 206,* 101, 1989.

35. **Schröder, S., Daggett, V., and Kollman, P.,** A comparison of the AM1 and PM3 semiempirical models for evaluating model compounds relevant to catalysis by serine proteases, *J. Amer. Chem. Soc., 113,* 8922, 1991.

36. **Kollman, P. A., and Hayes, D. M.,** Theoretical calculations on proton-transfer energetics: studies of methanol, imidazole, formic acid, and methanethiol as models for the serine and cysteine proteases, *J. Amer. Chem. Soc., 103,* 2955, 1981.

37. **Warshel, A., Naray-Szabo, G., Sussman, F., and Hwang, J.-K.,** How do serine proteases really work, *Biochemistry, 28,* 3629, 1989.

38. **Bernstein, F. C., Koetzle, T. F., Williams, G. J. B., Meyer, E. F. Jr, Brice, M. D., Rodgers, J. R., Kennard, O., Shimanouchi, T., and Tasumi, M.,** The protein data bank. A computer-based archival file for macromolecular structures, *J. Mol. Biol., 112,* 535, 1977; http://www.pdb.bnl.gov.

39. **Rieder S. C., and Rose, I. A.,** The mechanism of the triosephosphate isomerase reaction, *J. Mol. Biol., 234,* 1007, 1959.

40. **Herlihy, J. M., Maister, S. G., Albery, W. J., and Knowles, J. R.,** Energetics of triosephosphate isomerase: the fate of the $1(R)-^3H$ label of tritiated dihydroxyacetone phosphate in the isomerase reaction, *Biochemistry, 15,* 5601, 1976.

41. **de la Mare, S., Coulson, A. F. W., Knowles, J. R., Priddle, J. D., and Offord, R. E.,** Active site labelling of triose phosphate isomerase, *Biochem. J., 129,* 321, 1972.

42. **Waley, S. G., Mill, J. C., Rose, I. A., and O'Connell, E. L.,** Identification of site in triose phosphate isomerase labelled by glycidophosphate, *Nature (London), 227,* 181, 1970.

43. **Nickbarg, E. B., Davenport, R. C., Petsko, G. A., and Knowles, J. R.,** Triosephosphate isomerase: removal of a putatively electrophilic histidine residue results in a subtle change

in catalytic mechanism, *Biochemistry, 27,* 5948, 1988.

44. **Komives, E. A., Chang, L.C., Lolis, E., Tilton, R. F., Petsko, G. A., and Knowles, J. R.,** Electrophilic catalysis in triosephosphate isomerase: the role of histidine-95, *Biochemistry, 30,* 3011, 1991.

45. **Albery, W. J., and Knowles, J. R.,** Evolution of enzyme function and the development of catalytic efficiency, *Biochemistry, 15,* 5631, 1976.

46. **Albery, W. J., and Knowles, J. R.,** Efficiency and evolution of enzyme catalysis, *Angew. Chem., Int. Ed. Engl., 16,* 285, 1977.

47. **Toney, M. D., and Kirsch, J. F.,** Direct Brønsted analysis of the restoration of activity to a mutant enzyme by exogenous amines, *Science,* 243, 1485, 1989.

48. **Auld, D. S., and Bruice, T. C.,** Catalytic reactions involving azomethines (IX). General base catalysis of the transimination of 3-hydroxypyridine-4-aldehyde by alanine, *J. Amer. Chem. Soc., 89,* 2098, 1967.

49. **Dunathan, H. C.,** Stereochemical aspects of pyridoxal phosphate catalysis, *Adv. Enzymol., 35,* 79, 1971.

50. **Julin, D. A., and Kirsch, J. F.,** Kinetic isotope effect studies on aspartate aminotransferase: evidence for a concerted 1,3-prototropic shift mechanism for the cytoplasmic isozyme and L-aspartate and dichotomy of mechanism, *Biochemistry, 28,* 3825, 1989.

51. **Kuliopulos, A., Mullen, G. P., Xue, L., and Mildvan, A. S.,** Stereochemistry of the concerted enolization catalyzed by Δ^5-3-ketosteroid isomerase, *Biochemistry, 30,* 3169, 1991.

52. **Kuliopulos, A., Mildvan, A. S., Shortle, D., and Talalay, P.,** Kinetic and ultraviolet spectroscopic studies of active site mutants of Δ^5-3-ketosteroid isomerase, *Biochemistry, 28,* 149, 1989.

53. **Malhotra, S. K., and Ringold, H. J.,** Chemistry of conjugate anions and enols V. Stereochemistry, kinetics and mechanisms of the acid and enzymatic catalysed isomerisation of Δ^5-3-ketosteroids, *J. Amer. Chem. Soc., 87,* 3228, 1965.

54. **Xue, L., Talalay, P., and Mildvan, A. S.,** Studies of the mechanism of the Δ^5-3-ketosteroid isomerase reaction by substrate, solvent, and kinetic deuterium isotope effects on wild-type and mutant enzymes, *Biochemistry, 29,* 7491, 1990.

55. **Kollman, P. A.,** A theory of hydrogen-bond directionality, *J. Amer. Chem. Soc., 94,* 1837, 1972.

56. **Allen, L. C.,** A model for the hydrogen-bond, *Proc. Natl. Acad. Sci. USA, 72,* 4701, 1975.

57. **Allen, L. C.,** A simple model for the hydrogen-bonding, *J. Amer. Chem. Soc., 97,* 6921, 1975.

58. **Kollman, P. A., and Allen, L. C.,** The theory of the hydrogen-bond, *Chem. Revs.,* 72, 283, 1972.

59. **Shan, S-O., and Herschlag, D.,** The change in hydrogen-bond strength accompanying charge rearrangement: implication for enzymatic catalysis, *Proc. Natl. Acad. Sci. USA, 93,* 14474, 1996.

60. **Shan, S-O., and Herschlag, D.,** Energetic effects of multiple hydrogen-bonds. Implications for enzymatic catalysis, *J. Amer. Chem. Soc., 118,* 5515, 1996.

61. **Shan, S-O., and Herschlag, D.,** Energetics of hydrogen-bonds in model systems: implications for enzymatic catalysis, *Science, 272,* 97, 1996.

62. **Cleland, W. W.,** Low-barrier hydrogen-bonds and low fractionation factor bases in enzymatic reactions, *Biochemistry, 31,* 317, 1992.

63. **Gerlt, J. A., Kozarich, J. W., Kenyon, G. L., and Gassman, P. G.,** Electrophilic catalysis can explain the unexpected acidity of carbon acids in enzyme-catalysed reactions, *J. Amer. Chem. Soc., 113,* 9667, 1991.

64. **Cleland, W. W., and Kreevoy, M. M.,** Low-barrier hydrogen-bonds and enzymic catalysis, *Science, 264,* 1887, 1994.

65. **Wang, J. H.,** Directional character of proton transfer in enzyme catalysis, *Proc. Natl. Acad. Sci. USA, 66,* 874, 1970.

66. **Wang, J. H.,** Facilitated proton transfer in enzyme catalysis, *Science, 161,* 328, 1968.

67. **Wang, J. H., and Parker, L.,** Pretransition state protonation and the rate of chymotrypsin

catalysis, *Proc. Natl. Acad. Sci. USA, 58,* 2451, 1967.

68. **Wang, J. H., and Parker, L.,** On the mechanism of action at the acylation step of the α-chymotrypsin-catalysed hydrolysis of anilides, *J. Biol. Chem., 243,* 3729, 1968.

69. **Kirby, A. J.,** Efficiency of proton transfer catalysis in models and enzymes, *Accs. Chem. Res., 30,* 290, 1997.

70. **Kenyon, G. L., Gerlt, J. A., Petsko, G. A., and Kozarich, J. W.,** Mandelate racemase: structure-function studies of a pseudosymmetric enzyme, *Accs. Chem. Res., 28,* 178, 1995.

71. **Gerlt, J. A., and Gassman, P. G.,** Understanding the rates of certain enzyme-catalysed reactions: proton abstraction from carbon acids, acyl-transfer reactions, and displacement reactions of phosphodiesters, *Biochemistry, 32,* 11943, 1993.

72. **Babbitt, P. C., Mrachko, G. T., Hasson, M. S., Huisman, G. W., Kolter, R., Ringe, D., Petsko, G. A., Kenyon, G. L., and Gerlt, J. A.,** A functionally diverse enzyme superfamily that abstracts the α protons of carboxylic acids, *Science, 267,* 1159, 1995.

73. **Gerlt, J. A., and Gassman, P. G.,** Rapid enzyme catalyzed proton abstraction from C-acids - importance of late transition states in concerted mechanisms, *J. Amer. Chem. Soc., 115,* 11552, 1993.

74. **Gerlt, J. A., and Gassman, P. G.,** An explanation for rapid enzyme-catalyzed proton abstraction from carbon acids: importance of late transition states in concerted mechanisms, *J. Amer. Chem. Soc., 115,* 11552, 1993.

75. **Gerlt, J. A., and Gassman, P. G.,** Understanding enzyme-catalyzed proton abstraction from carbon acids: details of stepwise mechanisms for β-elimination reactions, *J. Amer. Chem. Soc., 114,* 5928, 1992.

76. **Guthrie, J. P., and Kluger, R.,** Electrostatic stabilization can explain the unexpected acidity of carbon acids in enzyme-catalyzed reactions, *J. Amer. Chem. Soc., 115,* 11569, 1993.

77. **Guthrie, J. P.,** Short strong hydrogen-bonds: can they explain enzymic catalysis? *Chemistry and Biology, 3,* 163, 1996.

78. **Alagona, G., Ghio, C., and Kollman, P. A.,** Do enzymes stabilise transition states by electrostatic interactions or pK$_a$ balance: the case of triose phosphate isomerase (TIM)?, *J. Amer. Chem. Soc., 117,* 9855, 1995.

79. **Frey, P. A.,** On low-barrier hydrogen-bonds and enzymic catalysis, *Science, 267,* 104, 1995.

80. **Cleland, W. W., and Kreevoy, M. M.,** On low-barrier hydrogen-bonds and enzymic catalysis, *Science, 267,* 104, 1995.

81. **Warshel, A., Papazyan, A., and Kollman, P. A.,** On low-barrier hydrogen-bonds and enzymic catalysis, *Science, 267,* 102, 1995.

82. **Emsley, J.,** Very strong hydrogen-bonding, *Chemical Society Reviews, 9,* 91, 1980.

83. **Kreevoy, M. M., and Liang, T.-M.,** Structures and isotopic fractionation factors of complexes, A1HA2-, *J. Amer. Chem. Soc., 99,* 5207, 1977.

84. **Kreevoy, M. M., and Liang, T.-M.,** Structures and isotopic fractionation factors of complexes, A1HA2-, *J. Amer. Chem. Soc., 102,* 3315, 1980.

85. **Frey, P. A., Whitt, S. A., and Tobin, J. B.,** A low-barrier hydrogen-bond in the catalytic triad of serine proteases, *Science, 264,* 1927, 1994.

86. **Gandour, R. D., Nabulsi, N. A. R., and Fronczek, F. R.,** Structural model of a short carboxyl-imidazole hydrogen-bond with a nearly centrally located proton: implications for the asp-his dyad in serine proteases, *J. Amer. Chem. Soc., 112,* 7816, 1990.

87. **Warshel, A.,** *Computer Modelling of Chemical Reactions in Enzymes and Solution,* John Wiley & Sons, New York, 1991.

88. **Åqvist, J., and Warshel, A.,** Simulation of enzyme reactions using valence bond force fields and other hybrid quantum/classical approaches, *Chem. Rev., 93,* 2523, 1993.

89. **Harrison, M. J., Burton, N. A., and Hillier, I. H.,** Catalytic mechanism of the enzyme papain: predictions with a hybrid quantum mechanical/molecular mechanical potential, *J. Amer. Chem. Soc., 119,* 12285, 1997.

90. **Harrison, M. J., Burton, N. A., Hillier, I. H., and Gould, I. R.,** Mechanism and transition state structure for papain catalysed amide hydrolysis, using a hybrid QM/MM potential, *Chem. Commun.,* 2769, 1996.

91. **Arad, D., Langridge, R., and Kollman, P. A.,** A simulation of the sulfur attack in the

catalytic pathway of papain using molecular mechanics and semiempirical quantum mechanics, *J. Amer. Chem. Soc., 112*, 491, 1990.

92. **Daggett, V., Schröder, S., and Kollman, P. A.**, Catalytic pathway of serine proteases - classical and quantum mechanical calculations, *J. Amer. Chem. Soc., 113*, 8926, 1991.

93. **Hwang, J.-K., and Warshel, A.**, Semiquantitative calculations of catalytic - free energies in genetically modified enzymes, *Biochemistry, 26*, 2669, 1987.

94. **Rao, S. N., Singh, U. C., Bash, P. A., and Kollman, P. A.**, Free-energy perturbation calculations on binding and catalysis after mutating asn-155 in subtilisin, *Nature (London), 328*, 551, 1987.

95. **Maegley, K. A., Admiraal, S. J., and Herschlag, D.**, Ras-catalyzed hydrolysis of GTP: a new perspective from model studies, *Proc. Natl. Acad. Sci. USA, 93*, 8160, 1996,

96. **Komiyama, M., and Bender, M. L.**, Do cleavages of amides by serine proteases occur through a stepwise pathway involving tetrahedral intermediates? *Proc. Natl. Acad. Sci. USA, 76*, 557, 1979.

97. **Hess, R. A., Hengge, A. C., and Cleland, W. W.**, Isotope effects on enzyme-catalyzed acyl transfer from *p*-nitrophenyl acetate: concerted mechanisms and increased hyperconjugation in the transition state, *J. Amer. Chem. Soc., 120*, 2703, 1998.

98. **Sowa, G. A., Hengge, A. C., and Cleland, W. W.**, Isotope effects support a concerted mechanism for ribonuclease A, *J. Amer. Chem. Soc., 119*, 2319, 1997.

99. **Davis, A. M., Regan, A. C., and Williams, A.**, Experimental charge measurement at leaving oxygen in the bovine ribonuclease A catalysed cyclisation of uridine 3'-phosphate aryl esters, *Biochemistry, 27*, 9042, 1988.

100. **Weiss, P. M., and Cleland, W. W.**, Alkaline phosphatase catalyzes the hydrolysis of glucose-6-phosphate *via* a dissociative mechanism, *J. Amer. Chem. Soc., 111*, 1928, 1989.

101. **Belasco, J. G., Albery, W. J., and Knowles, J. R.**, Double isotope fractionation: test for concertedness and for transition state dominance, *J. Amer. Chem. Soc., 105*, 2475, 1983.

102. **Hermes, J. D., Roeske, C. A., O'Leary, M. H., and Cleland, W. W.**, Use of multiple isotope effects to determine enzyme mechanisms and intrinsic isotope effects. Malic enzyme and glucose-6-phosphate dehydrogenase, *Biochemistry, 21*, 5106, 1982.

103. **Clark, J. D., O'Keefe, S. J., and Knowles, J. R.**, Malate synthase: proof of a stepwise Claisen condensation using the double-isotope fractionation test, *Biochemistry, 27*, 5961, 1988.

104. **Hanson, K. R., and Rose, I. A.**, Interpretations of enzyme reaction stereospecificity, *Accs. Chem. Res., 8*, 1, 1975.

105. **Belasco, J. G., Albery, W. J., and Knowles, J. R.**, Energetics of proline racemase: double fractionation experiment, a test for concertedness and for transition state dominance, *Biochemistry, 25*, 2552, 1986.

106. **Rétey, J., and Lynen, F.**, Biochemical function of biotin. IX. Steric process of the carboxylation of propionyl-CoA, *Bioch. Z., 342*, 256, 1965.

107. **Rétey, J.**, Enzymic reaction selectivity by negative catalysis or how do enzymes deal with highly reactive intermediates, *Angew. Chem., Int. Ed. Engl., 29*, 355, 1990.

108. **Bahnson, B. J., and Anderson, V. E.**, Crotonase-catalyzed β-elimination is concerted - a double isotope effect study, *Biochemistry, 30*, 5894, 1991.

109. **Hermes, J. D., Weiss, P. M., and Cleland, W. W.** Nitrogen-15 and deuterium isotope effects to determine the chemical mechanism of phenylalanine ammonia-lyase, *Biochemistry, 24*, 2959, 1985.

110. **Rendina, A. R., Hermes, J. D., and Cleland, W.W.**, Use of multiple isotope effects to study the mechanism of 6-phosphogluconate dehydrogenase, *Biochemistry, 23*, 6257, 1984.

111. **Hermes, J. D., Tipton, P. A., Fisher, M. A., O'Leary, M. H. , Morrison, J. F., and Cleland, W. W.**, Mechanism of enzymatic and acid-catalysed decarboxylations of prephenate, *Biochemistry, 23*, 6263, 1984.

112. **Fowler, A. V., and Zabin, I.**, The amino acid sequence of β-galactosidase of *Escherichia Coli, Proc. Natl. Acad. Sci. USA, 74*, 1507, 1977.

113. **Jacobson, R. H., Zhang, X. J., Dubose, R. F., and Matthews, B. W.**, 3-Dimensional structure of β-galactosidase from *Escherichia coli, Nature (London), 369*, 761, 1994.

114. **Knier, B. L. and Jencks, W. P.**, Mechanism of reactions of N-(methoxymethyl)-N,N-dimethylanilinium ions with nucleophilic reagents, *J. Amer. Chem. Soc., 102*, 6789, 1980.

115. **Richard, J. P.**, A consideration of the barrier for carbocation-nucleophile combination reactions, *Tetrahedron, 51*, 1535, 1995.

116. **Rosenberg, S., and Kirsch, J. F.**, Oxygen-18 leaving group isotope effects on the hydrolysis of nitrophenyl glycoside. II. Lysozyme and β-galactosidase: acid and alkaline hydrolysis, *Biochemistry, 20*, 3196, 1981.

117. **Dahlquist, F. W., Rand-Meir, T., and Raftery, M. A.**, Application of secondary α-deuterium kinetic isotope effects to studies of enzyme catalysis. Glycoside hydrolysis of lysozyme and β-galactosidase, *Biochemistry, 8*, 4214, 1968.

118. **Smith, L. E. H., Mohr, L. H., and Raftery, M. B.**, Mechanism for lysozyme-catalyzed hydrolysis, *J. Amer. Chem. Soc., 95*, 7497, 1973.

119. **Sinnott, M. L., Withers, S. G., and Viratelle, O. M.**, The necessity of magnesium cation for acid assistance of aglycone departure in catalysis by *Escherichia coli (lac Z)* β-galactosidase, *Biochem. J., 175*, 539, 1978.

120. **Sinnott, M. L., and Souchard, I. J. L.**, The mechanism of action of β-galactosidase. Effect of aglycone nature and α-deuterium substitution on the hydrolysis of aryl galactosidases, *Biochem. J., 133*, 89, 1973.

121. **Gopalan, V., Vanderjagdt, D. J., Libell, D. P., and Glen, R. H.**, Transglycosylation as a probe of the mechanism of action of mammalian cytosolic β-galactosidase, *J. Biol. Chem., 267*, 9629, 1992.

122. **Kempton, J. B., and Withers, S. G.**, Mechanism of *Agrobacterium* β-galactosidase - kinetic studies, *Biochemistry, 31*, 9961, 1991.

123. **Gebler, J. C., Aebersold, R., and Withers, S. G.**, Glu-537 and not glu-461 is the nucleophile in the active site of (lac z)β-galactosidase from *Escherichia coli.*, *J. Biol. Chem., 267*, 11126, 1992.

124. **Withers, S. G., and Street, I. P.**, Identification of a covalent α-D-glucopyranosyl enzyme intermediate formed on a β-glucosidase, *J. Amer. Chem. Soc., 110*, 8551, 1988.

125. **Street, I. P., Kempton, J. B., and Withers, S. G.**, Inactivation of a β-galactosidase through the accumulation of a stable 2-deoxy-2-fluoro-α-D-glucopyranosyl-enzyme intermediate - a detailed investigation, *Biochemistry, 31*, 9970, 1992.

126. **Herrisant, B., and Bairoch, A.**, New families in the classification of glycosyl hydrolases based on amino acid sequence similarities, *Biochem. J., 293*, 781, 1993.

127. **Brameld, K. A., and Goddard, W. A. III**, Substrate distortion to a boat conformation at subsite-1 is critical in the mechanism of family 18 chitinases, *J. Amer. Chem. Soc., 129*, 3571, 1998.

128. **Brameld, K. A., and Goddard, W. A. III**, The role of enzyme distortion in the single displacement mechanism of family 19 chitinases, *Proc. Natl. Acad. Sci. USA, 95*, 4276, 1998.

129. **Perrakis, A., Tews, I., Dauter, Z., Oppenheimer, A. B., Chet, I., Wilson, K. S., and Vorgiac, E.**, Crystal structure of a bacterial chitinase at 2.3Å, *Structure (London), 2*, 1169, 1994.

130. **Fruton, J. S.**, The active site of pepsin, *Accs. Chem. Res., 7*, 241, 1974.

131. **James, M. N. G., Hsu, I-N., and Delbaere, L. T. J.**, Mechanism of acid protease catalysis based on the crystal structure of penicillopepsin, *Nature, 267*, 808, 1977.

132. **Wlodawer, A., Miller, M., Jaskolski, M., Sathyanarayana, B. K., Baldwin, E., Weber, I. T., Selk, L. M., Clawson, L., Schneider, J., and Kent, S. B. H.**, Conserved folding in retroviral proteases: crystal structure of a synthetic HIV-1 protease, *Science, 245*, 616, 1989.

133. **Hubbard, C. D., and Stein, T. P.**, The pepsin catalysed hydrolysis of bis-p-nitrophenyl sulfite, *Bioch. Biophys. Res. Commun., 42*, 293, 1971.

134. **Erlanger, B. F., Vratsanos, S. M., Wassermann, N., and Cooper, A. G.**, Stereochemical investigation of the active center of pepsin using a new inactivator, *Bioch. Biophys. Res. Commun., 28*, 203, 1967.

135. **Chen, K. C. S., and Tang, J.**, Amino acid sequence around the epoxide-reactive residues in pepsin, *J. Biol. Chem., 247*, 2566, 1972.

136. **Hartsuck, J. A., and Tang, J.,** The carboxylate ion in the active center of pepsin, *J. Biol. Chem.*, *247*, 2575, 1972.

137. **Hamilton, G. A., Spona, J., and Crowell, L. D.,** The inactivation of pepsin by an equimolar amount of 1-diazo-4-phenylbutanone-2, *Bioch. Biophys. Res. Commun.*, *26*, 193, 1967.

138. **Reid, T. W., Stein, T. P., and Fahrney, D.,** The pepsin-catalysed hydrolysis of sulfite esters. II. Resolution of alkyl phenyl sulfites, *J. Amer. Chem. Soc.*, *89*, 7125, 1967.

139. **Erlanger, B. F., Vratsanos, S. M., Wassermann, N., and Cooper, A. G.,** A chemical investigation of the active centre of pepsin, *Bioch. Biophys. Res. Commun.*, *23*, 243, 1966.

140. **Chatfield, D. C., and Brooks, B. R.,** HIV-1 protease cleavage mechanism elucidated with molecular dynamics simulation, *J. Amer. Chem. Soc.*, *117*, 5561, 1995.

141. **Blundell, T. L., Cooper, J., Foundling, S. I., Jones, D. M., Atrash, B., and Szelke, M.,** On the rational design of renin inhibitors: X-ray studies of aspartic proteinases complexed with transition state analogues, *Biochemistry*, *26*, 5585, 1987.

142. **Jaskolski, M., Tomasselli, A. G., Sawyer, T. K., Staples, D. G., Heinrikson, R. L., Schneider, J., Kent, S. B. H., and Wlodawer, A.,** Structure at 2.5-Å resolution of chemically synthesised human immunodeficiency virus type 1 protease complexed with a hydroxyethylene-based inhibitor, *Biochemistry*, *30*, 1600, 1991.

143. **Foundling, S. I., Cooper, J., Watson, F. E., Cleasby, A., Pearl, L. H., Sibanda, B. L., Hemmings, A., Wood, S. P., Blundell, T. L., Valler, M. J., Norey, C. G., Kay, J., Boger, J., Dunn, B. M., Leckie, B. J., Jones, D. M., Atrash, B., Hallett, A., and Szelke, M.,** High resolution X-ray analyses of renin inhibitor-aspartic proteinase complexes, *Nature*, *327*, 349, 1987.

144. **Hyland, L. J., Tomaszek, T. A., and Meek, T. D.,** Human immunodeficiency virus-1 protease. 2. Use of rate studies and solvent kinetic isotope effects to elucidate details of chemical mechanism, *Biochemistry*, *30*, 8454, 1991.

145. **Hyland, L. J., Tomaszek, T. A., Roberts, G. D., Carr, S. A., Magaard, V. W., Bryan, H. L., Fakhoury, S. A., Moore, M. L., Minnich, M. D., Culp, J. S., Desjarlais, R. L., and Meek, T. D.,** Human immunodeficiency virus-1 protease 1. Initial velocity studies and kinetic characterisation of reaction intermediates by ^{18}O isotope exchange, *Biochemistry*, *30*, 8441, 1991.

146. **Adams, M. J.,** Oxido-reductases - pyridine nucleotide dependent enzymes, in *Enzyme Mechanisms*, Page, M. I., and Williams, A., eds., Royal Society of Chemistry, London, 1987, p. 477.

147. **Eklund, H., Plapp, B. V., Samana, J-P., and Branden, C-I.,** Binding of substrate in a ternary complex of horse liver alcohol dehydrogenase, *J. Biol. Chem.*, *257*, 14349, 1982.

148. **McInnes, I., Nonhebel, D. C., Orszulik, S. T., and Suckling, C. J.,** On the mechanism of hydrogen transfer by nicotinamide coenzymes and alcohol dehydrogenase, *J. Chem. Soc. Perkin Trans. 2*, 2777, 1983.

149. **Scharschmidt, M., Fisher, M. A., and Cleland, W. W.,** Variation of transition state structure as a function of the nucleophile in reactions catalyzed by dehydrogenases: 1. Liver alcohol dehydrogenase with benzyl alcohol and yeast alcohol dehydrogenase with benzaldehyde, *Biochemistry*, *23*, 547, 1984.

150. **Schmidt, J., Chen, J., De Traglia, M., Muikel, D., and McFarland, J. J.,** Solvent isotope effect on the liver alcohol dehydrogenase reaction, *J. Amer. Chem. Soc.*, *101*, 3634, 1979.

151. **Blanchard, J. S., and Cleland, W. W.,** Kinetic and chemical mechanism of yeast formate dehydrogenase, *Biochemistry*, *19*, 3543, 1980.

152. **Hermes, J. D., Morrical, S. W., O'Leary, M. H., and Cleland, W. W.,** Variation of transition state structure as a function of the nucleotide in reactions catalysed by dehydrogenase. II Formate dehydrogenase, *Biochemistry*, *23*, 5479, 1984.

153. **Silverman, R. B.,** Radical ideas about monoamine oxidase, *Accs. Chem. Res.*, *28*, 335, 1995.

154. **Miura, R., and Miyake, Y.,** The reaction mechanism of D-amino acid oxidase: concerted or not concerted? *Bioorganic Chemistry*, *16*, 97, 1988.

155. **Mehler, A. H., and Bloom, B.,** Interaction between rabbit muscle aldolase and

dihydoxyacetone phosphate, *J. Biol. Chem.*, *238*, 105, 1963.

156. **Di Iasio, A., Trombetta, G., and Grazi, E.,** Fructose-1,6-bisphosphate aldolase from liver: the absolute configuration of the intermediate carbinolamine, *FEBS Letters*, *73*, 244, 1977.

157. **Gamblin, S. J., Davies, G. J., Grimes, J. M., Jackson, R. M., Littlehead, J. A., and Watson, H. C.,** Activity and specificity of human aldolases, *J. Mol. Biol.*, *219*, 573, 1991.

158. **Cleland, W. W., Andrews, T. J., Gutteridge, S., Hartman, F. C., and Lorimer, G. H.,** Mechanism of rubisco: the carbamate as general base, *Chem. Rev.*, *98*, 549, 1998.

159. **Wilkinson, B., Zhu, M., Priestley, N. D., Nguyen, H. H. T., Morimoto, H., Williams, P. G., Chan, S. I., and Floss, H. G.,** A concerted mechanism for ethane hydroxylation by the particulate methane monooxygenase from *Methylococcus capsulatus* (Bath), *J. Amer. Chem. Soc.*, *118*, 921, 1996.

160. **Newcomb, M., Le Tadic-Biadatti, M-H., Chestney, D. L., Roberts, E. S., and Hollenberg, P. F.,** A nonsynchronous concerted mechanism for cytochrome P-450 catalyzed hydroxylation, *J. Amer. Chem. Soc.*, *117*, 12085, 1995.

161. **Ganem, B.,** The mechanism of the Claisen rearrangement: Déjà vu all over again, *Angew. Chem., Int. Ed. Engl.*, *35*, 936, 1996.

162. **Walsh, C. T., Liu, J., Rusnak, F., and Sakaitani, M.,** Molecular studies on enzymes in chorismate metabolism and the enterobactin biosynthesis pathway, *Chem. Rev.*, *90*, 1105, 1990.

163. **Lee, A. Y., Karplus, P. A., Ganem, B., and Clardy, J.,** Atomic structure of the buried catalytic pocket of *Escherichia coli* chorismate mutase, *J. Amer. Chem. Soc.*, *117*, 3627, 1995.

164. **Chook, Y. M., Gray, J. V., Ke, H., and Lipscomb, W. N.,** The monofunctional chorismate mutase from bacillus-subtilis - structure determination of chorismate mutase and its complexes with a transition state analog and prephenate, and implications for the mechanism of the enzymatic-reaction *J. Mol. Biol.*, *240*, 476, 1994.

165. **Chook, Y. M., Ke, H., and Lipscomb, W. N.,** Crystal-structures of the monofunctional chorismate mutase from bacillus-subtilis and its complex with a transition state analog, *Proc. Natl. Acad. Sci. USA*, *90*, 8600, 1993.

166. **Davidson, M. M., Gould, I. R., and Hillier, I. H.,** Claisen rearrangement - chorismate to prephenate by chorismate mutase, *J. Chem. Soc. Perkin Trans. 2*, *525*, 1996.

167. **Wiest, O., and Houk, K. N.,** On the transition state of the chorismate-prephenate rearrangement, *J. Org. Chem*, *59*, 7582, 1994.

168. **Wiest, O., and Houk, K. N.,** Stabilization of the transition state of the chorismate-prephenate rearrangement: an ab initio study of enzyme and antibody catalysis, *J. Amer. Chem. Soc.*, *117*, 11628, 1995.

169. **Gao, D., Pan, Y-K., Byun, K., and Gao, J.,** Theoretical evidence for a concerted mechanism of the oxirane cleavage and A-ring formation in oxidosqualene cyclization, *J. Amer. Chem. Soc.*, *120*, 4045, 1998.

Index